# NON-PERTURBATIVE RENORMALIZATION

*Published by*

World Scientific Publishing Co. Pte. Ltd.

5 Toh Tuck Link, Singapore 596224

*USA office:* 27 Warren Street, Suite 401-402, Hackensack, NJ 07601

*UK office:* 57 Shelton Street, Covent Garden, London WC2H 9HE

**British Library Cataloguing-in-Publication Data**
A catalogue record for this book is available from the British Library.

**NON-PERTURBATIVE RENORMALIZATION**

Copyright © 2008 by World Scientific Publishing Co. Pte. Ltd.

*All rights reserved. This book, or parts thereof, may not be reproduced in any form or by any means, electronic or mechanical, including photocopying, recording or any information storage and retrieval system now known or to be invented, without written permission from the Publisher.*

For photocopying of material in this volume, please pay a copying fee through the Copyright Clearance Center, Inc., 222 Rosewood Drive, Danvers, MA 01923, USA. In this case permission to photocopy is not required from the publisher.

ISBN-13 978-981-279-239-6
ISBN-10 981-279-239-2

Printed in Singapore.

# NON-PERTURBATIVE RENORMALIZATION

**Vieri Mastropietro**

Università di Roma "Tor Vergata", Italy

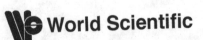

NEW JERSEY · LONDON · SINGAPORE · BEIJING · SHANGHAI · HONG KONG · TAIPEI · CHENNAI

# Preface

The notion of renormalization is at the core of several spectacular achievements of contemporary physics; originated in the context of Quantum Field Theory, where it appeared in order to solve the problem of the "ultraviolet divergences", it becomes later on (partly in a form known as "renormalization group") central in many other areas, like in the analysis of the critical properties close to phase transitions in classical statistical mechanics or in the theory of quantum liquids in condensed matter.

Renormalization is generally presented in a purely perturbative context (with no control of convergence of the series expansions), but in the last years, new mathematical techniques have been developed, allowing to put it on a firm mathematical basis. The aim of this book is to provide an introduction to rigorous non-perturbative renormalization in Quantum Field Theory, Statistical Physics and Condensed Matter. With respect to previous books on renormalization, the focus is mainly on fermionic (rather than bosonic) functional integrals, whose theory has been developed more recently and for which the structure of renormalization is not obscured by too many technicalities. Another important novelty is the implementation of Ward Identities based on local symmetries in the context of multiscale analysis, which allows the rigorous analysis of models with non trivial fixed points and anomalous behaviour. The book is devoted either to mathematicians and physicists aiming to enter into contact with the modern theory of renormalization; prerequisites are then limited to a minimum.

We start with an introduction to renormalization in physics and to the mathematical techniques for treating fermionic functional integrals, including multiscale decomposition techniques, tree expansions and determinant bounds. Such methods allow a unified treatment of models coming from Quantum Field Theory, Statistical Physics and Condensed matter. In particular, the first part of this book is devoted to constructive Quantum Field Theory, providing a mathematical construction of models at low dimensions and discussing the removal of the ultraviolet and infrared cut-off, the verification of the axioms and the validity of Ward Identities with the relative anomalies. The second part is devoted to lattice 2d Statistical Physics, analyzing in particular the theory of universality in perturbed Ising models and the computation of the non-universal critical indices in Vertex or Ashkin-Teller models. Finally in the third part, the theory of Quantum liquids like Luttinger or Fermi liquids is developed, considering models of interest in Condensed Matter like the Hubbard model in $1d$ or $2d$ or the Heisenberg spin chain.

Most of the material presented in this book grew out from common work with G. Benfatto and G. Gallavotti, and with the researchers composing the Roman school of rigorous renormalization, namely F. Bonetto, P. Falco, G. Gentile, G. Giuliani, A. Procacci and B. Scoppola. I have also benefitted from important discussions, which strongly influenced my point of view on renormalization, with J. Magnen and V. Rivasseau in Paris, with T. Spencer in Princeton and with K. Gawedzki in Lion.

Vieri Mastropietro

# Contents

Preface ................................................................. v

# Introduction to Renormalization     1

1. Basic Notions     3
    1.1 Relativistic quantum field theory ............... 3
        1.1.1 Quantum fields ........................... 3
        1.1.2 Functional integrals ..................... 5
        1.1.3 Perturbative renormalization ............. 8
    1.2 Classical statistical mechanics ................. 12
        1.2.1 Phase transitions ........................ 12
        1.2.2 Universality and non-universality ........ 14
    1.3 Condensed Matter ................................ 16
        1.3.1 Electrons in a crystal ................... 16
        1.3.2 The free Fermi gas ....................... 19
        1.3.3 Fermi liquids ............................ 21
        1.3.4 Luttinger liquids and BCS superconductors  23

2. Fermionic Functional Integrals     27
    2.1 Grassmann variables ............................. 27
    2.2 Grassmann measures .............................. 29
    2.3 Truncated expectations .......................... 31
    2.4 Properties of Grassmann integrals ............... 32
    2.5 Gallavotti-Nicoló tree expansion ................ 33
    2.6 Feynman graphs .................................. 39
    2.7 Determinant bounds for simple expectations ...... 42

| | | |
|---|---|---|
| 2.8 | The Brydges-Battle-Federbush representation | 45 |
| 2.9 | The Gawedzki-Kupiainen-Lesniewski formula | 51 |

# Quantum Field Theory     57

3. The Ultraviolet Problem in Massive QED2     59

| | | |
|---|---|---|
| 3.1 | Regularization and cut-offs | 59 |
| 3.2 | Integration of the bosons | 61 |
| 3.3 | Propagator decomposition | 63 |
| 3.4 | Renormalized expansion | 66 |
| 3.5 | Feynman graph expansion | 68 |
| 3.6 | Convergence of the renormalized expansion | 69 |
| 3.7 | Determinant bounds | 72 |
| 3.8 | The short memory property | 75 |
| 3.9 | Extraction of loop lines | 75 |
| 3.10 | The 2-point Schwinger function | 79 |
| 3.11 | The Yukawa model | 79 |

4. Infrared Problem and Anomalous Behavior     81

| | | |
|---|---|---|
| 4.1 | Anomalous dimension | 81 |
| 4.2 | Renormalization | 82 |
| 4.3 | Modification of the fermionic interaction | 83 |
| 4.4 | Bounds for the renormalized expansion | 89 |
| 4.5 | The beta function at lowest orders | 96 |
| 4.6 | Boundedness of the flow | 99 |
| 4.7 | The 2-point Schwinger function | 101 |

5. Ward Identities and Vanishing of the Beta Function     105

| | | |
|---|---|---|
| 5.1 | Schwinger functions and running couplings | 105 |
| 5.2 | Ward identities in presence of cut-offs | 107 |
| 5.3 | The correction identity | 109 |
| 5.4 | The Schwinger-Dyson equation | 115 |
| 5.5 | Analysis of the cut-off corrections | 118 |
| 5.6 | Vanishing of Beta function | 120 |
| 5.7 | Non-perturbative Adler-Bardeen theorem | 122 |
| 5.8 | Further remarks | 123 |

6. Thirring and Gross-Neveu Models     125

|   |   |   |
|---|---|---|
| 6.1 | The Thirring model | 125 |
| 6.2 | Removing the fermionic ultraviolet cut-off before the bosonic one | 126 |
| 6.3 | Removing the bosonic ultraviolet cut-off before the fermionic one | 128 |
| 6.4 | The Gross-Neveu model | 132 |

7. Axioms Verification and Wilson Fermions  133

|   |   |   |
|---|---|---|
| 7.1 | Osterwalder-Schrader axioms | 133 |
| 7.2 | Lattice regularization and fermion doubling | 135 |
| 7.3 | Integration of the doubled fermions | 137 |
| 7.4 | Lattice fermions | 138 |

8. Infraed QED4 with Large Photon Mass  143

|   |   |   |
|---|---|---|
| 8.1 | Regularization | 143 |
| 8.2 | Tree expansion | 145 |

# Lattice Statistical Mechanics  147

9. Universality in Generalized Ising Models  149

|   |   |   |
|---|---|---|
| 9.1 | The nearest neighbor Ising model | 149 |
| 9.2 | Heavy and light Majorana fermions | 153 |
| 9.3 | Generalized Ising models | 156 |
| 9.4 | Fermionic representation of the generalized Ising model | 157 |
| 9.5 | Integration of the $\chi$-variables | 159 |
| 9.6 | Integration of the light fermions | 160 |
| 9.7 | Correlation functions and the specific heat | 164 |

10. Nonuniversality in Vertex or Isotropic Ashkin-Teller Models  165

|   |   |   |
|---|---|---|
| 10.1 | Ashkin-Teller or Vertex models | 165 |
| 10.2 | Fermionic representation | 167 |
| 10.3 | Anomalous behaviour | 170 |
| 10.4 | Simmetry properties | 171 |
| 10.5 | Integration of the light fermions | 175 |
| 10.6 | The specific heat | 177 |

11. Universality-Nonuniversality Crossover in the Ashkin-Teller Model  181

| | | |
|---|---|---|
| 11.1 | The anisotropic AT model | 181 |
| 11.2 | Anomalous universality | 183 |
| 11.3 | Integration of the $\chi$ variables | 185 |
| 11.4 | Integration of the $\psi$ variables: first regime | 188 |
| 11.5 | Integration of the $\psi$ variables: second regime | 192 |
| 11.6 | Critical behaviour | 195 |

# Quantum Liquids   199

| | | |
|---|---|---|
| 12. | **Spinless Luttinger Liquids** | 201 |
| 12.1 | Fermions on a chain | 201 |
| 12.2 | Grassman representation | 203 |
| 12.3 | Luttinger liquid behavior | 204 |
| 12.4 | The ultraviolet integration | 207 |
| 12.5 | Quasi-particle fields | 209 |
| 12.6 | The flow of the running coupling constants | 213 |
| 12.7 | Density correlations | 215 |
| 12.8 | Quantum spin chains | 219 |
| 12.9 | Crystals and quasi-crystals | 222 |
| 13. | **The $1d$ Hubbard Model** | 225 |
| 13.1 | Spinning fermions | 225 |
| 13.2 | The effective potential | 226 |
| 13.3 | The flow of the running coupling constants | 228 |
| 13.4 | The auxiliary model | 231 |
| 13.5 | The effective renormalizations | 236 |
| 13.6 | Attractive interactions | 237 |
| 14. | **Fermi Liquids in Two Dimensions** | 239 |
| 14.1 | Interacting Fermions in $d=2$ | 239 |
| 14.2 | Multiscale integration | 242 |
| 14.3 | Bounds for the Feynman graphs | 245 |
| 14.4 | The sector decomposition | 246 |
| 14.5 | The sector lemma | 250 |
| 14.6 | Bounds for the tree expansion | 253 |
| 14.7 | Flow of runing coupling constants | 257 |
| 14.8 | Other results in $d=2$ | 260 |

| | | |
|---|---|---|
| 15. | **BCS Model with Long Range Interaction** | **263** |
| | 15.1 BCS model | 263 |
| | 15.2 Partial Hubbard-Stratonovich transformation | 266 |
| | 15.3 Corrections to the mean field | 268 |

| | | |
|---|---|---|
| **Appendix A** | **The Ising Model Fermionic Representation** | **273** |
| A.1 | The Grassmann representation of the 2d Ising model with open boundary conditions | 273 |
| A.2 | The Grassmann representation of the 2d Ising model with periodic boundary conditions | 283 |

*Bibliography*   287

# PART 1
# Introduction to Renormalization

# Chapter 1

# Basic Notions

## 1.1 Relativistic quantum field theory

### 1.1.1 *Quantum fields*

Aim of *Quantum Field Theory* (QFT) is to give a description of particles in agreement with the principles of Quantum Mechanics and Special Relativity (see Refs.[1],[2],[3] for complete expositions). A first attempt to get a coerent description of relativistic quantum particles was done by Dirac, who proposed an equation for the wave function of a relativistic fermion, the Dirac equation (with the convention $\hbar = c = 1$)

$$(i\bar{\gamma}_\mu \partial_\mu + m)\psi = 0 \tag{1.1}$$

where $x_\mu = (x_0, \vec{x})$, $\psi$ is a four dimensional spinor and $\bar{\gamma}_\mu$ are $4 \times 4$ matrices such that $\{\bar{\gamma}^\mu, \bar{\gamma}^\nu\} = 2g^{\mu,\nu}$, where $g^{\mu,\nu} = 0$ if $\mu \neq \nu$ and $1 = \bar{g}^{0,0} = -\bar{g}^{1,1} = -\bar{g}^{2,2} = -\bar{g}^{3,3}$.

The Dirac equation admits solutions with arbitrary negative energy so that apparently nothing prevent the particles to loose their energy indefinitely, for instance by interaction with the electromagnetic fields. This problem was solved by recasting the relativistic quantum mechanics as a many body problem, and assuming that the infinite states with negative energy levels are filled up in the vacuum state; according to Pauli exclusion principle, fermions cannot occupy the already filled states so that stability of states with positive energy is ensured. Moreover a photon can give a positive energy to a fermion in the sea, leaving an "hole" which appear as a fermion with opposite charge, a positron in the case of an electron, which was later one experimentally observed. The presence of infinities, the fact that the number of quantum particles can be created or annihilated and the analogy between the two apparently unrelated areas of relativistic quantum physics and quantum many body theory are features of the Dirac analysis which remained in all the subsequent developments.

As a single particle description appears to be not suitable for quantum relativistic particle for the possibility of creation or annihilation of pairs, one has to adopt a description in terms of *quantum fields*. They are operators acting on the Hilbert space of physical states and verifing a number of properties, see Ref.[4], among which is the *relativistic covariance* and the *microcausality*, which means that the fields must commute or anticommute if their support is space-like separated. If the fields are commuting they describe quantum particles called *bosons*, while if they are anticommuting they describe *fermions*. Fields with integer spins are bosons, and fields with semi-integer spins are fermions; this is the content of the *spin-statistics* theorem.

Fields describing *free* quantum particles can be explicitely constructed; important examples are the Dirac field, describing a relativistic fermion with spin 1/2, or the scalar field, describing a spin zero boson particle. The expectation values of time ordered product of field operators, called *Green functions*, can be exactly computed; they are expressed in terms of sum of products of the 2-point functions, a property called Wick rule. For instance the 2-point function for the Dirac field is given by

$$g(\vec{x}, x_0) = -i \int dk_0 d\vec{k} e^{ik_\mu x_\mu} \frac{\bar{\gamma}_\mu k_\mu - m}{-\vec{k}^2 + k_0^2 - m^2} \qquad (1.2)$$

The properties of non-interacting quantum fields are then essentially understood. On the other hand, one is mostly interested in what happens when particles interact, that is to *interacting fields*; however, except in very rare situation, their are impossible to construct explicitely.

The *Reconstruction theorem* says that, if we have a set of functions veryfing a certain number of properties called *Wightman axioms*, such functions *completely define* a relativistic quantum field theory, in the sense that they are the expectation values of relativistic quantum fields. It turns out that it is much more convenient, from a mathematical point of view, to pass to imaginary times with the replacement $x_0 \to ix_0$ (and $k_0 \to -ik_0$). The $n$-point functions with imaginary time are called Euclidean Green functions or *Schwinger functions*, and they can be also used to completely reconstruct a Quantum Field Theory in real time, if they verify a set of properties called *Osterwalder-Schrader* axioms [5].

The rule of the game are then essentially given; one has to find a set of Schwinger functions verifying the Osterwalder-Schrader axioms, and from them a QFT is completely determined. Of course one has, among all the possible QFT models, to find the ones really describing the interaction of quantum particles in the physical word; in the same way that in classical physics one can study the motion of a particles with a generic force, but of course only specific type of forces are realized in nature.

There is no indications in the axioms how to compute the expectations. What physicists have done is to try to get correlations verifying the axioms starting from suitable functional integrals; this procedure has been very successful, as the prediction obtained in this way have been verified with the maximum possible precision in the experiments, but of course other ways of deriving a set of Schwinger functions may be in principle possible.

### 1.1.2  Functional integrals

QFT models can be constructed starting from functional integrals, by generalizing the minimal action principle in classical mechanics. Such principle, due to the Hamilton, says that the trajectory of a particle follows the path minimizing a functional called the *action*, at least for small time intervals.

An extension of this principle holds also in quantum mechanics, and it is particularly suggestive; it says that the transition probability from one state to another is found summing over all the possible trajectories or paths, each of them weighted by a phase factor $e^{iS}$, where $S$ is the action. This principle can be still generalized to quantum fields, either bosonic or fermionic, considering functional integrals over all the possible field configurations weighted by $e^{iS}$ with $S = \int dx_0 d\vec{x} \mathcal{L}$, where $\mathcal{L}$ is the lagrangian density.

To give a mathematical meaning to such functional integrals is extremely difficult. One can expect a great semplification considering imaginary times with the replacement $x_0 \to ix_0$, so that one has to integrate over all field configurations weighted by the exponential $e^{-S}$ instead that with the oscillating factor $e^{iS}$. To have an idea of the simplification one can think to the integral $\int_{-\infty}^{\infty} dx e^{ix^2}$ with respect to $\int_{-\infty}^{\infty} dx e^{-x^2}$.

Historically among the first QFT which were considered is *Quantum Electrodynamics* in four dimensions (QED4), describing the interaction of a fermionic particle, like an electron, with a quantized electromagnetic field. The gauge invariance of classical Electrodynamics leads to well known difficulties in its quantization, and the standard method to overcome them consists to add to the action a gauge-fixing term. The Schwinger functions of QED can be written, if $\mathbf{x} = (x_0, \vec{x})$

$$< \psi_{\mathbf{x}_1};...;\psi_{\mathbf{x}_n}; A_{\mu_1,\mathbf{y}_1};...;A_{\mu_1,\mathbf{y}_1} > = \frac{\partial^n}{\partial \phi_{\mathbf{x}_1}...\partial \phi_{\mathbf{x}_n}} \frac{\partial^m}{\partial J_{\mathbf{y}_1}...\partial J_{\mathbf{y}_m}} \mathcal{W}(\phi, J)|_{\phi=0}$$
(1.3)

where $\mathcal{W}$ is the *formal* functional integral

$$\mathcal{W} = \log \int \mathcal{D}\psi \mathcal{D}A e^{-\int d\mathbf{x} \mathcal{L}(A,\psi) + \int d\mathbf{x}(\phi_{\mathbf{x}}\bar{\psi}_{\mathbf{x}} + \psi_{\mathbf{x}}\bar{\phi}_{\mathbf{x}} + J_\mu A_\mu)}$$

$$\mathcal{L}(A,\psi) = \frac{1}{2}F_{\mu,\nu}F_{\mu,\nu} + \frac{1}{2}(\partial_\mu A_\mu)^2 + \bar{\psi}_{\mathbf{x}}(\slashed{\partial} + m + e\gamma_\mu A_\mu)\psi_{\mathbf{x}} \quad (1.4)$$

and $\slashed{\partial} = \gamma_\mu \partial_\mu$, $\gamma_\mu$ are euclidean $\gamma$-matrices $\{\gamma_\mu, \gamma_\nu\} = 2\delta_{\mu,\nu}$, $e$ is the electric charge, $m$ is the electron mass, $F_{\mu,\nu} = \partial_\mu A_\nu - \partial_\nu A_\mu$, and $\frac{1}{2}(\partial_\mu A_\mu)^2$ is the gauge fixing term.

We will discuss the exact meaning of an expression like (1.4) in the following section; for the moment, we just say that $A_\mu$ is a gaussian variables describing the photon and $\psi, \bar{\psi}$ are spinor Grasmann variables describing the fermions. What it is important to stress at this point is that there are well defined rules allowing to express the Schwinger functions as a power series in $e$, whose $n - th$ order is given by the sum of a large number of terms admitting a graphical representation in terms of *Feynman graphs*. To a graph is associated, according to well defined rules, a value which is expressed by integrals over all momenta of products of the fermionic Euclidean propagator

$$g(\mathbf{k}) = \frac{\slashed{k} + im}{\mathbf{k}^2 + m^2} \quad (1.5)$$

and the photon propagators $v_{\mu,\nu}(\mathbf{k}) = \delta_{\mu,\nu}\mathbf{k}^{-2}$.

The Schwinger functions are related to physical observables (for instance one can compute from them the cross sections, which can be measured in accelerators); as there well defined rules to write the Schwinger functions as series, it is natural starting by truncating the perturbative series at some order (hoping that the contributions of the other orders is somewhat negligible), obtaining certain numbers and comparing them with the experiments.

However life is not so easy; while reasonable expressions are found at the very first orders in the series expansion, in going to higher orders one encounters Feynman graphs which are expressed by diverging integrals; this is the famous "problems of the infinities" in QFT. A closer look to the divergences of the integrals reveals that they are of two different types, called *ultraviolet* and *infrared*. The first are related to the fact that the integrands of the Feynman graphs do not decay fast enough at large monenta, while the second are due to the divergence of the integrand at zero momenta, and are present only with massless particles. The infrared divergences are connected to the low energy properties of the theory, while the infrared divergences are related to the high energy behaviour.

Even if the presence of the infinities seems to say that the above functional integrals are simply meaningless, we can forget for a moment such a problem and try instead to manipulate (1.4) *as* it would be a meaningful object. We can perform the *phase transformation*

$$\psi_{\mathbf{x}} \to e^{i\alpha_{\mathbf{x}}}\psi_{\mathbf{x}} \qquad \bar{\psi}_{\mathbf{x}} \to \bar{\psi}_{\mathbf{x}} e^{-i\alpha_{\mathbf{x}}} \tag{1.6}$$

so (1.12) becomes, assuming that the Jacobian of the transformation is 1

$$\mathcal{W} = \log \int \mathcal{D}\psi \mathcal{D}A e^{-\int d\mathbf{x} \mathcal{L}(A,\psi)}$$

$$e^{-\int d\mathbf{x}\alpha(\mathbf{x})\partial_\mu(\bar\psi_{\mathbf{x}}\gamma_\mu\psi_{\mathbf{x}})+\int d\mathbf{x}(e^{i\alpha_{\mathbf{x}}}\phi_{\mathbf{x}}\bar\psi_{\mathbf{x}}+e^{-i\alpha_{\mathbf{x}}}\psi_{\mathbf{x}}\bar\phi_{\mathbf{x}}+J_\mu A_\mu)} \tag{1.7}$$

and making a derivative with respect to $\alpha_{\mathbf{x}}, \phi_{\mathbf{y}}, \phi_{\mathbf{z}}$ one finds, if $j_{\mathbf{x},\mu} = \bar\psi_{\mathbf{x}}\gamma_\mu\psi_{\mathbf{x}}$

$$\partial_\mu \langle j_{\mathbf{x},\mu} \bar\psi_{\mathbf{y}}\psi_{\mathbf{z}} \rangle = \delta(\mathbf{x}-\mathbf{y})\langle \psi_{\mathbf{x}}\bar\psi_{\mathbf{z}}\rangle - \delta(\mathbf{x}-\mathbf{z})\langle \psi_{\mathbf{y}}\bar\psi_{\mathbf{z}}\rangle \tag{1.8}$$

or equivalently

$$i p_\mu \langle A_{\mu,\mathbf{p}} \psi_{\mathbf{k}} \bar\psi_{\mathbf{k}-\mathbf{p}} \rangle = ev(\mathbf{p})[\langle \psi_{\mathbf{k}-\mathbf{p}} \bar\psi_{\mathbf{k}-\mathbf{p}}\rangle - \langle \psi_{\mathbf{k}}\bar\psi_{\mathbf{k}}\rangle] \tag{1.9}$$

and similar ones with any number of fields. This means that the Schwinger functions are not indipendent one from the other but are related by an infinite set of identities, called *Ward Identities* (WI); such relations somewhat replace the gauge invariance of the classical theory. The WI can be also obtained by Feynman graph expansions, from the relation (in the massless case)

$$g(\mathbf{k}-\mathbf{p}) - g(\mathbf{k}) = \not{p} g(\mathbf{k}) g(\mathbf{k}+\mathbf{p}) \tag{1.10}$$

obvious consequence of $g(\mathbf{k}) = (i\not{k})^{-1}$. The WI (1.8) can be also derived from the conservation of the current $\partial_\mu j_{\mathbf{x},\mu} = 0$. Of course the Ward Identities (1.8), (1.9) are only formal as both the l.h.s.and the r.h.s. are infinite.

Other Ward Identities can be obtained, in the massless case $m = 0$, by the *chiral transformation* $\psi_{\mathbf{x}} \to e^{i\alpha\gamma_5}\psi_{\mathbf{x}}, \bar\psi_{\mathbf{x}} \to e^{-i\alpha\gamma_5}\bar\psi_{\mathbf{x}}$; it is found, if $j^5_{\mathbf{x},\mu} = \bar\psi_{\mathbf{x}}\gamma_\mu\psi_{\mathbf{x}}$

$$\partial_\mu \langle j^5_{\mathbf{x},\mu} \bar\psi_{\mathbf{y}}\psi_{\mathbf{z}} \rangle = \delta(\mathbf{x}-\mathbf{y})\gamma_5 \langle \psi_{\mathbf{x}}\bar\psi_{\mathbf{z}}\rangle - \delta(\mathbf{x}-\mathbf{z})\gamma_5\langle \psi_{\mathbf{y}}\bar\psi_{\mathbf{z}}\rangle \tag{1.11}$$

The WI (1.11) can be also derived from the equation of conservation of the axial current $\partial_\mu j^5_{\mathbf{x},\mu} = 0$.

Several other QFT models have been introduced in addition to QED, considering also interactions involving only boson or fermions. An important model is the *Nambu-Jona Lasinio* model, known as *Thirring model* in $d = 2$, describing the self interaction of a massive fermion via a current-current interaction; the Schwinger functions are obtained by the derivatives of the following functional integral

$$\mathcal{W} = \log \int \mathcal{D}\psi e^{-\int d\mathbf{x}\mathcal{L}(\psi) + \int d\mathbf{x}(\phi_\mathbf{x}\bar{\psi}_\mathbf{x} + \psi_\mathbf{x}\phi_\mathbf{x})}\Big|_{\phi=0} \quad (1.12)$$

with Lagrangian

$$\mathcal{L}(\psi) = \bar{\psi}_\mathbf{x}(\gamma_\mu \partial_\mu + m)\psi_\mathbf{x} + \lambda(\bar{\psi}_\mathbf{x}\gamma_\mu\psi_\mathbf{x})(\bar{\psi}_\mathbf{x}\gamma_\mu\psi_\mathbf{x}) \quad (1.13)$$

Also for such models the formal WI (1.8), (1.11) are valid. A variant of the model consist to add a colour index to the fermions, $\psi_{\mathbf{x},i}$, and $i = 1, .., N$; in such a case the model is called *Gross-Neveu model*.

Of course in real applications the dimension has to be 4, but it can be convenient consider QFT at lower dimensions, in which the analytical difficulties are much simpler and one can learn informations on general properties.

### 1.1.3 *Perturbative renormalization*

Let us return to the problem of infinities in the Feynman graph expansion of the functional integrals, which makes the theory apparently meaningless. The infrared divergences can be avoided assuming that all the particles have a mass; this is a good starting point, despite in nature neutrinos or photons have no mass and in any case the mass of electrons is very small with respect to the other quantities, in adimensional units.

The idea of Tomonaga, Schwinger and Feynman was to try to absorb the "ultraviolet divergences" in the parameters appearing in the lagrangian $\mathcal{L}$, according to a procedure called *renormalization*. The physical idea behind such procedure is that the parameters appearing in the Lagrangian (called *bare* parameters), like the electron mass or charge, are not the ones really observed, the *dressed* or physical parameters; in QED, for instance, their values are deeply modified by the interaction with the electromagnetic field.

Hence the Schwinger functions, which are written initially as functions of the bare parameters, must be re-expressed in terms of the dressed ones in order to compare them with real experiments. One has to consider the bare QED Lagrangian

$$\mathcal{L}_B = \frac{1}{2}Z_3 F_{\mu,\nu}F_{\mu,\nu} + Z_2 \bar{\psi}_\mathbf{x}\gamma_\mu\partial_\mu\psi_\mathbf{x} + Z_4 m\bar{\psi}_\mathbf{x}\psi_\mathbf{x} + eZ_1\bar{\psi}_\mathbf{x}\gamma_\mu\partial_\mu\psi_\mathbf{x} A_\mu \quad (1.14)$$

where $Z_2$ and $Z_3$ are the fermionic and bosonic bare wave function normalization,

$$e\frac{Z_1}{Z_2\sqrt{Z_3}} = e_0 \qquad (1.15)$$

is the bare electric charge and $mZ_4$ the bare mass.

The idea is then to choose the bare parameters so that the dressed ones have the observed values, that is the charge is $e$, the mass $m$ and the normalizations are equal to 1; in order to accomplish this goal it is convenient to write $\mathcal{L}_B$ as $\mathcal{L} + \delta\mathcal{L}$, with $\mathcal{L}$ was given by (1.4) and

$$\delta\mathcal{L}(A,\psi) = \frac{1}{2}\delta z_3 F_{\mu,\nu}F_{\mu,\nu} + \delta z_2 \bar\psi_{\mathbf{x}}\gamma_\mu\partial_\mu\psi_{\mathbf{x}} + \delta z_4 m\bar\psi_{\mathbf{x}}\psi_{\mathbf{x}} + e\delta z_1 \bar\psi_{\mathbf{x}}\gamma_\mu\partial_\mu\psi_{\mathbf{x}}A_\mu \qquad (1.16)$$

where $Z_3 = \sqrt{(1+\delta z_3)}$, $Z_2 = \sqrt{(1+\delta z_2)}$, $Z_1 = (1+\delta z_1)$ and $Z_4 = 1+\delta z_4$; the $\delta z_i$ are called counterterms. One considers then the (formal) generating function

$$\log\int \mathcal{D}\psi\mathcal{D}A e^{-\int d\mathbf{x}(\mathcal{L}+\delta\mathcal{L}) + \int d\mathbf{x}(\phi_{\mathbf{x}}\bar\psi_{\mathbf{x}} + \psi_{\mathbf{x}}\bar\phi_{\mathbf{x}} + J_\mu A_\mu)} \qquad (1.17)$$

which can be still written as sum of Feynman graphs, containing diverging expressions; in order to manage such divergences the Feynamnn graphs need to be *regularized*. There are several possible regularizations, each one with their proper advantages and disadvantages; at a purely perturbative level, the best regularizations are the one preserving the symmetries of the classical theory (and respecting (1.10)), like the *dimensional regularization*. One writes the *counterterterms* $\delta z_i$ as power series in $e$

$$\delta z_i(\Lambda) = e\delta z_{i,1}(\Lambda) + e^2 \delta z_{i,2}(\Lambda) + ... \qquad (1.18)$$

where $\Lambda$ is a parameter depending on the regularization, and try to choose the coefficients $\delta z_i(\Lambda)$ so the Schwinger functions are expressed by power series (in $e$) which are order by order *finite* removing the regularization ($\Lambda \to \infty$): this means that the infinities in the expansion are exactly compensated by the counterterms and that the resulting theory remain order by order finite when the regularization is removed.

Of course it is not obvious at all that such a procedure is successfull; there is apparently no reason *a priori* for which by adding a $\delta\mathcal{L}$ of the form (1.16) (that is exactly of the form of $\mathcal{L}$) one can exactly compensate all the infinities arising in the graph expansion. Indeed this can be checked explicitly at lowest orders but it was only through a great analytical effort that it was proved that, by choosing properly the $\delta z_i$ (diverging as $\Lambda \to \infty$), the theory is finite *at any order*. This property is called *perturbative renormalizability* is it is verified only by a few QFT models; it was proved for QED by Weinberg, Bogolubov, Zimmermann, Hepp and others.

At the end, it turns out that the *renormalized* Schwinger functions can be written as an expansion in $e$ order by order finite; all the infinities are absorbed in bare parameters which are of course diverging removing the cut-offs, but this is not important as they are by definition unobservable. On the contrary, the physical quantities, expressed in terms of the dressed parameters, are in excellent agreement with experiments.

Moreover, by using regularizations respecting (1.10), like the dimensional one, the Ward Identities (1.8) are valid for the *renormalized* Schwinger functions, as identities valid order by order in the perturbative expansion; one says that the WI are *preserved* by the renormalization procedure. The validity of the WI implies that the counterterms are not indipendent one from the other; for instance it holds the identity

$$Z_2 = Z_1 \qquad (1.19)$$

as order by order statement. It says that $e_0 = eZ_3^{-\frac{1}{2}}$, that is the renormalization of the charge depends only from the photon renormalization.

While the WI based on the total phase transformation are preserved by the renormalization, quite surprisingly the WI based on a chiral transformation are instead *not preseverd*; the analogue equation for the axial current (1.11) is not true in the renormalized theory but it appears an extra term

$$\frac{\alpha}{4\pi} \langle F_{\mu,\nu} F_{\mu,\nu}; \bar{\psi}_\mathbf{y} \psi_\mathbf{z} \rangle \qquad (1.20)$$

where $\alpha = e^2/4\pi$. The presence of the additional term in the Ward Identity is called a *quantum anomaly*. Note also that the anomaly coefficient is linear in $\alpha$, that is there are no higher orders corrections; this property is called *anomaly renormalization* or Adler-Bardeen theorem, and it is again a perturbative order by order statement. In QED2, anomaly also appears which are linear in the charge.

As we said, a peculiarity of $QED4$ is that the counterterms one needs to eliminate the singularities are exactly of the form of the parameters appearing in the original lagrangian; this property is true only for a small number of models, said *renormalizable*. For instance it is not true in the Gross Neveu or Thirring model in $d = 4$, while it is true if such models are considered in $d = 2$. Some models are indeed *superrinormalizable*; there is no need of counterterms (there are no divergences) or the divergence are only in a a finite number of Feynman graphs so that the series in (1.18) is just a finite sum; this is the case of QED2.

In addition to QED, the Weak and the Strong interactions (responsible respectively, for instance, of the decay of a neutron and of the formation of atomic nuclei) have been described in terms on renormalizable QFT models in the celebrated *Standard Model*. Actually such a theory provides an understanding of all the fundamental forces in nature (except gravity) and it has been tested in experiments with a precision never reached by any other physical theory. It provides an explanation of an enormous range of phenomena, from the properties of sub-atomic particles to the properties of stars. It is however important to recall that such computations are purely *perturbative*, that is obtained by arbitrary truncation series expansions, whose convergence cannot be proved (most of them are probably not convergent at all); hence there is no mathematical proof that a consistent QFT corresponding to such theory really exists.

A different point of view on renormalization, mainly due to Wilson Ref.[6], has also emerged in more recent years, and it is know as Renormalization Group or effective action approach, see Refs.[7],[8]. One introduces an ultraviolet momentum cut-off at some large momentum scale, that is the energies greater than some value $\Lambda$ are forbidden. One then starts from an action at scale $\Lambda$ and integrate iteratively the fields of decreasing energy scale, obtaining a sequence of effective actions. The bare parameters appearing in the action at the scale $\Lambda$ are chosen so that the effective action at energy corresponding to our experiments have the correct values. This method has some disadvantage, as it is more complex to prove the validity of the WI in the renormalized theory; for finite values of the cut-offs, the momentum regularizations violate (1.10) and produce additional terms in the WI, so that one has to show that WI are finally restored removing the cut-offs.

On the other hand, this approach has the advantage to be suitable in principle for going beyond a purely perturbative approach, see Refs.[3], [9],[10],[11], and indeed using it the existence of several nontrivial QFT models, mainly in $d = 2$, can be rigorously established at a non-perturbative level, as we will see in the first part of this book. While such models are still far from the realistic ones, they are important to show that QFT based on functional integrals and the renormalization procedure can really provide a coherent mathematical understanding of all fundamental interactions (except possibly gravity).

## 1.2 Classical statistical mechanics

### 1.2.1 *Phase transitions*

The aim of statistical mechanics is to compute the macroscopic properties of systems composed by a huge number of atoms or molecules, given only a knowledge of the microscopic forces between the components (for detailed expositions, see Refs.[12],[13]). More exactly, starting from the hamiltonian of a system (taking into account of all the interaction between atoms or molecules) one can compute, according to the postulates of statistical physics, thermodinamic quantities like the pressure or the entropy. Typical phenomena one is interested in statistical mechanics are the *phase transitions*; a well known example in everyday life is for intance the transition from water to vapour. Another important example of phase transition, somewhat simpler to study, happens when a magnetic field is applied to a magnet; at temperatures lower than $T_c$ the magnets developes a magnetization which persists even when the magnetization is turned off (ferromagnetic phase), while above $T_c$ there is no spontaneous magnetization (paramagnetic phase).

Experimentally in corrispondence of a phase transition the physical observables (like the magnetization or the specific heat) have some singularity, typically of the form $O(|T - T_c|^{-\alpha})$, with $\alpha$ called a *critical index*.

Can phase transitions be explained in the statistical mechanics framework? Singularities in the thermodinamic functions, signaling phase transitions, can possibly appear only in the so-called thermodynamic limit, consisting in taking the limit of infinite volume and particle number taking the density fixed. For a long time it was though that such singularities do not appear, and that phase transition cannot be understood starting from the forces between the molecules componing a material. It was Onsager the first who showed a microscopic model exhibiting a phase transition: the 2-dimensional, nearest neighbor *Ising model*, in which the thermodinamic functions can be explicitely computed.

The Ising model is a paradigmatic model for statistical mechanics. It describes a magnet as a lattice made up of molecules with a magnetic dipole wich either points in some direction or in the opposite one (of course the magnetic moment of a molecule should be a vector pointing in any direction, hence such a description is rather crude). To each point $\mathbf{x}$ of the lattice a spin $\sigma_\mathbf{x} = \pm 1$ is then associated, and the energy interaction, due to the dipole force, of two spins is $J_{\mathbf{x},\mathbf{x}'}$ if the two spins have the same value, and

$-J_{\mathbf{x},\mathbf{x}'}$ if they have different values, so that the hamiltonian is

$$H_I = -\sum_{\mathbf{x},\mathbf{x}'} J_{\mathbf{x},\mathbf{x}'} \sigma_{\mathbf{x}} \sigma_{\mathbf{x}'} \qquad (1.21)$$

Typically $J_{\mathbf{x},\mathbf{x}'}$ is chosen as a short-range interaction; particularly important is the case when $J_{\mathbf{x},\mathbf{x}'}$ is different from zero only if $\mathbf{x}, \mathbf{x}'$ are nearest neighbor, as in this case several semplifications in the mathematical analysis appear.

The partition function at inverse temperature $\beta$ is given by

$$Z = \sum_{\{\sigma_{\mathbf{x}} = \pm\}_{\mathbf{x} \in \Lambda}} e^{-\beta H_I} \qquad (1.22)$$

and, if $\beta = (\kappa T)^{-1}$, $\kappa$ is the Boltzmann constant and $N$ the number of point in $\Lambda$,

$$f(\beta) = -\beta \lim_{N \to \infty} N^{-1} \log Z \qquad (1.23)$$

is the *free energy* for site; the limit $N \to \infty$ is called thermodinamic limit. The *specific heat* is given by

$$C_v = -\frac{\partial}{\partial T} \frac{\partial}{\partial \beta} (\beta^{-1} f(\beta))] \qquad (1.24)$$

and in a similar way are defined all the other thermodynamic functions.

Even with so many simplifying features, the computation of the thermodynamic functions corresponding to (9.1) is quite difficult. Explicit values for the critical indices can be obtained quite easily in the so called *mean field approximation*, but such values are in general quantitatively not correct.

If one considers only nearest neighbor interactions the Ising model con be solved in $d = 1$ (where it does not exhibit phase transitions) and in $d = 2$, through the remarkable exact solution found by Onsager; in $d = 2$ the hamiltonian of the nearest-neighbor Ising model is given by, if $\mathbf{x} = (x_0, x)$

$$H_{n.n.} = \sum_{\mathbf{x},\mathbf{x}' \in \Lambda} [\sigma_{x,x_0} \sigma_{x,x_0+1} + \sigma_{x,x_0} \sigma_{x+1,x_0}] \qquad (1.25)$$

and there is a phase transition at the critical temperature

$$\tanh \beta_c = \sqrt{2} - 1 \qquad (1.26)$$

The thermodinamic quantities corresponding to (1.25) are singular at $\beta = \beta_c$; for instance the specific heat is given by

$$C_v = -C_1 \log |\beta - \beta_c| + C_1 \qquad (1.27)$$

with $C_1, C_2$ constants. The above result should be compared with what is found in the mean field approximation, in which the specific heat is discontinuous. Other thermodynamic variables can be considered; a remarkable one is the spontaneous magnetization, from which one can see that the system has at $\beta = \beta_c$ a phase transition from a paramagnetic to a ferromagnetic phase; the spontaneous magnetization vanishes as $O(|\beta - \beta_c|^{\frac{1}{2}})$.

The solution of the $2d$ nearest-neighbor Ising model was followed by the solutions of other lattice spin models, see [14], like the *Ice models*, with a physical meaning within the theory of the idrogen bond, or the *Vertex models*, and a lot of important informations were achieved from them, which in several cases were also experimentally verified.

It should be noted however that the exact solvability requires a quite special structure, and it is immediately destroyed even by apparently innocuous modifications. For instance in the case of the Ising model, if one includes also a next to nearest interaction (there is physically no reason for which only nearest neighbor spins should interact) the exact solvability is immediately lost. Moreover, while the Ising model in $1d$ or $2d$ is solvable, there is no exact solution for the $3d$ Ising model.

### 1.2.2 *Universality and non-universality*

A crucial role in theory of phase transition is played by the principle of *universality*. Let us consider an hamiltonian of the form

$$H = H_0 + \lambda V \tag{1.28}$$

where $H_0$ is some simple hamiltonian, whose thermodynamic quantities can be computed, $V$ is a complicated perturbation and $\lambda$ is a parameter measuring its strenght.

The natural question is if the critical properties are modified or not by the presence of $\lambda V$. According to the *universality hypothesis*, the singularities in the thermodinamic functions, in particular the critical indices should be insensitive to perturbations as long as symmetry and some form of locality are retained.

The most natural model in which universality can be investigated is the $d = 2$ Ising model; we can choose $H_0$ as (1.25) and $V$ is some complex term involving many spin interaction. What the universality hypothesis says in this case is that, while the thermodinamical quantities (and the critical temperature) depend in general from the perturbation, the critical indices would be identical to the one of the Ising model.

The importance of such hypothesis is clear; if universality holds, one can use extremely simplified and higly idealized models instead of more realistic but extremely complicated ones, and the critical properties should be the same; models with the same critical behaviour are called to be in the same "universality class". Indeed universality seems verified in experiments: for instance carbon dioxide, xenon and the $d = 2$ Ising model appears to be in the same universality class.

On the other hand, a too naive extension of the notion of univesality can be incorrect. The 8-vertex model is equivalent to two Ising models coupled by a quartic interaction, but for such a model universality does not hold; the critical indices of such a model can be explicitely computed ond one sees that they are different with respect to the ones of the Ising model.

How it is possible to check if the universality holds in a model? How it is possible to compute critical indices, when exact solutions are lacking? The more promising technique is the Wilson Renormalization Group, based an an iterative integration leading to a sequence of effective theories. A very important achievement of this method was the computation of the non-universal critical indices in the $3d$ Ising model which, contrary to the $2d$ case, is not solvable. The indices (different from the ones computed in the $d = 2$ case) can be written in terms of a series expansion and are in remarkable agreement with experimental data, see Ref. [6]; unfortunately, a proof of the convergence of such series is still lacking.

A simpler applications of such ideas, which can be instead performed with a full mathematical rigor, can be done for the computation of the critical indices in $2d$ lattice spin models, like non nearest neighbor Ising model, or Vertex or Askhin-Teller models. In such cases in fact one can exploit the remarkable mapping of the Ising model is a system of free fermions in $d = 2$ dimensions, very similar to the ones for $d = 2$ QFT seen before; the mass of the fermions corresponds to $|T - T_c|$ so that the critical point corresponds to massless fermions. Consequently, models which are perturbations of the Ising model can be mapped in fermionic interacting systems; the lack of solvability is reflected in the fact that the Lagrangian is not quadratic.

It turns out then that many thermodynamic quantities (like the specific heat) of several spin lattice models can be written as a fermionic functional integral of the form

$$\int \mathcal{D}\psi e^{-S_0(\psi) - S_1(\psi)} \tag{1.29}$$

where $S_0(\psi)$ is the lattice regularization of the euclidean action of Dirac fermions in $d = 2$ and $S_1(\psi)$ corrsponds to the interacting part. The lattice functional integrals introduced in QFT in $d = 1 + 1$ somewhat artificaly to cure the ultraviolet divergences, naturally appear in statistical physiscs as perturbations of the Ising model. Note however that in QFT the continuum limit has to be taken while in statistical physics the lattice is fixed; one has only the infrared problem to face.

The representation (1.29) is very useful as it allows to apply the methods of renormalization developed in QFT to several 2d lattice spin models. In the second part of this book we will see that such methods allow to give a proof of universality for certain classes of perturbed Ising model, in the sense that the behaviour of physical quantities is the same as of the Ising model up to a renormalization of the critical temperature. The same methods allow also the rigorous computations of several non universal critical indices in solvable or non solvable models, like Vertex or Askin-Teller models, equivalent to Ising models coupled by quartic interactions.

## 1.3 Condensed Matter

### 1.3.1 *Electrons in a crystal*

Condensed matter is concerned with the average properties of a system composed by a large number of quantum particles (see Refs. [15],[16]). A crystal can be described as a lattice of atoms in which the valence electrons are lost by the atoms (which become ions) and move freely in the metal; they are responsible of the conduction properties of the crystal. The conduction electrons, whose number is enormous, interact either with the ions and between each other, in a way which depend from the relative positions; the final effect of all such interactions is of course terribly complicated and the macroscopic properties of the crystal, like its conductivity or specific heat, depend from it.

We recall that, according to quantum mechanics, particles are described by a complex, square integrable *wave function* $\Psi(\vec{x}_1, ..., \vec{x}_N)$ with $|\Psi|^2$ representing the probability density of finding $N$ particles at positions $\vec{x}_1, \vec{x}_2, ..., \vec{x}_N$, which we will assume in a $d$-dimensional square box with side $L$ and periodic boundary conditions. The time evolution of the wavefunction is driven by the Schroedinger equation

$$-i\frac{\partial}{\partial t}\psi = H_N \psi \qquad (1.30)$$

where $H_N$ is the *Hamiltonian operator*, and the choice of such operator is determined by the physical system one wants to describe.

In order to understand the properties of the conduction electrons in a metal one should determine an an antisymmetric wave function verifing the Schroedinger equation (1.30) with an hamiltonian of the form

$$H_N = \sum_{i=1}^{N}[-\frac{\partial^2_{\vec{x}_i}}{2m} + u\,c(\vec{x}_i)] + \sum_{i<j}\lambda\,v(\vec{x}_i - \vec{x}_j) \qquad (1.31)$$

in which the first term represents the non relativistic kinetic energy of the electrons ($m$ is the mass), $u\,c(\vec{x})$ is a periodic potential due to the ions in the lattice ($c(\vec{x}) = c(\vec{x} + \vec{R})$ with $\vec{R} = (n_1 a_1, .., n_d a_d)$, $n_i \in \mathbb{Z}$) and $\lambda\,v(\vec{x} - \vec{y})$ is a two body interaction potential, which is described by short range potential to take into account, phenomenologically, the electrostatic screening. Finally $\lambda$ and $u$ are couplings which measure the "strength" of the corresponding interaction. Much more complicated and "realistic" Hamiltonians could be considered; for instance one can add an interaction with a stochastic field to take into account impurities in the lattice, or with a boson field to take into account the dynamics of the ions, and so on. Note also that one can study not only three dimensional Fermi systems ($d = 3$), but also $d = 2$ or $d = 1$ systems; they can describe the conduction electrons of crystals so anisotropic to be considered as bidimensional or one dimensional system.

The *Second quantization* formalism introduces some technical simplification in the analysis of interacting systems. One introduces the Hilbert space of states of a system of $N > 1$ fermions as the space $\mathcal{H}_N$ of all the complex square integrable antisymmetric functions $\Psi(\vec{x}_1, ..., \vec{x}_N)$. Let be $\{\phi_{\vec{k}}(\vec{x})\}_{\vec{k} \in R^d}$ be a *basis* for $\mathcal{H}_1$ (the one particle Hilbert space of all the complex square integrable functions $\Psi(\vec{x}_1)$), where $\vec{k}$ is an index called *quantum number*.

Usually the set of $\phi_{\vec{k}}(\vec{x})$ is chosen as the *eigenfunctions* of the single particle Hamiltonian

$$-\frac{\partial^2_{\vec{x}}}{2m} + u\,c(\vec{x}) \qquad (1.32)$$

with eigenvalue $\varepsilon(\vec{k})$. If $u = 0$ then $\phi_{\vec{k}}(\vec{x}) = \frac{1}{L^{d/2}}e^{i\vec{k}\vec{x}}$ with $\vec{k}$ representing the *momentum*, and $\varepsilon(\vec{k}) = \frac{|\vec{k}|^2}{2m}$; due to periodic boundary conditions $\vec{k}$ has the form $\vec{k} = \frac{2\pi}{L}\vec{n}$, $\vec{n} = n_1, ..n_d$ with $n_i$ integer and $-[L/2] \leq n_i \leq [(L-1)/2]$.

If we call $|\vec{k}_1, ..\vec{k}_N\rangle$ the normalized antisymmetrization of

$$\phi_{\vec{k}_1}(\vec{x}_1)\phi_{\vec{k}_2}(\vec{x}_2)...\phi_{\vec{k}_N}(\vec{x}_N) \qquad (1.33)$$

we have that the set of all possible $|\vec{k}_1, ..\vec{k}_N\rangle$ is a *basis* for $\mathcal{H}_N$; $|\vec{k}_1, ..\vec{k}_N\rangle$ describes a state in which the $N$ fermions have quantum numbers $\vec{k}_1, .., \vec{k}_N$.

One can introduce the *creation or annihilation operators* $a^+_{\vec{k}}, a^-_{\vec{k}}$: they are *anticommuting operators*

$$\{a^+_{\vec{k}}, a^-_{\vec{k}'}\} = \delta_{\vec{k},\vec{k}'} \qquad \{a^+_{\vec{k}}, a^+_{\vec{k}'}\} = \{a^-_{\vec{k}}, a^-_{\vec{k}'}\} = 0 \qquad (1.34)$$

such that $a^+_{\vec{k}}|\vec{k}_1,..\vec{k}_N\rangle = |\vec{k}, \vec{k}_1,..\vec{k}_N\rangle$ if $\vec{k} \neq \vec{k}_i, i = 1,...N$ and 0 otherwise; $a^-_{\vec{k}}$ is the *adjoint* of $a^+_{\vec{k}}$. The state $|0\rangle$ such that $a^-_{\vec{k}}|0\rangle = 0$ for all $\vec{k}$ is called *vacuum state* and it represents a state with zero particles. The *Fock space* is defined as the direct sum of the Hilbert spaces with any number of particles, and all the elements of the Fock space can be generated by superposing linearly products of creation operators acting over the vacuum state.

In terms of $a^+_{\vec{x},\sigma} = L^{-d/2} \sum_{\mathbf{k}} \phi_{\vec{k}}(\vec{x}) a^+_{\vec{k},\sigma}$ and of its adjoint $a^-_{\vec{x},\sigma}$, the Hamiltonian can be written as an operator on the Fock space

$$H = \sum_\sigma [\int_V d\vec{x} a^+_{\vec{x},\sigma} \frac{-\partial^2_{\vec{x}}}{2m} a^-_{\vec{x},\sigma} + \qquad (1.35)$$

$$+ u \int_V d\vec{x} c(\vec{x}) a^+_{\sigma,\vec{x}} a^-_{\vec{x},\sigma}] + \sum_{\sigma,\sigma'} \lambda \int_V d\vec{x} \int_V d\vec{y} v(\vec{x} - \vec{y}) a^+_{\vec{x},\sigma} a^-_{\vec{x},\sigma} a^+_{\vec{y},\sigma'} a^-_{\vec{y},\sigma'}$$

In many cases, one gets a good description of the effects of the crystal lattice on the conduction electron considering the so called tight-binding approximation, in which electrons occupy sites of a lattice and can can jump from one lattice site to another one. The Hamiltonian, (called *Hubbard hamiltonian* for local repulsive interactions), is given by

$$H = \sum_{\vec{x} \in \Lambda} \sum_{\sigma=\uparrow\downarrow} a^+_{\vec{x},\sigma}(-\frac{\Delta}{2} - \mu) a^-_{\vec{x},\sigma} + \lambda \sum_{\vec{x},\vec{y}\in\Lambda} v(\vec{x}-\vec{y}) a^+_{\vec{x},\sigma} a^-_{\vec{x},\sigma} a^+_{\vec{y},\sigma'} a^-_{\vec{y},\sigma'} \qquad (1.36)$$

where $\Lambda \subset \mathbb{Z}^d$ is a square sublattice of $\mathbb{Z}^d$ with side $L$ and $\Delta$ is the discrete Laplacean.

As in classical statistical mechanics, one introduces the grand canonical partition function $Z = \text{Tr} e^{-\beta(H-\mu N)}$, where $\mu$ is the chemical potential, $N$ is the particle number operator $N = \sum_\sigma \int d\vec{x} a^+_{\vec{x},\sigma} a^-_{\vec{x},\sigma}$ and Tr is the trace operation over the Fock space. Many macroscopic observables can be expressed in terms of $Z$, like the specific heat. The thermodynamical *average* of an observable $O$, typically expressed by a monomial in the $a^\pm$ operators, is given by

$$<O> = Z^{-1} \text{Tr}[e^{-\beta(H-\mu N)} O] \qquad (1.37)$$

The Schwinger functions are defined as, if $\mathbf{x} = (\vec{x}, x_0)$ and $x_{0,1} \geq x_{0,2} \geq \ldots x_{0,s}$, $s$ even

$$S_s(\mathbf{x}_1, \mathbf{x}_2, \ldots, \mathbf{x}_s) = \qquad (1.38)$$
$$\frac{\text{Tr} e^{-(\beta - x_{0,1})(H - \mu N)} a_{\vec{x}_1}^{\varepsilon_1} e^{-(x_{0,1} - x_{0,2})(H - \mu N)} a_{\vec{x}_2}^{\varepsilon_2} \ldots e^{-x_{0,s}(H - \mu N)}}{\text{Tr} e^{-\beta(H - \mu N)}}$$

with $\varepsilon_i = \pm$ and $-\beta/2 \leq t_i \leq \beta/2$.

The physical observables of interest at temperature $\beta^{-1}$ can be obtained from the Schwinger functions. For instance the averaged number of electrons with momentum $\vec{k}$ is given, in the infinite volume limit, by

$$< a_{\vec{k},\sigma}^+ a_{\vec{k},\sigma}^- > = \int d\vec{x} e^{i\vec{k}\vec{x}} S(\vec{x}, 0^+; 0, 0) \qquad (1.39)$$

Important physical quantities which can be obtained from the higher order Schwinger functions are the *response functions*, which measure the density of the system to a perturbation; for intance the density-density response function is given by, $x_0 > y_0$

$$\Omega(\mathbf{x}, \mathbf{y}) = \qquad (1.40)$$
$$\sum_{\sigma,\sigma'} \frac{\text{Tr} e^{-(\beta - x_0)(H - \mu N)} a_{\vec{x},\sigma}^+ a_{\vec{x},\sigma}^- e^{-(x_0 - y_0)(H - \mu N)} a_{\vec{y},\sigma'}^+ a_{\vec{y},\sigma'}^- e^{-y_0(H - \mu N)}}{\text{Tr} e^{-\beta(H - \mu N)}}$$

### 1.3.2 The free Fermi gas

Computing the physical observables corresponding to the complete Hamiltonian (1.36) is a very difficult task. If there is no interaction $\lambda = 0$ one obtains a model called *free Fermi gas* which can be analytically investigated and which is very succesfull in understanding the properties of the conduction electrons in metals.

It holds that $|\vec{k}_1, \sigma_1, ..\vec{k}_N \sigma_N\rangle$ are eigenfunctions of $H$ with eigenvalue $\sum_{\vec{k},\sigma} \varepsilon(\vec{k}) n_{\vec{k},\sigma}$, where $n_{\vec{k},\sigma} = 0, 1$ is the eigenvalue of $a_{\vec{k},\sigma}^+ a_{\vec{k},\sigma}^-$; $n_{\vec{k},\sigma} = 1$ if in the state there is a fermion with momentum $\mathbf{k}$ and spin $\sigma$ and it is zero otherwise.

The eigenfunction $|\Omega\rangle$ of $H$ with lowest energy is called *ground state*, and it determines the low temperatures properties of the system. In order to find the ground state $|\Omega\rangle$, one has to minimize $\sum_{\vec{k},\sigma} \varepsilon(\vec{k}) n_{\vec{k},\sigma}$ with the constraint that $n_{\vec{k},\sigma}$ can take only the values 0 or 1 and $\sum_{\vec{k},\sigma} n_{\vec{k},\sigma} = N$; if there are many solutions to this problem one says that the ground state is *degenerate*. In the case $u = 0$, in the limit $L \to \infty$ the ground state is such that $n_{\vec{k},\sigma} = 1$ if $\vec{k}$ is in a *sphere* of radius $k_F = (3\pi^2 \rho)^{\frac{1}{3}}$, if $\rho$ is the density.

The boundary of the sphere with radius $k_F$ in the space of momenta is called *Fermi surface* and it is a key notion in the theory of Fermi systems; if $d = 2$ it is replaced by a circle and in $d = 1$ by two points.

If $u \neq 0$ the Fermi surface is still given by the set $\vec{k} : \varepsilon(\vec{k}) = \varepsilon_F$ with $\varepsilon_F$ determined by the condition $\sum_{\mathbf{k}:\varepsilon(\vec{k}) \leq \varepsilon_F} 1 = N$. However in this case the Fermi surface is not anymore a sphere in $d = 3$, but it is in general polyhedron of a very complex shape.

The averaged number of electrons with momentum $\vec{k}$ is given, in the infinite volume limit, by

$$< a^+_{\vec{k},\sigma} a^-_{\vec{k},\sigma} > = (1 + e^{\beta(\varepsilon(\vec{k})-\mu)})^{-1} \tag{1.41}$$

at zero temperature it reduces to $\vartheta(\varepsilon(\vec{k}) - \varepsilon_F)$ ($\mu = \varepsilon_F$ at $T = 0$), *i.e.* it has a *discontinuity* at the Fermi surface, while at high temperature it is very close to the Maxwell distribution $\simeq e^{-\beta(\varepsilon(\vec{k})-\mu)}$.

The two point Schwinger function $g(\mathbf{x}_1 - \mathbf{x}_2)$ is given by, using that $e^{H x_0} \psi^\pm_{\vec{k}} e^{-H x_0} = e^{\pm(\varepsilon(\vec{k})-\mu)x_0} \psi^\pm_{\vec{k}}$ and calling $t = x_{0,1} - x_{0,2}$

$$g(\vec{k},t) = e^{(\varepsilon(\vec{k})-\mu)t} \frac{\text{Tr} e^{-\beta(H-\mu N)} T(a^-_{\vec{k}} a^+_{\vec{k}})}{\text{Tr} e^{-\beta(H-\mu N)}}$$

$$= \frac{e^{-(\varepsilon(\vec{k})-\mu)t}}{1+e^{\beta(\varepsilon(\vec{k})-\mu)}} [\vartheta(t) - e^{-\beta(\varepsilon(\vec{k})-\mu)} \vartheta(-t)] \tag{1.42}$$

Note that $g(\vec{k},t) = -g(\vec{k},t+\beta)$; we can then write $g(\vec{k},t)$ in Fourier series in the following form

$$g(\vec{k},t) = \frac{2\pi}{\beta} \sum_{k_0 = 2\pi(n_0+1/2)\beta^{-1}} e^{-ik_0 t} \widehat{g}(\mathbf{k}) \tag{1.43}$$

and

$$\widehat{g}(\mathbf{k}) = \int_{-\frac{\beta}{2}}^{\frac{\beta}{2}} dt\, e^{ik_0 t} g(t,\vec{k}) = \frac{1}{-ik_0 + \varepsilon(\vec{k}) - \mu} \tag{1.44}$$

At finite temperature $\beta < \infty$ is not singular; only in the limit $\beta \to \infty$ it can be singular when $k_0 = 0$ and at the Fermi surface $\varepsilon(p_F(\vartheta)) = \mu$. Assume that the Fermi surface is sufficiently regular, and that it is possible to parametrize the Fermi surface $\varepsilon(\vec{k}) = \mu$ as $\vec{p}_F(\vec{\vartheta})$, where $\vec{\vartheta}$ is a angle (in $d = 2$) or a couple of angles (in $d = 3$); close to the singularity $\widehat{g}(\mathbf{k})$ has the form

$$\widehat{g}(\mathbf{k}) = \frac{1}{-ik_0 + \vec{v}_F^{(0)}(\vec{\vartheta}) \cdot (\vec{k} - \vec{p}_F^{(0)}(\vec{\vartheta})) + R(\vec{k})} \tag{1.45}$$

where $\vec{v}_F^{(0)}(\vec{\vartheta}) = (\partial \varepsilon_0 / \partial \vec{k})|_{\vec{k}=\vec{p}_F(\vartheta)}$ is the *free Fermi velocity*. Moreover, near the Fermi surface, $|R(\vec{k})| \leq C|\vec{k} - \vec{p}_F^{(0)}(\vartheta)|^2$, for some positive constant $C$.

In presence of the interaction the Schwinger functions cannot be exactly computed as in the free case. The Schwinger functions (1.38) can be written as a fermionic functional integral; in the case of the Hubbard model, for instance, the Schwinger functions can be written as

$$S(\mathbf{x}_1, \mathbf{x}_2, ..., \mathbf{x}_N) = \frac{\partial^N}{\partial \phi_{\mathbf{x}_1}^{\varepsilon_i} ... \partial \phi_{\mathbf{x}_N}^{\varepsilon_i}} \mathcal{W}(\varphi)|_{\varphi=0} \qquad (1.46)$$

where, if $\psi_{\mathbf{x}}^{\pm}$ are *Grassmann variables*, $\mathbf{x} = (x_0, \vec{x})$ and $\vec{x}$ are points on a square lattice $\Lambda$ and, if $v(\mathbf{x} - \mathbf{y}) = v(\vec{x} - \vec{y})\delta(x_0 - y_0)$

$$e^{\mathcal{W}(\phi)} = \int P(d\psi) \qquad (1.47)$$

$$e^{-\lambda \sum_{\sigma, \sigma'} \int_{-\frac{\beta}{2}}^{\frac{\beta}{2}} dx_0 dy_0 \sum_{\vec{x}, \vec{y} \in \Lambda} v(\mathbf{x}-\mathbf{y}) \psi_{\mathbf{x},\sigma}^+ \psi_{\mathbf{x},\sigma}^- \psi_{\mathbf{y},\sigma'}^+ \psi_{\mathbf{y},\sigma'}^- + \sum_{\sigma} \int d\vec{x} \phi_{\mathbf{x},\sigma}^+ \psi_{\mathbf{x},\sigma}^- + \phi_{\mathbf{x},\sigma}^- \psi_{\mathbf{x},\sigma}^+}$$

and, if $\mathcal{D}_L = \{\vec{k} = \frac{2\pi}{L}(n_1, n_2, ..., n_d) : -[L/2] \leq n_1, n_2, ..., n_d \leq [(L-1)/2]\}$ and $\psi_{\mathbf{x}}^{\pm} = \frac{1}{\beta L^d} \sum_{\mathbf{k} \in \mathcal{D}_{\beta, L}} e^{\pm i \mathbf{k} \mathbf{x}} \psi_{\mathbf{k},\sigma}^{\pm}$, the "fermionic measure" is given by

$$P(d\psi) = \mathcal{D}(\psi) \exp[\frac{(2\pi)^{d+1}}{(L^d \beta)} \sum_{\mathbf{k}} \psi_{\mathbf{k},\sigma}^+(-ik_0 + 2 - \sum_{i=1}^{d} \cos k_i - \mu) \psi_{\mathbf{k},\sigma}^-] \quad (1.48)$$

The above functional integrals are very similar to the ones of QFT seens before. The lattice is a natural ultraviolet cut-off, and the temperature plays the role of an infrared cut-off; no divergences are then present at finite volume and non zero temperature, and indeed at high temperature the functional integrals can be expressed by convergent series; the interacting Schwinger function has more or less the same properties of the free one. Things are however much different at very low or vanishing temperatures; in such case an explicit computation shows the Feynman integrals can be diverging or so large to make the power series not converging. This is a signal that the interacting Schwinger and the free one are not perturbatively close even if the coupling $\lambda$ is small.

### 1.3.3 Fermi liquids

If there is no interaction between the particles, the properties on the $N$-particle system can be deduced from the single particle ones. When an interaction is switched on, there is no reason a priori to expect this: the interaction between an (essentially infinite) number of particles can induce the emergence of a radically new behavior with respect to the free case. Nevertheless, it is an experimental matter of fact that the properties of the conduction electrons in a number of metals are well described by the free Fermi gas model, up to a redefinition (or *renormalization*) of the parameters. In other words, a description in terms of non interacting electrons is still valid for many interacting systems, modulo a renormalization of the the parameters like the mass or the wave function renormalization. A fermionic system with such property is called a *Landau Fermi liquid*, from Landau who introduced in the 50's such a notion.

For a number of years, physicists were very happy of the fortunate circonstance that, at least for temperatures not too low, metals were well described in terms of Fermi liquids. At very low temperatures the Fermi liquid description breaks down, and some sort of phase transition toward more complex states appear: before it, in their normal phase, the Landau Fermi liquid description was quite successfull in many metals. However in more recent years, the discovery of high $T_c$ superconductivity focus the attemption on a number of material which, in their normal phase, apparently do not behave as Fermi liquids. This leads people, see for instance Ref. [17], to reconsider the notion of Fermi liquid and to try to understand, starting from functional integrals of the form (1.47), how such behavior emerges (or does not emerge) from a microscopic model.

One calls *Landau Fermi liquid* a system whose interacting Schwinger functions are similar to the free one, up to a renormalization of the physical parameters. In a Fermi liquid the 2-point Schwinger function has the form

$$\widehat{S}(\mathbf{k}) = \frac{1}{Z(\vec{\vartheta})} \frac{1}{-ik_0 + \vec{v}_F(\vec{\vartheta}) \cdot (\vec{k} - \vec{p}_F(\vec{\vartheta})) + R(\mathbf{k})} \tag{1.49}$$

where $Z(\vec{\vartheta}) - 1, \vec{v}_F(\vec{\vartheta}) - \vec{v}_F^{(0)}(\vec{\vartheta}), \vec{p}_F(\vec{\vartheta}) - \vec{p}_F^{(0)}(\vec{\vartheta})$ essentially independent from the temperature, and in addition

$$|R(\mathbf{k})| \leq C\big[|\vec{k} - \vec{p}_F(\vec{\vartheta})|^2 + k_0^2 + |\vec{k} - \vec{p}_F(\vartheta)||k_0|\big] \tag{1.50}$$

According to the above definition, the Schwinger functions of the interacting system are very similar to the Schwinger function of a free Fermi gas (1.45), and as a consequence the physical properties of the interacting system (which can be deduced from the Schwinger functions) are qualitatively very similar to the ones of the free Fermi gas, up to a renormalization of the parameters.

Not all systems are Fermi liquids; surely $d = 1$ interacting fermionic systems are not Fermi liquids, and the same is true at higher dimensions when the Fermi surface has some cusps or flat pieces, like in the Hubbard model in the half-filled band case. Moreover at very low temperature Fermi liquid behavior is generically absent as a consequence of phase transitions, for instance toward a superconducting state.

In general, to prove that a certain model has a Fermi liquid behavior is not an easy task; starting from the functional integral (1.47), one has to prove the convervence of very complicate expansions in Feynman graphs. The issue of convergence is not just a mathematical curiosity; indeed in the debate on high $T_c$ superconductivity it has been proposed that the apparent discrepancy between theoretical prediction and experimental data was due to the fact computation at lowest order give wrong results for the lack of convergence of the series.

In the third part of this book, we will see how the renormalization mehods allow to give a proof of Fermi liquid behavior in the 2d interacting fermionc systems, up to exponentially small temperatures $T \leq O(e^{\frac{-k}{\lambda}})$ and for free Fermi surfaces verifying suitable convexity properties (including for instance the Hubbard model not at half filling).

### 1.3.4 Luttinger liquids and BCS superconductors

What happens at lower temperatures, that is from exponentially small temperatures up to $T = 0$? The answer depends critically on the dimension.

In one dimension, systems have generically a non Fermi liquid behavior, that is their Schwinger function cannot be written as in (2.19) even above exponentially small temperatures. In order to describe $1d$ systems, the notion of *Luttinger liquids* has been introduced; such systems have with the Luttinger model the same relation that the Landau Fermi liquids have with the free Fermi gas.

The Luttinger model has unique peculiarity (in many body theory) to be interacting and exactly solvable in a strong sense, that is all its Schwinger function can be explicitely computed, see ref.[18].

The model describes a system of two kinds of interacting fermions in $d = 1$ described by a field $\psi^{\pm}_{\vec{x},\omega}$, $\omega = \pm$, with hamiltonian

$$H = \int d\vec{x} \sum_{\omega=\pm} a^+_{\vec{x},\omega}(i\omega\partial_x - p_F)a^-_{\vec{x},\omega} + \lambda \int d\vec{x}d\vec{y}v(\vec{x}-\vec{y})\rho_+(\vec{x})\rho_-(\vec{y}) \quad (1.51)$$

with $\rho_\omega(\vec{x}) = a^+_{\vec{x},\omega}a^-_{\vec{x},\omega}$ and $v(\vec{x}-\vec{y})$ is a short range interaction. The single particle energy $\varepsilon(\vec{k}) = \omega\vec{k}$ is not bounded from below, and, as in Dirac theory, one has to fill all the states with negative energy; this means that the operators $H$ and $\rho_\omega$ can be regarded as operators acting on the Hilbert space $\mathcal{H}$ constructed by completing the space given by the span of the vectors obtained applying finitely many creation or annihilaton operators over the state $|0>$ defined as $|0>= \prod_{\vec{k}\leq 0} a^+_{\vec{k},+} a^+_{\vec{k},-}|vac>$.

In the Hilbert space, the operators $\rho_\omega(\vec{p})$ verify a bosonic comutation relation

$$[\rho_\omega(\varepsilon\vec{p}), \rho_\omega(-\varepsilon\vec{p})] = -\varepsilon\omega\vec{p}L/2\pi \qquad (1.52)$$

and the hamiltonian can be diagonalized in terms of bosonic operators. In other words the Luttinger model, describing interacting fermions, can be mapped in a system of non-interacting bosons; this property is called *bosonization*.

The Schwinger functions can be then exactly computed; one sees that the 2-point functions behaves for large distances at $T=0$ as

$$S(\mathbf{x}) \simeq_{\mathbf{x}\to\infty} \frac{1}{i\omega x + x_0} \frac{1}{(x^2+x_0^2)^\eta} \qquad (1.53)$$

with $\eta(\lambda) = a\lambda^2 + O(\lambda^3)$ is a *critical index*; it is easy to verify that the Fourier trasform diverges at the Fermi points as $O((|k_0|+||\mathbf{k}|-p_F|)^{-1+\eta})$, that is, contrary to what happens in Landau Fermi liquids, the interaction changes qualitatively the singularity; it is still a power law but a different $\lambda$-dependent exponent.

Proceeding as for Landau Fermi liquids, one can introduce then the notion of Luttinger liquid for systems which behave qualitativey as the Luttinger model. The we say that a system is a *Luttinger liquid* is the Schwinger function has the form

$$\widehat{S}(\mathbf{k}) = \frac{[k_0^2 + v_F^2(\lambda)(|\vec{k}|-p_F(\lambda))^2]^{2\eta}}{-ik_0 + v_F(\lambda)[|\vec{k}|-p_F(\lambda)]}[1+A_\lambda(\mathbf{k})] \qquad (1.54)$$

where $p_F(\lambda) = k_F + O(\lambda)$ and $A_\lambda(\mathbf{k})$. As a consequence the physical properties are different with respect to a Fermi liquid; for instance the averaged number of electrons with momentum $\vec{k}$ is continuous at $\beta = \infty$

$$<a^+_{\vec{k},\omega} a^-_{\vec{k},\omega}> \simeq const + O(||\vec{k}|-p_F|^{2\eta}) \qquad (1.55)$$

In general bosonization requires linear dispersion relation and a Dirac sea of fermions with negative energy, all features making the Luttinger model a quite unrealistic description for the conduction electrons in metals. The only way to establish Luttinger liquid behavior in more realistic models like the Hubbard model is to analyze the fermionic function integrals (1.48).

In the third part of this book, will show that the methods of non-perturbative renormalization allow an essential complete understaing for Luttinger liquids in $d = 1$, even in realistic models, like the repulsive Hubbard model; the attractive case,in which no Luttinger liquid behavior is present, is much less understood.

The behavior of system in $d = 2, 3$ is more difficult to analyze. If the temperature is low enough, it is expected that Fermi liquid behavior breaks down, as a consequence of quantum instabilities present in the systems. The most famous of such instabilities is given by the phenomenon of superconductivity. According to the theory of Baardeen, Cooper and Schrieffer (BCS theory) the interaction between fermions leads to the formation of a gap in the energy spectrum, below the critical temperature; it is found, under certain approximations, that for $T$ small enough

$$\lim_{L\to\infty} \widehat{S}(\mathbf{k}) = \frac{-ik_0 - \varepsilon(\vec{k}) + \mu}{k_0^2 + (\varepsilon(\vec{k}) - \mu)^2 + \Delta_\lambda^2} \qquad (1.56)$$

where $\Delta$ is exponentially small in $\lambda$. The physical properties predicted by (1.56) are completely different with respect to the free case: the occupation number is continuous, there is an energy gap in the spectrum, the specific heat is $O(e^{-\Delta_\lambda T})$ and the phenomenon of superconductivity appears. At the moment, the theory of superconductivity, and the derivation of (1.56), are based on a *mean field* approximation, and a mathematical derivation is still lacking, despite it is reasonable to hope that the renormalization methods will allow to understand such important phenomena in the near future.

## Chapter 2

# Fermionic Functional Integrals

## 2.1 Grassmann variables

In the previous chapter we have seen several physical quantities which are expressed in terms of fermionic functional integrals; it is time to define them and to see how they are computed.

A finite dimensional *Grassman algebra*, see also Refs.[12],[19], is a set of *Grassman variables* $\psi_\alpha$, with $\alpha$ an index belonging to some finite set $A = (1,..,2n)$ which are anticommuting, that is

$$\{\psi_\alpha, \psi_{\alpha'}\} = \psi_\alpha \psi_{\alpha'} + \psi_{\alpha'} \psi_\alpha = 0 \tag{2.1}$$

and commuting with numbers.

A *Grassmann integral* $\int d\psi_\alpha$ is a linear operation defined as

$$\int d\psi_\alpha = 0 \qquad \int d\psi_\alpha \psi_\alpha = 1 \tag{2.2}$$

The Grassmann integral of any analytic function can be obtained by linearity; for instance

$$\int d\psi_\alpha e^{\psi_\alpha} = \int d\psi_\alpha (1 + \psi_\alpha) = 1 \tag{2.3}$$

$d\psi_\alpha$ is also a Grassmann variable, anticommuting with $\psi_\alpha$, so that $\int \psi_\alpha d\psi_\alpha = -1$. A slight generalization of (2.3) is

$$\int \prod_\alpha d\psi_\alpha e^{\frac{1}{2}\sum_{\alpha,\beta} \psi_\alpha A_{\alpha,\beta} \psi_\beta} = \text{Pf}A \tag{2.4}$$

where $A$ is an even antisymmetric $2n$-matrix and Pf$A$ denotes the *Pfaffian*. It holds

$$\int d\psi_{2n}...d\psi_1 \exp \frac{1}{2} \sum_{\alpha,\beta} \psi_\alpha A_{\alpha,\beta} \psi_\beta = \int d\psi_{2n}...d\psi_1 \prod_{\alpha<\beta}(1 + A_{\alpha,\beta}\psi_\alpha\psi_\beta) =$$

$$\frac{1}{2^n n!} \sum_p (-1)^p A_{p_1,p_2} A_{p_3,p_4}...A_{p_{2n-1},p_{2n}} \equiv \text{Pf}A \tag{2.5}$$

where the sum is over all the permutations.

Suppose that we can split the variables $\psi$ in two sets $\psi_1,...,\psi_{2n} = \psi_1^+,...,\psi_n^+,\psi_1^-,...,\psi_n^-$ and that $A$ has the form

$$A = \begin{pmatrix} 0 & -K \\ K^t & 0 \end{pmatrix} \qquad (2.6)$$

then

$$\text{Pf} A = \int d\psi_1^+...d\psi_n^+ \psi_1^-...d\psi_n^- e^{\frac{1}{2}\sum_{i,j} \psi_\alpha^+ A_{i,j} \psi_j^-} = \det K \qquad (2.7)$$

Moreover by defining

$$\langle O \rangle = \frac{\int d\psi_1...d\psi_{2n} e^{\frac{1}{2}\sum_{\alpha,\beta} \psi_\alpha A_{\alpha,\beta} \psi_\beta} e^{\sum_\alpha \psi_\alpha h_\alpha}}{\int d\psi_1...d\psi_{2n} e^{\frac{1}{2}\sum_{\alpha,\beta} \psi_\alpha A_{\alpha,\beta} \psi_\beta}} \qquad (2.8)$$

we get

$$\langle e^{\sum_\alpha \psi_\alpha \eta_\alpha} \rangle = e^{\frac{1}{2}\sum_{\alpha,\beta} h_\alpha (A^{-1})_{\alpha,\beta} h_\beta} \qquad (2.9)$$

and differentiating

$$\langle h_1 h_2...h_{2p} \rangle = \text{Pf}(A^{-1})_{1,...,2p} \qquad (2.10)$$

where $(A^{-1})_{1,...,2p}$ means the restriction of $A^{-1}$ to the subspace $1,..,2p$.

If $A$ is of the form (2.6) and

$$h_1, h_2, ..., h_{2p} = \eta_1^+, \eta_2^+ ..., \eta_p^+, \eta_1^-, \eta_2^-, ..., \eta_p^- \qquad (2.11)$$

we have

$$\langle \eta_1^+ \eta_2^+ ... \eta_p^+ \eta_1^- \eta_2^- ... \eta_p^- \rangle = \text{Det}(K^{-1})_{1,...,p} \qquad (2.12)$$

In the very simple case in which $n = 1$

$$\int d\psi_\alpha^+ d\psi_\alpha^- e^{-a\psi_\alpha^+ \psi_\alpha^-} = -a \int d\psi_\alpha^+ d\psi_\alpha^- \psi_\alpha^+ \psi_\alpha^- = a \qquad (2.13)$$

and

$$\frac{\int d\psi_\alpha^+ d\psi_\alpha^- \psi_\alpha^+ \psi_\alpha^- e^{-a\psi_\alpha^+ \psi_\alpha^-}}{\int d\psi_\alpha^+ d\psi_\alpha^- e^{-a\psi_\alpha^+ \psi_\alpha^-}} = \frac{1}{a} \qquad (2.14)$$

The above formula is strongly reminiscent of the formulas for gaussian integrals

$$\frac{\int_{-\infty}^{\infty} dx\, x^2 e^{-ax^2}}{\int_{-\infty}^{\infty} dx\, e^{-ax^2}} = \frac{1}{2a} \qquad (2.15)$$

and this analogy motivates the name "integrals" given to the above operations.

The well known relation $(\text{Pf}A)^2 = \det A$ can be quite easily deduced by the above Grassmann integrals; it can be written as

$$\int \prod_\alpha d\psi_\alpha^- d\psi_\alpha^+ e^{\psi_\alpha^- A_{\alpha,\beta} \psi_\beta^+} = \int \prod_\alpha d\psi_\alpha^{(1)} e^{\frac{1}{2}\psi_\alpha^{(1)} A_{\alpha,\beta} \psi_\beta^{(1)}} \int \prod_\alpha d\psi_\alpha^{(2)} e^{\frac{1}{2}\psi_\alpha^{(2)} A_{\alpha,\beta} \psi_\beta^{(2)}} \tag{2.16}$$

which can be proved by the change of variables

$$\psi_\alpha^+ = \frac{1}{\sqrt{2}}(\psi_\alpha^{(1)} + i\psi_\alpha^{(2)}) \quad \psi_\alpha^- = \frac{1}{\sqrt{2}}(\psi_\alpha^{(1)} - i\psi_\alpha^{(2)}) \tag{2.17}$$

in $\int d\psi_\alpha^- d\psi_\alpha^+ e^{\psi_\alpha^- A_{\alpha,\beta} \psi_\beta^+}$. Then $d\psi_\alpha^- d\psi_\alpha^+ = i d\psi_\alpha^{(1)} d\psi_\alpha^{(2)}$ and

$$\psi_\alpha A_{\alpha,\beta} \psi_\beta^+ = \frac{1}{2}\psi_\alpha^{(1)} A_{\alpha,\beta} \psi_\beta^{(1)} + \frac{1}{2}\psi_\alpha^{(2)} A_{\alpha,\beta} \psi_\beta^{(2)} \tag{2.18}$$

as

$$\psi_\alpha^{(1)} A_{\alpha,\beta} \psi_\beta^{(2)} - \psi_\alpha^{(2)} A_{\alpha,\beta} \psi_\beta^{(1)} = \psi_\alpha^{(1)} A_{\alpha,\beta} \psi_\beta^{(2)} - \psi_\beta^{(1)} A_{\beta,\alpha} \psi_\alpha^{(2)} = 0 \tag{2.19}$$

## 2.2 Grassmann measures

Pursuing further the analogy with Gaussian integrals, we can consider a "fermionic measure". Let be $\mathcal{D}$ the set of $\mathbf{k} = (k_1, .., k_d)$ with $k_i = \frac{2\pi}{L_i}(n_i + \frac{\sigma_i}{2})$, with $n_i$ an integer such that $-L_i/2a_i \leq n_i \leq L_i/2a_i - 1$ while $\sigma_i = 0, 1$ (depending if periodic or antiperiodic conditions are imposed). A fermionic measure $P(d\psi)$ is an expression of the form, if $V = \prod_i L_i$, $\mathcal{N} = \prod_{\mathbf{k}}(V g_{\mathbf{k}})^{-1}$

$$P(d\psi) = \frac{1}{\mathcal{N}} \prod_{\mathbf{k} \in \mathcal{D}} d\widehat{\psi}_{\mathbf{k}}^+ d\widehat{\psi}_{\mathbf{k}}^- e^{-\frac{1}{V}\sum_{\mathbf{k} \in \mathcal{D}} \widehat{\psi}_{\mathbf{k}}^+ g_{\mathbf{k}}^{-1} \widehat{\psi}_{\mathbf{k}}^-} ; \tag{2.20}$$

with $g_{\mathbf{k}}$ $2\pi/a_i$-periodic and finite. It is easy to verify that, by (2.12)

$$\int P(d\psi) = 1 , \quad \int P(d\psi) \widehat{\psi}_{\mathbf{k}}^- \widehat{\psi}_{\mathbf{k}'}^+ = \delta_{\mathbf{k},\mathbf{k}'} g_{\mathbf{k}} . \tag{2.21}$$

Indeed

$$\int P(d\psi) \widehat{\psi}_{\mathbf{k}'}^- \widehat{\psi}_{\mathbf{k}'}^+ = \int \frac{1}{\mathcal{N}} \prod_{\mathbf{k} \in \mathcal{D}} [d\widehat{\psi}_{\mathbf{k}}^+ d\widehat{\psi}_{\mathbf{k}}^- (1 - \frac{1}{V}\widehat{\psi}_{\mathbf{k}}^+ g_{\mathbf{k}}^{-1} \widehat{\psi}_{\mathbf{k}}^-)] \widehat{\psi}_{\mathbf{k}'}^- \widehat{\psi}_{\mathbf{k}'}^+ = g_{\mathbf{k}'} \tag{2.22}$$

In general $P(d\psi)$ will be called a *Gaussian fermionic integration measure* with covariance $g$: for any analytic function $F$ defined on the Grassman algebra the *simple expectation* is defined as

$$\int P(d\psi) F(\psi) = \varepsilon(F) . \tag{2.23}$$

If $F(\psi)$ is a monomial, $\mathcal{E}(F)$ is given by the following formula, called *Wick rule* and following from (2.12):

$$\int P(d\widehat{\psi})\, \widehat{\psi}^-_{\mathbf{k}_1}...\widehat{\psi}^-_{\mathbf{k}_n} \widehat{\psi}^+_{\mathbf{k}'_1},...,\widehat{\psi}^+_{\mathbf{k}'_m} = \delta_{n,m} \sum_\pi (-1)^{p_\pi} \prod_{i=1}^n \delta_{\mathbf{k}_i,\mathbf{k}'_{\pi(j)}} g_{\mathbf{k}_i}, \quad (2.24)$$

where the sum is over all the permutations $\pi = \{\pi(1),\ldots,\pi(n)\}$ of the indices $\{1,\ldots,n\}$ with parity $p_\pi$ with respect to the fundamental permutation.

We will consider also the case of Grassmann measures of the form

$$P(d\psi) = \frac{1}{\mathcal{N}} \prod_{\mathbf{k}\in\mathcal{D}} \prod_\sigma d\widehat{\psi}^+_{\mathbf{k},\sigma} d\widehat{\psi}^-_{\mathbf{k},\sigma}\, e^{-\frac{1}{V}\sum_{\sigma,\sigma'} \sum_{\mathbf{k}\in\mathcal{D}} \widehat{\psi}^+_{\mathbf{k},\sigma} [g_\mathbf{k}^{-1}]_{\sigma,\sigma'} \widehat{\psi}^-_{\mathbf{k},\sigma}} ; \quad (2.25)$$

and

$$\int P(d\psi) = 1, \qquad \int P(d\psi)\, \widehat{\psi}^-_{\mathbf{k},\sigma'} \widehat{\psi}^+_{\mathbf{k}',\sigma'} = \delta_{\mathbf{k},\mathbf{k}'} [g_\mathbf{k}]_{\sigma,\sigma'} . \quad (2.26)$$

and the Wick rule is modified in an obvious way.

If $\mathbf{x} = (x_0,...,x_d)$ and $x_i$ are points of the form $r_i = m_i a_i$, $m_i = -L_i/2a_i,..., L_i/2a_i - 1$, that is lattice points with step $\frac{N_i}{M_i}$, the Grassman fields $\psi^\pm_\mathbf{x}$ defined as

$$\psi^\pm_\mathbf{x} = \frac{1}{V} \sum_{\mathbf{k}\in\mathcal{D}} \widehat{\psi}^\pm_\mathbf{k} e^{\pm i\mathbf{k}\cdot\mathbf{x}}, \quad (2.27)$$

and

$$\int P(d\psi)\psi^-_\mathbf{x}\psi^+_\mathbf{y} = \frac{1}{V} \sum_{\mathbf{k}\in\mathcal{D}} e^{-i\mathbf{k}\cdot(\mathbf{x}-\mathbf{y})} \widehat{g}(\mathbf{k}) \equiv g(\mathbf{x}-\mathbf{y}), \quad (2.28)$$

Of course the properties for the Grassman variables seen above extend trivially to the Grassman fields; for instance the Wick rule for Grassman fields is

$$\int P(d\psi)\, \psi^-_{\mathbf{x}_1}...\psi^-_{\mathbf{x}_n}\psi^+_{\mathbf{y}_1},...,\psi^+_{\mathbf{y}_m} \equiv \mathcal{E}(\psi^-_{\mathbf{x}_1}...\psi^-_{\mathbf{x}_n}\psi^+_{\mathbf{y}_1},...,\psi^+_{\mathbf{y}_m}) =$$

$$\delta_{n,m} \sum_\pi (-1)^{p_\pi} \prod_{i=1}^n g(\mathbf{x}_i - \mathbf{y}_{\pi(j)}) \quad (2.29)$$

or equivalently

$$\mathcal{E}(\psi^-_{\mathbf{x}_1}...\psi^-_{\mathbf{x}_n}\psi^+_{\mathbf{y}_1},...,\psi^+_{\mathbf{y}_n}) = \det G, \quad (2.30)$$

where

$$G = \begin{pmatrix} g(\mathbf{x}_1-\mathbf{y}_1) & g(\mathbf{x}_1-\mathbf{y}_2) & \ldots & g(\mathbf{x}_1-\mathbf{y}_n) \\ g(\mathbf{x}_2-\mathbf{y}_1) & g(\mathbf{x}_2-\mathbf{y}_2) & \ldots & g(\mathbf{x}_2-\mathbf{y}_n) \\ g(\mathbf{x}_3-\mathbf{y}_1) & g(\mathbf{x}_3-\mathbf{y}_2) & \ldots & g(\mathbf{x}_3-\mathbf{y}_n) \\ \ldots & \ldots & \ldots & \\ g(\mathbf{x}_n-\mathbf{y}_1) & g(\mathbf{x}_n-\mathbf{y}_2) & \ldots & g(\mathbf{x}_n-\mathbf{y}_n) \end{pmatrix}$$

It is very useful to represent the Wick rule (2.30) in a graphical way. Given a monomial $\tilde\psi_B = \prod_{\alpha \in B} \psi_{\mathbf{x}(\alpha)}^{\sigma(\alpha)}$, we can represent each each $\psi_{\mathbf{x}(\alpha)}^+$ as a line line exiting from the point $\mathbf{x}(f)$, while each $\psi_{\mathbf{x}(f)}^-$, as a line entering $\mathbf{x}(\alpha)$. Let $\mathcal{G}$ be the set of graphs obtained by contracting such lines in all possible ways so that only lines with opposite $\sigma(\alpha)$ are contracted: given $\alpha, \beta \in B$, denote by $(\alpha\beta)$ the line joining $\mathbf{x}(\alpha)$ and $\mathbf{x}(\beta)$ and by $G$ an element of $\mathcal{G}$, i.e. a graph in $\mathcal{G}$. Then the Wick rule can be stated as

$$\mathcal{E}(\tilde\psi_B) = \sum_{G \in \mathcal{G}} \prod_{(\alpha\beta) \in G} (-1)^{\pi_G} g(\mathbf{x}(\alpha) - \mathbf{x}(\beta)) , \qquad (2.31)$$

where $\pi_G$ is a sign which depends on the graph $G$.

## 2.3 Truncated expectations

The *truncated expectations* are defined as

$$\mathcal{E}^T(X_1, \ldots, X_p; n_1, \ldots, n_p) = \frac{\partial^{n_1 + \ldots + n_p}}{\partial \lambda_1^{n_1} \ldots \partial \lambda_p^{n_p}} \log \int P(d\psi) \, e^{\lambda_1 X_1(\psi) + \ldots + \lambda_p X_p(\psi)} \bigg|_{\lambda=0}, \qquad (2.32)$$

where $X_1, \ldots, X_p$ are monomials of Grassmann fields and $\lambda = \{\lambda_1, \ldots, \lambda_p\}$. It is easy to check that $\mathcal{E}^T$ is a linear operation verifying

$$\mathcal{E}^T(c_1 X_1 + \ldots + c_p X_p; n) = \sum_{n_1 + \ldots + n_p = n} \frac{n!}{n_1! \ldots n_p!} c_1^{n_1} \ldots c_p^{n_p} \mathcal{E}^T(X_1, \ldots, X_p; n_1, \ldots, n_p), \qquad (2.33)$$

and that the following relations are true:

$$\mathcal{E}^T(X; 1) = \mathcal{E}(X) \quad \mathcal{E}^T(X; 0) = 0$$
$$\mathcal{E}^T(X, \ldots, X; n_1, \ldots, n_p) = \mathcal{E}^T(X; n_1 + \ldots + n_p) \qquad (2.34)$$

As a particular case of (2.32) one has

$$\mathcal{E}^T(X; n) = \frac{\partial^n}{\partial \lambda^n} \log \int P(d\psi) e^{\lambda X(\psi)}|_{\lambda=0} \qquad (2.35)$$

so that

$$\log \int P(d\psi) e^{X(\psi)} = \sum_{n=0}^{\infty} \frac{1}{n!} \frac{\partial^n}{\partial \lambda^n} \log \int P(d\psi) \, e^{\lambda X(\psi)}|_{\lambda=0} = \sum_{n=0}^{\infty} \frac{1}{n!} \mathcal{E}^T(X; n) \qquad (2.36)$$

Also the truncated expectation

$$\mathcal{E}^T\left(\tilde\psi_{B_1}, \ldots, \tilde\psi_{B_p}; n_1, \ldots, n_p\right) \qquad (2.37)$$

can be graphically represented. Draw in the plane $n_1$ boxes $H_{11}, \ldots, H_{1n_1}$, such that each of them contains all points representing the indices belonging to $B_1$, $n_2$ boxes $H_{21}, \ldots, H_{2n_2}$, such that each of them contains all points representing the indices belonging to $B_2$, and so on: we call *clusters* such boxes. Then consider all possible graphs $G$ obtained by contracting as before all the lines emerging from the points in such a way that no line is left uncontracted and with the property that if the clusters were considered as points then $G$ would be connected. If we denote the lines as before we have

$$\mathcal{E}^T\left(\tilde{\psi}_{B_1}, \ldots, \tilde{\psi}_{B_p}; n_1, \ldots, n_p\right) = \sum_{G \in \mathcal{G}_0} \prod_{(\alpha\beta) \in \tau} (-1)^{\pi_G} g(\mathbf{x}(\alpha) - \mathbf{x}(\beta)),$$
(2.38)

where $\mathcal{G}_0$ denotes the set of all graphs obtained following the just given prescription; again $\pi_G$ is a sign depending on $G$.

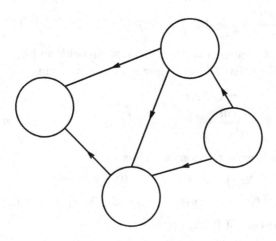

Fig. 2.1  A Feynman diagram $G \in \mathcal{G}_0$ in (2.38)

## 2.4 Properties of Grassmann integrals

The fermionic measures verifies a number of important properties which will be heavily used in the following. They are:

(1) *Addition property.* Given two integrations $P(\mathrm{d}\psi_1)$ and $P(\mathrm{d}\psi_2)$, with covariance $g_1$ and $g_2$ respectively, then, for any function $F$ which can be written as sum over monomials of Grassman variables, *i.e.* $F = F(\psi)$, with $\psi = \psi_1 + \psi_2$, one has

$$\int P(\mathrm{d}\psi_1) \int P(\mathrm{d}\psi_2) F(\psi_1 + \psi_2) = \int P(\mathrm{d}\psi) F(\psi) , \qquad (2.39)$$

where $P(\mathrm{d}\psi)$ has covariance $g \equiv g_1 + g_2$.

It is sufficient to prove it for $F(\psi) = \psi^- \psi^+$, then one uses the anticommutation rules. One has

$$\int P(\mathrm{d}\psi_1) \int P(\mathrm{d}\psi_2) \left(\psi_1^- + \psi_2^-\right) \left(\psi_1^+ + \psi_2^+\right) \qquad (2.40)$$

$$= \int P(\mathrm{d}\psi_1) \psi_1^- \psi_1^+ \int P(\mathrm{d}\psi_2) + \int P(\mathrm{d}\psi_1) \int P(\mathrm{d}\psi_2) \psi_2^- \psi_2^+ = g_1 + g_2$$

where (11.35) has been used.

(2) *Invariance of exponentials.* From the definition of truncated expectations, it follows that, if $\phi$ is an "external field", *i.e.* a not integrated field, then

$$\int P(\mathrm{d}\psi) e^{X(\psi+\phi)} = \exp\left[\sum_{n=0}^{\infty} \frac{1}{n!} \mathcal{E}^T \left(X(\cdot+\phi); n\right)\right] \equiv e^{X'(\phi)} , \qquad (2.41)$$

which says that integrating an exponential one still gets an exponential, whose argument is expressed by the sum of truncated expectations.

(3) *Change of integration.* If $P_g(\mathrm{d}\psi)$ denotes the integration with covariance $g$, then, for any analytic function $F(\psi)$, one has

$$\frac{1}{\mathcal{N}_\nu} \int P_g(\mathrm{d}\psi) e^{-\nu \psi^+ \psi^-} F(\psi) = \int P_{\tilde{g}}(\mathrm{d}\psi) F(\psi) , \qquad \tilde{g}^{-1} = g^{-1} + \nu , \qquad (2.42)$$

where

$$\mathcal{N}_\nu = \frac{g^{-1} + \nu}{g^{-1}} = 1 + g\nu = \int P_g(\mathrm{d}\psi) e^{-\nu \psi^+ \psi^-} \qquad (2.43)$$

The proof is very easy from the definitions.

## 2.5 Gallavotti-Nicoló tree expansion

The functional integrals which appears in the physical applications have generally the form

$$\int P(\mathrm{d}\psi) e^{\mathcal{V}(\psi)} \qquad (2.44)$$

where $\mathcal{V}(\psi)$ is a sum of monomials in $\psi$. Such functional integrals can be represented as a power series using (2.36) and (2.38), that is by writing (2.44) as series of truncated expectations which are computed by the Wick rule. In physical applications, the propagator $g_\mathbf{k}$ is unbounded or it has a weak decay properties for large $\mathbf{k}$, so that typically such expansions are difficult to control.

A different way of computing (2.44) is obtained when the propagator can be written as, with $-j, N$ positive integers

$$\widehat{g}_\mathbf{k} = \sum_{k=j}^{N} \widehat{g}_\mathbf{k}^{(k)} \tag{2.45}$$

with $\widehat{g}_\mathbf{k}^{(k)}$ either bounded and fast decaying.

By defining $\widehat{g}_\mathbf{k}^{(j,N-1)} = \sum_{k=j}^{N-1} \widehat{g}_\mathbf{k}^{(k)}$, we can use the *addition property* to write

$$\int P(d\psi) e^{\mathcal{V}(\psi)} = \int P(d\psi^{[j,N-1]}) \int P(d\psi^{(N)}) e^{\mathcal{V}(\psi^{(j,N-1)}+\psi^{(N)})} \tag{2.46}$$

where $P(d\psi^{(j,N-1)})$ and $P(d\psi^{(N)})$ are the Grassman integrations with propagators $\widehat{g}_\mathbf{k}^{(j,N-1)}$ and $\widehat{g}_\mathbf{k}^{(N)}$ respectively.

By using the *invariance of exponentials* property, the r.h.s. of (2.46) can be written as

$$e^{E^{(N-1)}} \int P(d\psi^{(j,N-1)}) e^{\mathcal{V}^{(N-1)}(\psi^{(j,N-1)})} \tag{2.47}$$

where the *effective potential* $\mathcal{V}^{(N-1)}$ is given by

$$E^{(N-1)} + \mathcal{V}^{(N-1)}(\psi^{(j,N-1)}) = \log \int P(d\psi^{(N)}) e^{\mathcal{V}(\psi^{(j,N-1)}+\psi^{(N)})} =$$

$$\sum_{n=0}^{\infty} \frac{1}{n!} \mathcal{E}_N^T(\mathcal{V}; n) \tag{2.48}$$

which can be graphically represented as in the fig 2.2. The above procedure can be iterated and

$$\int P(d\psi^{(j,N-1)}) e^{\mathcal{V}^{(N-1)}(\psi^{(j,N-1)})} =$$

$$\int P(d\psi^{(j,N-2)}) \int P(d\psi^{(N-1)}) e^{\mathcal{V}^{(N-1)}(\psi^{(\leq N-2)}+\psi^{(N-1)})} =$$

$$e^{E^{(N-2)}} \int P(d\psi^{(j,N-2)}) e^{\mathcal{V}^{(N-2)}(\psi^{(j,N-2)})} \tag{2.49}$$

where

$$E^{(N-2)} + \mathcal{V}^{(N-2)}(\psi^{(j,N-2)}) = \sum_{n=0}^{\infty} \frac{1}{n!} \mathcal{E}_{N-1}^T(\mathcal{V}^{(N-1)}; n) \tag{2.50}$$

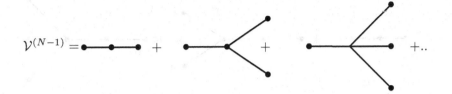

Fig. 2.2  Graphic representation of the expansion (2.47)

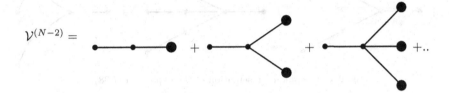

Fig. 2.3  Graphic representation of the expansion (2.50)

which can graphically represented as in fig.2.3.

On the other hand $\mathcal{V}^{(N-2)}$ can written as

$$\mathcal{V}^{(N-2)}(\psi^{[j,N-2]}) = \sum_{n=0}^{\infty} \frac{1}{n!} \mathcal{E}_{N-1}^T \left( \sum_{n'=0}^{\infty} \frac{1}{n'!} (\mathcal{V}; n'); n \right) \qquad (2.51)$$

and more explicitly as

$$\mathcal{V}^{(N-2)}(\psi^{[j,N-2]}) = \mathcal{E}_{N-1}^T(\mathcal{E}_N^T(\mathcal{V})) + \frac{1}{2!}\mathcal{E}_{N-1}^T(\mathcal{E}_N^T(\mathcal{V},\mathcal{V}))$$

$$+ \frac{1}{3!}\mathcal{E}_{N-1}^T(\mathcal{E}_N^T(\mathcal{V},\mathcal{V},\mathcal{V})) + \frac{1}{2!}\mathcal{E}_{N-1}^T(\mathcal{E}_N^T(\mathcal{V})\mathcal{E}_N^T(\mathcal{V}))$$

$$+ \frac{1}{3!}\mathcal{E}_{N-1}^T(\mathcal{E}_N^T(\mathcal{V})\mathcal{E}_N^T(\mathcal{V})\mathcal{E}_N^T(\mathcal{V}))$$

$$+ \frac{1}{3!2!}\mathcal{E}_{N-1}^T(\mathcal{E}_N^T(\mathcal{V})\mathcal{E}_N^T(\mathcal{V}\mathcal{V})) + \cdots \qquad (2.52)$$

which can be of course represented as in Fig.2.4.

Now we can iterate further the above procedure, by integrating all the fields $\psi^{(N)}, \psi^{(N-1)}, \psi^{(N-2)}, \ldots, \psi^{(k+1)}$, so obtaining a contribution to $\mathcal{V}^{(k)}$ recursively defined as, if $k > h$

$$\mathcal{V}^{(h)} = \sum_{n=0}^{\infty} \frac{1}{n!} \mathcal{E}_{h+1}^T(\mathcal{V}^{(h+1)}; n) \qquad (2.53)$$

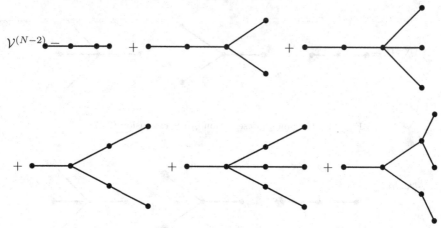

Fig. 2.4 Graphic representation of the expansion (7.38)

where $\mathcal{V}^{(h+1)}$ in turn can be expressed in terms of $\mathcal{V}^{(h+2)}$ as (2.53) with $h$ replaced with $h+1$, and so on until $\mathcal{V}^{(h)}$ is expressed in terms of $\mathcal{V}^{(N)} \equiv \mathcal{V}$.

It follows then that $\mathcal{V}^{(h)}$ can be represented as the sum over *labeled trees*, called Gallavotti-Nicoló trees and introduced in Refs. [8],[20]. In order to define the labeled trees, we introduce first the *unlabeled trees*, defined as the family of all trees which can be constructed by joining a point $r$, the *root*, with an ordered set of $n \geq 1$ points, the *endpoints* of the *unlabeled tree*, so that $r$ is not a branching point. $n$ will be called the *order* of the unlabeled tree and the branching points will be called the *non trivial vertices*. The unlabeled trees are partially ordered from the root to the endpoints in the natural way; we shall use the symbol $<$ to denote the partial order.

Two unlabeled trees are identified if they can be superposed by a suitable continuous deformation, so that the endpoints with the same index coincide.

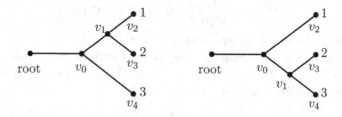

Fig. 2.5 Two unequivalent unlabeled trees of order 3

The number of trees is controlled through the following result.

**Lemma 2.1.** *The number of unlabeled trees with $n$ points is bounded by $C^n$ for some constant $C$.*

*Proof.* The number of (rooted) unlabeled trees is bounded by the number of one-dimensional random walks $W$ with $2n$ steps. This can be proved as follows.

We can imagine to move along the tree by remaining to the left of the lines and starting from the root line. We move forward until an endpoint is reached: in this case we turn backwards until we meet a nontrivial vertex; then we turn once more forward and so on, until we come back to the root line. See Fig. 2.5: + means that we move from left to right along the line, while − means that we move from right to left.

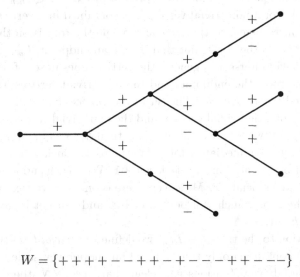

$$W = \{+ + + + - - + + - + - - - - + + - - - -\}$$

Fig. 2.6 A rooted tree and the corresponding walk $w$

Each time we move forward along a line we associate to it a sign +, while we associate to it a sign − when we move backwards. So the tree can be characterized by a collections of $2n$ signs $\pm$ which define a walk $W = \{\pm \pm \ldots \pm\}$. Note that not all one-dimensional random walks with $2n$ steps correspond to unlabeled trees: we call *compatible* the random walks for which this happens. For instance the first sign is always a + and the last one is always a $-$: moreover the overall number of $+$ signs has to be equal to the overall number of signs $-$: note that the correspondence between unlabeled trees and one-dimensional compatible random walks is 1-to-1. By neglecting all the constraints we can bound the number of collections of $2n$ signs, hence the number of unlabeled trees with $n$ nodes, by $2^{2n}$, that is the overall number of random walks with $2n$ steps. So we can choose $C = 4$ and the assertion follows. ∎

A *labeled tree* can be obtained from an unlabeled tree by assigning labels $h_v$ to its vertices $v$.

We define the set $\mathcal{T}_{h,n}$ the corresponding set of labeled trees with $n$ endpoints in the following way. We associate a label $h \leq N-1$ with the root and we introduce a family of vertical lines, labeled by an an integer taking values in $[h, N+1]$, and we represent any tree $\tau \in \mathcal{T}_{h,n}$ so that, if $v$ is an endpoint or a non trivial vertex, it is contained in a vertical line with index $h_v > h$, to be called the *scale* of $v$, while the root is on the line with index $h$. There is the constraint that, if $v$ is an endpoint, $h_v > h+1$.

The tree will intersect in general the vertical lines in set of points different from the root, the endpoints and the non trivial vertices; these points will be called *trivial vertices*. The set of the *vertices* of $\tau$ will be the union of the endpoints, the trivial vertices and the non trivial vertices. Note that, if $v_1$ and $v_2$ are two vertices and $v_1 < v_2$, then $h_{v_1} < h_{v_2}$. Moreover, there is only one vertex immediately following the root, which will be denoted $v_0$ and can not be an endpoint; its scale is $h+1$. With each endpoint $v$ of scale $h_v = N+1$ we associate $\mathcal{V}$. Moreover, there is only one vertex immediately following the root, which will be denoted $v_0$ and can not be an endpoint; its scale is $h+1$.

In addition to the trees $\tau \in \mathcal{T}_{h,n}$, we define the *trivial trees* (not belonging to $\mathcal{T}_{h,1}$) containing only the root and an endpoint of scale $h+1$.

With the above definitions it is clear that, if $h \leq N$, the effective potential can be written as sum over trees:

$$\mathcal{V}^{(h)}(\psi^{(j,h)}) + E_h = \sum_{n=1}^{\infty} \sum_{\tau \in \mathcal{T}_{h,n}} \mathcal{V}^{(h)}(\tau, \psi^{(j,h)}) \qquad (2.54)$$

where $\mathcal{V}^{(h)}(\tau, \psi^{(j,h)})$ is defined iteratively as follows. If $\tau$ is not trivial and $v_0$ is the first vertex of $\tau$ and $\tau_1, \ldots, \tau_s$ (with $s = s_{v_0}$) are the subtrees of $\tau$ with root $v_0$, then

$$\mathcal{V}^{(h)}(\tau, \psi^{(j,h)}) = \frac{1}{s!} \mathcal{E}^T_{h+1} \left( \mathcal{V}^{(h+1)}(\tau_1, \psi^{(j,h+1)}), \ldots, \mathcal{V}^{(h+1)}(\tau_s, \psi^{(j,h+1)}) \right) \qquad (2.55)$$

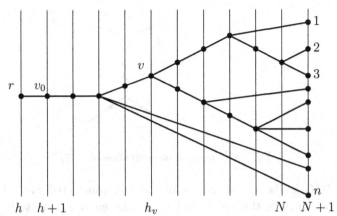

Fig. 2.7  A tree appearing in the graphic representation of $\mathcal{V}^{(h)}$.

and if $\tau_i$ is a trivial tree necessarily $h = N - 1$ and $\mathcal{V}^{(N)}(\tau_i, \psi^{(j,N)}) = \mathcal{V}(\psi^{(j,N)})$.

## 2.6  Feynman graphs

Let us consider, for fixing ideas, a grassman integral of the form

$$\bar{V}(\phi) = \log \int P(d\psi)\, e^{\mathcal{V}(\psi+\phi)} = \sum_{n=0}^{\infty} \frac{1}{n!} \mathcal{E}^T(\mathcal{V}; n) \qquad (2.56)$$

with

$$\mathcal{V}(\psi) = \lambda \int d\mathbf{x} \int d\mathbf{y}\, v(\mathbf{x}-\mathbf{y}) \psi^+_\mathbf{x} \psi^-_\mathbf{x} \psi^+_\mathbf{y} \psi^-_\mathbf{y} \qquad (2.57)$$

In order to compute $\bar{V}(\phi)$ we can use the Wick rule (2.38) and the relative graphical representation. It is then clear the expansion (2.56) can be written as sum over *Feynman graphs* in the following way.

A Feynman graph at order $n$ contributing to (2.56) is obtained taking $n$ elements represented as in Fig.2.8 and joining (contracting) the lines with consistent orientation so that all the $n$ vertices are connected. Calling $\ell$ the contracted lines of the graph, if $\mathcal{G}_n$ is the set of all possible Feynman graphs of order $n$, for any graph $G \in \mathcal{G}_n$ we can associate a *value* Val$(G)$ as, if $n = m + s$

$$\text{Val}(G) = (-1)^\pi \int d\mathbf{x}_1...d\mathbf{x}_n d\mathbf{y}_1...d\mathbf{y}_n [\prod_{i=1}^n \lambda v(\mathbf{x}_i - \mathbf{y}_i)][\prod_\ell g(\mathbf{x}_\ell - \mathbf{y}_\ell)] \prod_{i \in A(G)} \phi^{\sigma_i}_{\mathbf{x}_i} \qquad (2.58)$$

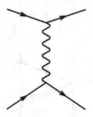

Fig. 2.8   Graphical representation of (2.57)

where $A(G)$ is the set of indices of the non contracted lines, $\ell$ are the contracted lines of the graph and $\mathbf{x}_\ell, \mathbf{y}_\ell$ the coordinates at the edge of the line, and $(-1)^\pi$ is the sign associated to the graph. With the above definitions

$$\bar{\mathcal{V}}(\phi) = \sum_{n=0}^{\infty} \sum_{G \in \mathcal{G}_n} \mathrm{Val}(G) \qquad (2.59)$$

In a similar way also the tree expansion seen in the previous section can be written as a sum of Feynman graphs

$$\mathcal{V}^{(k)}(\psi^{(\leq k)}) = \sum_{n=1}^{\infty} \sum_{\tau \in \mathcal{T}_{h,n}} \sum_{G \in \mathcal{G}(\tau)} \mathrm{Val}(G) \qquad (2.60)$$

where $\mathcal{G}(\tau)$ is the set of Feynman graphs defined in the following way. Given a tree $\tau$, the tree structure provides an arrangement of endpoints into a hierarchy of *clusters* contained into each other.

A Feynman graph at order $n$ is obtained associating to each end-point a graph element as in Fig.2.8, and joining (contracting) the lines with consistent orientation so that all the $n$ vertices are connected. Calling $\ell$ the contracted lines of the graph, if $\mathcal{G}_n(\tau)$ is the set of all possible Feynman graphs of order $n$, for any graph $G \in \mathcal{G}_n$ we can associate a *value* $\mathrm{Val}(G)$ as

$$\mathrm{Val}(G) = \qquad (2.61)$$
$$(-1)^\pi \int d\mathbf{x}_1...d\mathbf{x}_n d\mathbf{y}_1...d\mathbf{y}_n [\prod_{i=1}^{n} \lambda v(\mathbf{x}_i - \mathbf{y}_i)][\prod_{\ell} g^{(h_\ell)}(\mathbf{x}_\ell - \mathbf{y}_\ell)] \prod_{i \in A(G)} \psi_{\mathbf{x}_i}^{(\leq h)\sigma_i}$$

where $h_\ell$ is the scale of the cluster such that $\ell$ is internal to it and not to any smaller cluster. An example of Feynman diagram, with the tree and the cluster structure associated to it, is given in Fig. 2.10.

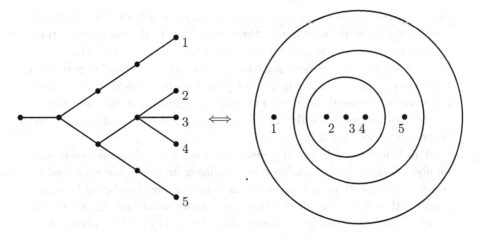

Fig. 2.9 A tree of order 5 and the corresponding clusters. Only the clusters corresponding to the nontrivial vertices are explicitly drawn.

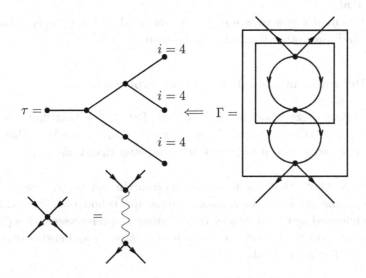

Fig. 2.10 An example of Feynman graph $G$ with its clusters. All the endpoints are supposed to be of type $\lambda$.

The value of the Feynman graph in the above picture is given by (if $v(\mathbf{x}-\mathbf{y}) = \delta(\mathbf{x}-\mathbf{y})$ for semplicity)

$$\lambda^2 \int d\mathbf{x}_1 d\mathbf{x}_2 d\mathbf{x}_3 g^{(h)}(\mathbf{x}_1-\mathbf{x}_2) g^{(h)}(\mathbf{x}_1-\mathbf{x}_2) g^{(k)}(\mathbf{x}_2-\mathbf{x}_3) g^{(k)}(\mathbf{x}_2-\mathbf{x}_3) \quad (2.62)$$

The expansions (2.59) or (2.60) allow to write the functional integrals we are interested in as sum of Feynman graphs which can be explicitely computed. In the physical applications we will see in the following sections we will see that several problems arises in practice. First of all, several Feynman graphs can be diverging when the cut-offs introduced to regularize the theory are removed. It is necessary to introduce different expansions, based on a multiscale analysis but more complex than the one seen in the previos section, in order to extract the dangerous contributions and renormalize the theory.

This is however still not sufficient; when the integrals is over an infinite number of Grassmann variables the Schwinger functions are expressed by series in which each term is given by the sum of a huge number of Feynman graphs. Hence, even if all the Feynman graphs would be finite, one has also to worry about the convergence of the series in (2.56). Unfortunately, even if each Feynman graph is bounded by $C^n \lambda^n \frac{1}{n!}$ (the best bound we can hope for) their number is $O(n!^2)$ so that each addend in (2.56) is bounded by $\widetilde{C}^n \lambda^n n!$.

A Feynman graph expansion cannot be used to get non-perturbative results and some other method must be used.

## 2.7 Determinant bounds for simple expectations

An important observation, due to Caianiello Ref. [21], is that there are important cancellations between Feynman graphs which make the behaviour of the series expansion at large orders much better than what seems at first sight.

We saw above that the fermionic expectation can be written as sum over Feynman graphs; if the fermionic propagator is bounded, each addend can be bounded by $C^n$ so that by the Feynman graph representation (2.31) one gets at most the bound, for a suitable constant $C_1$ and noting that the number of Feynman graphs is $O(n!)$

$$|\mathcal{E}(\psi^-_{\mathbf{x}_1}...\psi^-_{\mathbf{x}_n}\psi^+_{\mathbf{y}_1},...,\psi^+_{\mathbf{y}_n})| \leq n! C_1^n \qquad (2.63)$$

On the other hand, the above bound does not take into account the relative signs between graphs; by using them the above bound is dramatically improved.

We have seen in fact that the fermionic expectations can be written in a form of a determinant (see (2.30))

$$\mathcal{E}(\psi^-_{\mathbf{x}_1}...\psi^-_{\mathbf{x}_n}\psi^+_{\mathbf{y}_1},...,\psi^+_{\mathbf{y}_n}) = det G, \qquad (2.64)$$

where $G_{i,j} = q(\mathbf{x}_i - \mathbf{y}_j)$.

In physical applications, the fermionic propagator can be typically written in the following way

$$g(\mathbf{x} - \mathbf{y}) = \int d\mathbf{z} A(\mathbf{x} - \mathbf{z}) \bar{B}(\mathbf{y} - \mathbf{z}) \tag{2.65}$$

where $A, B$ are vectors in $L^2$.

The above representation is very useful as it allows to bound the fermionic expectation by the *Gram inequality*.

**Lemma 2.2.** *Let $\mathcal{H}$ an Hilbert space and $\{\mathbf{f}_j\}_{j=1}^m$, $\{\mathbf{g}_j\}_{j=1}^m \in \mathcal{H}$. Then if $H$ is an $m \times m$ matrix such that $H_{i,j} = (\mathbf{f}_i, \mathbf{g}_j)$ then*

$$|\det H| \leq \prod_{j=1}^m \|\mathbf{f}_j\| \|\mathbf{g}_j\| \tag{2.66}$$

*Proof.* We may assume that $g_1, .., g_m$ are linearly indipendent, otherwise the determinant is vanishing and the inequality is trivially satisfied. Moreover, if is $\mathcal{G}$ the span of $g_1, .., g_m$, we may also assume that each $f_j$ is in $\mathcal{G}$ as

$$\det(\mathbf{g}_i, \mathbf{f}_j) = \det(\mathbf{g}_i, \mathcal{P}\mathbf{f}_j) \tag{2.67}$$

where $\mathcal{P}$ is the orthogonal projection in $\mathcal{G}$, and $\prod_{j=1}^m \|\mathbf{g}_j\| \|\mathcal{P}\mathbf{f}_j\| \leq \prod_{j=1}^m \|\mathbf{g}_j\| \|\mathbf{f}_j\|$. Finally we may assume that $f_1, .., f_m$ are linearly indepedent, as otherwise the determinant vanishes.

If $\{\mathbf{g}_j\}_{j=1}^m$ is an orthogonal basis in $\mathcal{G}$ (so that $\{\mathbf{e}_j\}_{j=1}^m$, with $\mathbf{e}_j = \|\mathbf{g}_j\|^{-1}\mathbf{g}_j$, is an orthonormal basis) then

$$|\det(\mathbf{g}_i, \mathbf{f}_j)| = |\det(\mathbf{e}_i, \mathbf{f}_j)| \prod_{j=1}^m \|\mathbf{g}_j\| \leq \prod_{j=1}^m \|\mathbf{g}_j\| \|\mathbf{f}_j\| \tag{2.68}$$

Now consider the case in which the only conditions on the vectors $\{\mathbf{g}_j\}_{j=1}^m$ is that they are linearly independent. Set $\tilde{\mathbf{g}}_j = \|\mathbf{g}_j\|^{-1}\mathbf{g}_j$, so that $\|\tilde{\mathbf{g}}_j\|^2 = 1$, and define inductively the orthonormal basis in $\mathcal{G}$

$$\mathbf{e}_1 \equiv \tilde{\mathbf{g}}_1 , \tag{2.69}$$

$$\mathbf{e}_2 \equiv \frac{\tilde{\mathbf{g}}_2 - (\tilde{\mathbf{g}}_2, \tilde{\mathbf{g}}_1)\tilde{\mathbf{g}}_1}{1 - (\tilde{\mathbf{g}}_2, \tilde{\mathbf{g}}_1)^2} \tag{2.70}$$

and so on, in such a way that one has $(\mathbf{e}_i, \mathbf{e}_j) = \delta_{i,j}$.

If $c_2 = 1 - (\tilde{\mathbf{g}}_2, \tilde{\mathbf{g}}_1)^2$, with $0 \leq c_2 \leq 1$, one has

$$\tilde{\mathbf{g}}_2 = c_2 \mathbf{e}_2 + c_2(\tilde{\mathbf{g}}_2, \mathbf{g}_1)\tilde{\mathbf{g}}_1 . \tag{2.71}$$

Note that, by computing $\det(\widetilde{\mathbf{g}}_i, \mathbf{f}_j)$, no difference is made by replacing the vector $\widetilde{\mathbf{g}}_2$ with $c_2\mathbf{e}_2$; in fact the contribution from the remaining part of (2.71) is vanishing.

We can proceed analogously for the terms with $j = 3, \ldots, m$ using the Gram-Schmidt orthogonalization procedure, with $0 \leq c_j \leq 1$, and we get

$$\det(\widetilde{\mathbf{g}}_i, \mathbf{f}_j) = \prod_{j=1}^{m} c_j \det(\mathbf{e}_i, \mathbf{f}_j) \qquad (2.72)$$

so that

$$|\det(\mathbf{g}_i, \mathbf{f}_j)| = |\det(\widetilde{\mathbf{g}}_i, \mathbf{f}_j)| \prod_{j=1}^{m} \|\mathbf{g}_j\| = |\det(\mathbf{e}_i, \mathbf{f}_j)| \prod_{j=1}^{m} c_j \|\mathbf{g}_j\| \quad (2.73)$$

$$= \prod_{j=1}^{m} c_j \|\mathbf{g}_j\| \|\mathbf{f}_j\| \leq \prod_{j=1}^{m} \|\mathbf{g}_j\| \|\mathbf{f}_j\|$$

$$(2.74)$$

and (2.66) follows. ∎

We can apply the above Gram inequality to get a bound for the fermionic expectations. If $A(\mathbf{x} - \cdot)$ and $B(\mathbf{y} - \cdot)$ are vectors in the Hilbert space $\mathcal{H}$, with inner product

$$(A(\mathbf{x} - \cdot), B(\mathbf{y} - \cdot)) = \int d\mathbf{z} A(\mathbf{x} - \mathbf{z}) \bar{B}(\mathbf{y} - \mathbf{z}) \qquad (2.75)$$

and $\|\cdot\|$ the norm induced by that inner product, then given $\{A(\mathbf{x}_j - \cdot)\}_{j=1}^{n}$ and $\{B(\mathbf{y}_j - \cdot)\}_{j=1}^{n}$ we can write, by (2.65), the matrix elements of $G$ as

$$g(\mathbf{x}_i - \mathbf{y}_j) = (A(\mathbf{x}_i - \cdot), B(\mathbf{y}_j - \cdot)) \qquad (2.76)$$

so that by (2.66)

$$|\mathcal{E}(\psi^-_{\mathbf{x}_1}...\psi^-_{\mathbf{x}_n}\psi^+_{\mathbf{y}_1},\ldots,\psi^+_{\mathbf{y}_n})| \leq \prod_{j=1}^{n} \|A(\mathbf{x}_i - \cdot)\| \|B(\mathbf{x}_i - \cdot)\| \qquad (2.77)$$

As an obvious consequence of the above bound, assuming that $\|A(\mathbf{x}_i - \cdot)\| \leq C_0$, $\|B(\mathbf{x}_i - \cdot)\| \leq C_0$, one obtains

$$|\mathcal{E}(\psi^-_{\mathbf{x}_1}...\psi^-_{\mathbf{x}_n}\psi^+_{\mathbf{y}_1},\ldots,\psi^+_{\mathbf{y}_n})| \leq C_0^{2n} \qquad (2.78)$$

Comparing (2.63) and (2.78) we see that the Feynman graph bound is very rough at large $n$; by bounding each Feynman graph by its absolute value one misses important cancellations which have the effect that the apparent $n!$-behaviour is not present. In order to get nonperturbative results, the determinant structure must be preserved as much as possible in order to not generate spurious $n!$ in the bounds.

## 2.8 The Brydges-Battle-Federbush representation

The appplication of such ideas to the evaluation of (2.56) is not immediate, as there appears truncated expectations. Of course we could write the truncated expectations in terms of simple expectaions, but in this way we cannot use the decay properties of the propagators in order to perform the coordinate integrations appearing in the expansions. What we need is to extract the minimal number of propagators in order to perform the integrations, leaving all the other fields "packed" in the determinant. A possible way to achieve this is obtained through the Brydges-Battle-Federbusch representation, introduced in Refs. [22],[23].

We start from the fermionic expectation

$$\mathcal{E}\left(\widetilde{\psi}(P_1)...\widetilde{\psi}(P_s)\right) = \det G \qquad (2.79)$$

where $\sum_{j=1}^{s}|P_j^-| = \sum_{j=1}^{s}|P_j^+|$ (otherwise (2.79) is vanishing), $G$ is the matrix with entries $G_{ji,ji'} = g(\mathbf{x}_{j,i} - \mathbf{y}_{j',i'})$, $j, j' = 1, .., s$ and $i = 1, .., |P_j^+|$, $i' = 1, .., |P_j^-|$. We can write, by (2.7)

$$\mathcal{E}\left(\widetilde{\psi}(P_1)...\widetilde{\psi}(P_s)\right) = \int \mathcal{D}\eta \, e^{-V(X)} \qquad (2.80)$$

where $\eta_{i,j}^{\pm}$ are Grassmann variables, $\mathcal{D}\eta = \prod_{j=1}^{s}\prod_{i=1}^{|P_j^-|} d\eta_{j,i}^{-} \prod_{i=1}^{|P_j^+|} d\eta_{j,i'}^{+}$ and, if $X = \{1, .., s\}$

$$V(X) = \sum_{j,j'=1}^{s} \sum_{i=1}^{|P_j^-|} \sum_{i'=1}^{|P_{j'}^+|} \eta_{j',i'}^{+} G_{ji,ji'} \eta_{j,i}^{-} = \sum_{1 \le j,j' \le k} V_{j,j'} \qquad (2.81)$$

where

$$V_{jj'} = \sum_{i=1}^{|P_j^-|} \sum_{i'=1}^{|P_{j'}^+|} \eta_{j',i'}^{+} g(\mathbf{x}_{j,i} - \mathbf{y}_{j',i'}) \eta_{j,i}^{-} \qquad (2.82)$$

If $X_1 \equiv \{1\}$, we introduce $W_X(X_1, t_1)$ with $t_1 \in [0,1]$ defined as

$$W_X(X_1; t_1) = \sum_{\ell} t_1(\ell) V_\ell \qquad (2.83)$$

where $\ell = (jj')$ is a pair of elements $j, j' \in X$ and the function $t_1(\ell) = t_1$ if $\ell \sim \partial X_1$, that is if $\ell = (jj')$ "intesects the boundary" of $X_1$, i.e. it connects a point belonging to $P_1$ to a point contained inside some $P_{j'}$ with $j' \ne 1$, and $t_1(\ell) = 1$ otherwise.

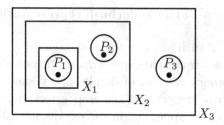

Fig. 2.11  An example of the sets $X_k$ for $k=1,2,3$. One has $X_1=\{1\}$, $X_2=\{1,2\}$ and $X_3=\{1,2,3\}$

More explicitly

$$W_X(X_1;t_1) = \sum_{j=2}^{s} t_1 V_{1j} + V_{11} + \sum_{1<j'\leq j} V_{j'j} = \quad (2.84)$$

$$t_1(\sum_{j=2}^{s} V_{1j} + V_{11} + \sum_{1<j'\leq j} V_{j'j}) + (1-t_1)(V_{11} + \sum_{1<j'\leq j} V_{j'j}) =$$

$$t_1 V(X) + (1-t_1)\left[V(X_1) + V(X \setminus X_1)\right]$$

so that when $t_1 = 0$ the set $X_1$ and $X \setminus X_1$ are decoupled and $W_X(X_1;1) = V(X)$. We can write

$$e^{-V(X)} = \int_0^1 dt_1 \left[\frac{\partial}{\partial t_1} e^{-W_X(X_1;t_1)}\right] + e^{-W_X(X_1;0)} \quad (2.85)$$

$$= -\sum_{\ell_1 \sim \partial X_1} V_{\ell_1} \int_0^1 dt_1 \, e^{-W_X(X_1;t_1)} + e^{-W_X(X_1;0)}$$

We have expressed $e^{-V(X)}$ as sum of two terms, in the first a bond $\ell_1$ between 1 and the elements in $X$ is exhibited, in the second one $X_1 = \{1\}$ is decoupled from the rest of $X$.

We introduce now the following quantities

$$W_X(X_1,\ldots,X_r;t_1,\ldots,t_r) = \sum_\ell \prod_{k=1}^{r} t_k(\ell) \, V_\ell \quad (2.86)$$

where $X_k$ are subsets of $X$ with $|X_k| = k$, inductively defined as $X_1 = \{1\}$ and $X_{k+1} \supset X_k$; $\ell = (jj')$ is a pair of elements $j, j' \in X$ and the sum in (2.86) is over all the possible pairings $(jj')$, and finally the functions $t_k(\ell) = t_k$ if $\ell \sim \partial X_k$, that is if $\ell = (jj')$ "intesects the boundary" of $X_k$, i.e. it connects a point belonging to some $P_j$ with $j \in X_k$ to a point contained inside some $P_{j'}$ with $j' \notin X_k$, and $t_k(\ell) = 1$ otherwise.

We choose $X_2 = \{1, \text{point connected by } \ell_1 \text{ with } 1\}$ and we write, according to the above definition

$$W_X(X_1, X_2; t_1, t_2) = (1 - t_2)\left[W_{X_2}(X_1; t_1) + V(X \setminus X_2)\right] + t_2 W_X(X_1; t_1) \tag{2.87}$$

so that when $t_2 = 0$ $X_2$ and $X \setminus X_2$ are decoupled. By substituing under the first term in (2.86), under the sum over $\ell_1$

$$e^{-W_X(X_1;t_1)} = \int_0^1 dt_2 \left[\frac{\partial}{\partial t_2} e^{-W_X(X_1,X_2;t_1,t_2)}\right] +$$

$$e^{-W_X(X_1,X_2;t_1,0)} \tag{2.88}$$

$$= -\sum_{\ell_2 \sim \partial X_2} V_{\ell_2} \int_0^1 dt_2 \, t_1(\ell_2) \, e^{-W_X(X_1,X_2;t_1,t_2)} + e^{-W_X(X_1,X_2;t_1,0)}$$

so that

$$e^{-V(X)} = \sum_{\ell_1 \sim \partial X_1} \sum_{\ell_2 \sim \partial X_2} \int_0^1 dt_1 \int_0^1 dt_2 \, (-1)^2 \, V_{\ell_1} V_{\ell_2} \, t_1(\ell_2) \, e^{-W_X(X_1,X_2;t_1,t_2)}$$

$$+ \sum_{\ell_1 \sim \partial X_1} \int_0^1 dt_1 \, (-1) \, V_{\ell_1} \, e^{-W_X(X_1,X_2;t_1,0)} + e^{-W_X(X_1;0)} \tag{2.89}$$

where in the first term a bound crossing the boundary of $X_2$ is exhibited. We can define $X_3 = \{X_2, \text{point connected by } \ell_2 \text{ with } X_2\}$ and so on, continuing until $X_r = X$ for some $r$, leaving decoupled terms alone; noting that

$$W_X(X_1, \ldots, X_r; t_1, \ldots, t_{r-1}, 0) = W_{X_r}(X_1, \ldots, X_{r-1}; t_1, \ldots, t_{r-1}) + V(X \setminus X_r) \tag{2.90}$$

we get

$$e^{-V(X)} = \sum_{r=1}^{s-1} K(X_r) e^{-V(X/X_r)} \tag{2.91}$$

where $K(X_1) = e^{-V(X_1)}$ otherwise

$$K(X_r) = \sum_{\ell_1 \sim \partial X_1} \cdots \sum_{\ell_r \sim \partial X_{r-1}} \int_0^1 dt_1 \cdots \int_0^1 dt_{r-1} \, (-1)^r \, V_{\ell_1} \ldots V_{\ell_r}$$

$$\left(\prod_{k=1}^{r-1} t_1(\ell_{k+1}) \ldots t_k(\ell_{k+1})\right) e^{-W_{X_r}(X_1,\ldots,X_{r-1};t_1,\ldots,t_{r-1})} \tag{2.92}$$

The above quantity can be also rewritten as

$$K(X_r) = \sum_{X_2 \ldots X_{r-1}} \sum_{T \text{ on } X_r} \prod_{\ell \in T} V_\ell \qquad (2.93)$$

$$\int_0^1 dt_1 \ldots \int_0^1 dt_{r-1} \left( \prod_{\ell \in T} \frac{\prod_{k=1}^{r-1} t_k(\ell)}{t_{n(\ell)}} \right) e^{-W_{X_r}(X_1,\ldots,X_{r-1};t_1,\ldots,t_{r-1})}$$

where $T$ is a set of lines $\ell$ such that for each $k = 1, \ldots, r$, it contains one line incident with $(j, i)$, where $j \in X_k$ and $i \in \{1, \ldots, |P_j^\pm|\}$, each line $\ell \in T$ intersects at least one boundary $\partial X_k$ and the lines $\ell_1, \ell_2, \ldots$ are ordered so that $\ell_1 \sim \partial X_1, \ell_2 \sim \partial X_2, \ldots$.

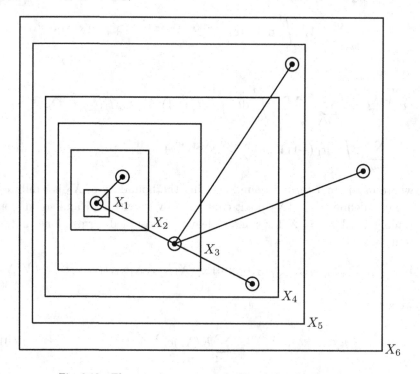

Fig. 2.12  The sets $X_1,\ldots,X_6$ and the lines belonging to $T$

For each $\ell \in T$, $T$ called *anchored tree*, one defines two indices $n(\ell)$ and $n'(\ell)$ such that

$$n(\ell) = \max\{k : \ell \sim \partial X_k\},$$
$$n'(\ell) = \min\{k : \ell \sim \partial X_k\}. \qquad (2.94)$$

In (2.94) we can sum first over the trees $T$, then over the sets $X_k$,

$$\sum_{X_2\ldots X_{r-1}} \sum_{T \text{ on } X_r} = \sum_{T \text{ on } X_r} \sum_{\substack{X_2\ldots X_{r-1} \\ \text{fixed } T}} \tag{2.95}$$

where "fixed $T$" recalls that the sets $X_2,\ldots,X_r$ have to be compatible with the tree $T$.

Moreover we can write, by (2.86),

$$W_{X_r}(X_1,\ldots,X_{r-1};t_1,\ldots,t_{r-1}) = \tag{2.96}$$

$$\sum_{\ell \in X_r} t_1(\ell)\ldots t_{r-1}(\ell)V_\ell = \sum_{\ell \in X_r} t_{n'(\ell)}\ldots t_{n(\ell)-1}V_\ell$$

and set in (2.94)

$$\frac{\prod_{k=1}^{r-1} t_r(\ell)}{t_{n(\ell)}} = t_{n'(\ell)}\ldots t_{n(\ell)-1} \tag{2.97}$$

so obtaining

$$K(X_r) = \sum_{T \text{ on } X_r} \sum_{\substack{X_2\ldots X_{r-1} \\ \text{fixed } T}} \prod_{\ell \in T} V_\ell \int_0^1 dt_1 \ldots \int_0^1 dt_{l-1} \tag{2.98}$$

$$\prod_{\ell \in T} \left(t_{n'(\ell)}\ldots t_{n(\ell)-1}\right) e^{-\sum_{\ell \in X_r} t_{n'(\ell)}\ldots t_{n(\ell)-1}V_\ell} .$$

The above procedure can be repeated for the factors $e^{-V(X\setminus X_r)}$ until we get the following representation

$$\mathcal{E}\left(\prod_{j=1}^s \tilde\psi(P_j)\right) = \int \mathcal{D}\eta \sum_m (-1)^m \sum_{Q_1,\ldots,Q_m} K(Q_1)\ldots K(Q_m) . \tag{2.99}$$

where $\sum_{Q_1,\ldots,Q_m}$ is a sum over all the partitions of $\{1,..,s\}$ into non-empty disjoint sets.

On the other hand the expectations can be expressed in terms of truncated expectations, through the relation

$$\mathcal{E}\left(\prod_{j=1}^s \tilde\psi(P_j)\right) = \sum_m \sum_{Q_1,\ldots,Q_m} (-1)^\pi \mathcal{E}^T(\prod_{i \in Q_1} \tilde\psi(P_i))\ldots \mathcal{E}^T(\prod_{i \in Q_m} \tilde\psi(P_i)) \tag{2.100}$$

where $\sum_{Q_1,\ldots,Q_m}$ is a sum over all the partitions of $\{1,..,s\}$ into non-empty disjoint sets, $(-1)^\pi$ is the sign of the permutation; hence so that, by comparing (2.100) with (2.99) we find the following expression for the truncated expectations, up to a sign:

$$\mathcal{E}^T\left(\widetilde{\psi}(Q_1),\ldots,\widetilde{\psi}(Q_m)\right) = \int \mathcal{D}\eta \sum_{\substack{T \text{ on } X_m}} \sum_{\substack{X_2\cdots X_{m-1} \\ \text{fixed } T}} \prod_{\ell \in T} V_\ell \qquad (2.101)$$

$$\int_0^1 dt_1 \cdots \int_0^1 dt_{m-1} \prod_{\ell \in T} \left(t_{n'(\ell)} \cdots t_{n(\ell)-1}\right) e^{-\sum_{\ell \in X} t_{n'(\ell)} \cdots t_{n(\ell)-1} V(\ell)}$$

A remarkable property of (2.101) is the following result, see Ref.[22].

**Lemma 2.3.** *In (2.101) one has*

$$\sum_{\substack{X_2\cdots X_{m-1} \\ \text{fixed } T}} \int_0^1 dt_1 \cdots \int_0^1 dt_{m-1} \prod_{\ell \in T} \left(t_{n'(\ell)} \cdots t_{n(\ell)-1}\right) = 1 \qquad (2.102)$$

*for any anchored tree $T$. Moreover in (2.102)*

$$dP_T(\mathbf{t}) \equiv \sum_{\substack{X_2\cdots X_{p-1} \\ \text{fixed } T}} \prod_{\ell \in T} \left(t_{n'(\ell)} \cdots t_{n(\ell)-1}\right) \prod_{q=1}^{m-1} dt \qquad (2.103)$$

*can be interpreted as a probability measure in the variable $\mathbf{t} = (t_1,\ldots,t_{m-1})$.*

*Proof.* Let us denote by $b_k$ the number of lines $\ell \in T$ exiting from points $\mathbf{x}(j,i)$, with $j \in X_k$. By construction the parameter $t_k$ inside the integral in the left hand side of (2.102) appears to the power $b_k - 1$, as all the lines intersecting $\partial X_k$ contribute to $t_k$, except the one connecting $X_k$ with the point whose union with $X_k$ gives the set $X_{k+1}$ (this is clear by using the notations introduced after (2.86)). Then

$$\prod_{\ell \in T} \left(t_{n'(\ell)} \cdots t_{n(\ell)-1}\right) = \prod_{k=1}^{m-2} t_k^{b_k-1} \qquad (2.104)$$

and in (2.102) one has $m-1$ independent integrations

$$\int_0^1 dt_{m-1} \prod_{k=1}^{m-2} \left(\int_0^1 dt_k\, t_k^{b_k-1}\right) = \prod_{k=1}^{m-2} \frac{1}{b_k} \qquad (2.105)$$

which is a well defined expression as $b_k \geq 1$ for $k = 1,\ldots,m-2$. Moreover we can write

$$\sum_{\substack{X_2\cdots X_{m-1} \\ \text{fixed } T}} = \sum_{\substack{X_2 \\ \text{fixed } X_1}} \sum_{\substack{X_3 \\ \text{fixed } X_1, X_2}} \cdots \sum_{\substack{X_{m-1} \\ \text{fixed } X_1,\ldots,X_{m-2}}} \qquad (2.106)$$

where the number of possible choices in summing over $X_k$, once $X_1,\ldots,X_{k-1}$ have been fixed, is exactly $b_{k-1}$: if $b_{k-1}$ lines exit from $X_{k-1}$ then $X_k$ is obtained by adding to $X_{k-1}$ one of the $b_{k-1}$ points connected to $X_{k-1}$ through one of the lines of the tree. Then

$$\sum_{\substack{X_2\ldots X_{m-1}\\ \text{fixed } T}} 1 = b_1\ldots b_{m-2} \tag{2.107}$$

and, at the end,

$$\sum_{\substack{X_2\ldots X_{m-1}\\ \text{fixed } T}} \int_0^1 dt_1 \ldots \int_0^1 dt_{m-1} \prod_{\ell\in T}\left(t_{n'(\ell)}\ldots t_{n(\ell)-1}\right) = \prod_{k=1}^{m-2} \frac{b_k}{b_k} \tag{2.108}$$

which yields (2.102). ∎

## 2.9 The Gawedzki-Kupiainen-Lesniewski formula

The conclusion of the analysis in the previous section is that, calling

$$V(\mathbf{t}) \equiv \sum_{\ell\in X} t_{n'(\ell)}\ldots t_{n(\ell)-1}\, V_\ell\,, \tag{2.109}$$

the truncated expectation can be represented as

$$\mathcal{E}^T\left(\widetilde{\psi}(P_1),\ldots,\widetilde{\psi}(P_s)\right) = \int \mathcal{D}\eta \sum_T \prod_{(jj')\in T}\left(V_{jj'} + V_{jj'}\right) \int dP_T(\mathbf{t})\, e^{-V(\mathbf{t})} \tag{2.110}$$

where $T$ are trees anchored on some points $x_i^{(q)}$, $q = 1,\ldots,s$ and $i = 1,\ldots,|P_q|$. By (2.82), we know that each $V_{jj'}$ contains Grassman variables $\sum_{i=1}^{|P|_j} \eta_{j,i}^-$ and $\sum_{i'=1}^{|P|_{j'}} \eta_{j',i'}^+$; for each addend in the sum (2.110) we call $\eta_1$ the Grassman variables appearing in $\prod_{(jj')\in T}(V_{jj'} + V_{jj'})$ (whose number is $2(s-1)$), and $\eta_2$ the remaining Grassman variables, (whose number is $\sum_j |P|_j - 2(s-1)$); moreover we can write

$$V(\mathbf{t}) = V_1(\mathbf{t}) + V_2(\mathbf{t}) \tag{2.111}$$

where $V_2$ contains no variables $\eta_1$; hence by the definition of Grassman integration we get

$$\int \mathcal{D}\eta_1 \mathcal{D}\eta_2 \prod_{(jj')\in T}(V_{jj'} + V_{jj'}) \int dP_T(\mathbf{t})\, e^{-V(\mathbf{t})} =$$

$$\prod_{\ell\in T} g_\ell \int dP_T(\mathbf{t}) \int \mathcal{D}\eta_2\, e^{-V_2(\mathbf{t})}. \tag{2.112}$$

By definition

$$\int \mathcal{D}\eta_2 e^{-V_2(\mathbf{t})} = \det G^T(\mathbf{t}) \qquad (2.113)$$

where $G^T(\mathbf{t})$ is a square $[1/2 \sum_j |P|_j - (s-1)] \times [1/2 \sum_j |P|_j - (s-1)]$ matrix whose elements are

$$G^T_{(j,i)(j',i')} = t_{n'(jj')} \ldots t_{n(jj')-1} g(\mathbf{x}_{j,i} - \mathbf{x}_{j',i'}) \qquad (2.114)$$

and $\mathbf{x}_{j,i}, \mathbf{x}_{j',i'}$ are all the points of the fields in the truncated expectation except the one in the propagators in $T$.

Also $\det G^T(\mathbf{t})$ is a Gram determinant; indeed one can define a family of vectors in $\mathbb{R}^s$ inductively as

$$\mathbf{u}_1 = \mathbf{v}_1,$$
$$\mathbf{u}_j = t_{j-1}\mathbf{u}_{j-1} + \mathbf{v}_j\sqrt{1 - t_{j-1}^2}, j = 2, \ldots, m \qquad (2.115)$$

where $\{\mathbf{v}_i\}_{i=1}^m$ is an orthonormal basis and the sets $X_k$ have been relabeled so that $X_1 = \{1\}$, $X_2 = \{1,2\}$, ..., $X_m = \{1, 2, \ldots, m\}$, hence

$$t_{n'(jj')} \ldots t_{n(jj')-1} = t_j \ldots t_{j'-1} \qquad (2.116)$$

for a line $(jj')$. By the definitions (2.115) one has

$$\mathbf{u}_j \cdot \mathbf{u}_{j'} = t_j \ldots t_{j'-1} \qquad (2.117)$$

The conclusion of this analysis is the following formula for the trunctaed expectations, obtained in Refs. [24],[25] (similar formulas appeared in Ref.[26]), if $s > 1$

$$\mathcal{E}^T\left(\tilde{\psi}(P_1), \ldots, \tilde{\psi}(P_s)\right) = \sum_T \left(\prod_{\ell \in T} g_\ell\right) \int dP_T(\mathbf{t}) \det G^T(\mathbf{t}), \qquad (2.118)$$

(1) $T$ is a set of lines forming an *anchored tree* between the clusters of points $P_1, \ldots, P_s$, i.e. $T$ is a set of lines which becomes a tree if one identifies all the points in the same cluster,
(2) $\mathbf{t} = \{t_{i,i'} \in [0,1], 1 \le i, i' \le s\}$ and $dP_T(\mathbf{t})$ is a probability measure with support on a set of $\mathbf{t}$ such that $t_{i,i'} = \mathbf{u}_i \cdot \mathbf{u}_{i'}$ for some family of vectors $\mathbf{u}_i \in \mathbb{R}^s$ of unit norm.
(3) $G^T(\mathbf{t})$ is a $(n-s+1) \times (n-s+1)$ matrix, whose elements are given by

$$[G^T(\mathbf{t})]_{(j,i),(j',i')} = t_{j,j'} g(\mathbf{x}(j,i) - \mathbf{x}(j',i')) \qquad (2.119)$$

where $1 \le j, j' \le s$ and $1 \le i \le |P_j|$, $1 \le i' \le |P_{j'}|$, such that the lines $\ell = \mathbf{x}(j,i) - \mathbf{x}(j',i')$ do not belong to $T$.

It is important to stress the crucial difference with respect the expression of the truncated expecation in terms of Feynman graph and the above one. In the first case one sums over all the possible ways of contracting *all* the half lines associated to the fields $\widetilde{\psi}(P_1), \ldots \widetilde{\psi}(P_s)$; a single contribution is given as in Fig. 2.1.

On the other hand in (2.118) one contracts only the half-lines associated to the fields $\widetilde{\psi}(P_1), \ldots \widetilde{\psi}(P_s)$ which are necessary to produce a tree connecting the clusters $\widetilde{\psi}(P_1), \ldots \widetilde{\psi}(P_s)$; the propagator of all the other fields are in the determinant in (2.118).

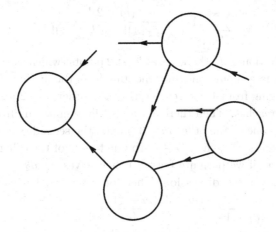

Fig. 2.13  A term contributing to the truncated expectation according to the expansion (2.118)

In order to bound the trunctated expectations, we need a bound on the number of anchored trees $T$, which is provided by the following lemma.

**Lemma 2.4.** *Given $m$ set of indices $Q_1, \ldots, Q_m$ such that*

$$\{\mathbf{x}(f) : f \in Q_q\} = \{\mathbf{x}_1^{(q)}, \ldots, \mathbf{x}_{|Q_q|}^{(q)}\}, \qquad q = 1, \ldots, m, \qquad (2.120)$$

*and $\sum_{q=1}^m |Q_q^+| = \sum_{q=1}^m |Q_q^-| = n$, then the number of trees $T$ anchored on $\{Q_1, \ldots, Q_m\}$ is bounded by*

$$\sum_T 1 \leq m!\, C^n \qquad (2.121)$$

*for some constant $C$.*

*Proof.* The proof goes through the following steps.
(1) First suppose that each set $Q_q$ is a point: we shall see at the end what

happens if the sets contain several points. We can write

$$\sum_T 1 = \sum_{\{d_q\}} \sum_{\substack{T \\ \text{fixed }\{d_q\}}} 1, \qquad (2.122)$$

where in the right hand side the first sum is over all the possible configurations $\{d_q\}$, if we denote by $d_q$ the number of lines emerging from (*i.e.* entering or exiting from) $Q_q \equiv q$, while the second sum is over all the trees compatible with a fixed configuration $\{d_q\}$.

(2) The second sum in the right hand side of (2.122) can be exactly computed and it gives

$$\sum_{\substack{T \\ \text{fixed }\{d_q\}}} 1 = \frac{(m-2)!}{(d_1-1)!\ldots(d_m-1)!}. \qquad (2.123)$$

In fact, by definition of $T$, there are at least 2 points (which we can call 1 and $m$) such that there is only one line emerging from them: then $d_1 = d_m = 1$. The line emerging from 1 can reach one of the other $m-2$ points: we call 2 the point it reaches. Then there are $d_2 - 1$ lines emerging from 2 leading the first one to one of the other $m-3$ points, the second one to one of the other $m-4$ points, ..., the $(d_2-1)$-th one to one of the other $m-d_2-1$ points; moreover if we permute between themselves the $d_2 - 1$ lines ther is no change in the above discussion. Therefore so far we have obtained

$$\frac{(m-2)}{(d_1-1)!} \cdot \frac{(m-3)(m-4)\ldots(m-d_2-1)}{(d_2-1)!} \qquad (2.124)$$

possible contributions. By iterating until the $m$-th point is reached we find (2.123).

(3) The first sum in the r.h.s. of (2.122) can be bounded by

$$\sum_{\{d_q\}} 1 \leq C^m, \qquad (2.125)$$

where one can choose $C = 2$. In fact one has two constraints $\sum_{q=1}^m d_q = 2(m-1)$ and $1 \leq d_q \leq m-1 \ \forall \ i = 1, \ldots, m$, as the tree $T$ has $m-1$ lines, each lines emerge from two points and each point is connected with no less than 1 point and no more than with all the others. Then, if we set $M = 2(m-1)$ and ignore for simplicity the second constraint on $\{d_q\}$, we have

$$\sum_{\{d_q\}} 1 \leq \int_0^M dx_1 \int_0^{M-x_1} dx_2 \ldots \int_0^{M-\sum_{q=1}^{m-1} 1} dx_m \qquad (2.126)$$

$$\leq \frac{1}{m!} M^m = \frac{1}{m!}[2(m-1)]^m \leq 2^m \frac{m^m}{m!}$$

and as $e^{-m} \leq m^m/m! \leq 1$, then (2.125) immediately follows with $C = 2$.

(4) Now we take into account that, for each $q = 1, \ldots, m$, $Q_q$ is a collection of points. Fixed $T$ on $\mathcal{Q}$, the number of anchored trees is

$$\prod_{q=1}^{m} \frac{|Q_q|!}{(|Q_q| - d_q)!} \qquad (2.127)$$

as we have to consider the $|Q_q|!$ permutations of the $|Q_q|$ elements of the set $Q_q$ and divide by the $(|Q_q| - d_q)!$ permutations of the elements of $Q_q$ which no line emerges from. So, by using that $[(d_q - 1)!]^{-1} \leq [d_q!]^{-1} 2^{d_q}$ and $\prod_{q=1}^{m} 2^{d_q} = 2^{2(m-1)} \leq 4^m$, we obtain

$$\sum_{\substack{T \\ \text{fixed } \{d_q\}}} 1 \leq \widetilde{C}^{2(n+m)} (m-2)! , \qquad (2.128)$$

where one can take $\widetilde{C} = 2^2$. ∎

# PART 2
# Quantum Field Theory

## Chapter 3

# The Ultraviolet Problem in Massive QED2

## 3.1 Regularization and cut-offs

Now that we have developed several technical tools to analyze functional integrals we are finally ready to consider physical models; a natural starting point is Quantum Electrodynamics in $d = 2$, both for its semplicity and its physical interest. Its formal generating function is given by (1.4) but such expression needs to be regularized to have a meaning, introducing suitable cut-offs which must be removed at the end.

A possible regularization for the generating function of QED2 with massive photon fields and in the Feynman gauge is

$$e^{\mathcal{W}_{h,N}(J^A,\phi)} = \int P(d\psi)P(dA)e^{\int d\mathbf{x}[e\bar{\psi}_\mathbf{x}(A_{\mu,\mathbf{x}}\gamma_\mu)\psi_\mathbf{x} + J^A_{\mu,\mathbf{x}}A_{\mu,\mathbf{x}} + \phi_\mathbf{x}\bar{\psi}_\mathbf{x} + \bar{\phi}_\mathbf{x}\psi_\mathbf{x}]}$$
(3.1)

where $\phi, J^A$ are external fields and:

-)in $\Lambda = [0, L] \times [0, L]$ a lattice $\Lambda_a$ is introduced whose sites are given by the space-time points $\mathbf{x} = (x, x_0) = (na, n_0 a)$ with $L/a$ integer and $n, n_0 = -L/2a, 1, \ldots, L/2a - 1$. We also consider the set $\mathcal{D}$ of space-time momenta $\mathbf{k} = (k, k_0)$ with $k = (m + \frac{1}{2})\frac{2\pi}{L}$ and $k_0 = (m_0 + \frac{1}{2})\frac{2\pi}{L}$ with $m, m_0 = 0, 1, \ldots, L/a - 1$. To simplify the notations we write $\int d\mathbf{x} = a^2 \sum_{\mathbf{x} \in \Lambda_a}$ and $\int \frac{d\mathbf{k}}{(2\pi)^2} = \frac{1}{L^2} \sum_{\mathbf{k} \in \mathcal{D}}$.

-)$\psi_\mathbf{x}, \bar{\psi}_\mathbf{x}, \mathbf{x} \in \Lambda_a$ are a finite set of *Grassmann spinors* and $P(d\psi)$ is the fermionic integration with propagator

$$g^{(h,N)}(\mathbf{x} - \mathbf{y}) = \int \frac{d\mathbf{k}}{(2\pi)^2} \frac{-i \slashed{k} + m}{\mathbf{k}^2 + m^2} e^{-i\mathbf{k}(\mathbf{x}-\mathbf{y})} \chi_{h,N}(\mathbf{k}) =$$

$$\int \frac{d\mathbf{k}}{(2\pi)^2} \hat{g}^{(h,N)}(\mathbf{k}) e^{-i\mathbf{k}(\mathbf{x}-\mathbf{y})}$$
(3.2)

where $\slashed{k} = \mathbf{k}_\mu \gamma_\mu$. More explicitely $\psi = (\psi_+^-, \psi_-^-)$, $\psi^+ = (\psi_+^+, \psi_-^-)$, $\bar{\psi} =$

$\psi^+\gamma_0$ and the $\gamma$ matrices can be chosen as

$$\gamma_0 = \begin{pmatrix} 0 & 1 \\ 1 & 0 \end{pmatrix}, \quad \gamma^1 = \begin{pmatrix} 0 & -i \\ i & 0 \end{pmatrix}$$

We will denote a spinor element by $\psi_\omega^\pm$ with $\omega = \pm$. The function $\chi_{h,N}(\mathbf{k})$ is a smooth $C^\infty(\mathbb{R}^+)$ compact-support cutoff function selecting momenta $\gamma^h \leq |\mathbf{k}| \leq \gamma^N$ with $\gamma > 1$ and , $-h, N$ positive integers; $\gamma^h$ and $\gamma^N$ are respectively the infrared and ultraviolet fermionic momentum cut-off. We will call $\lim_{h \to -\infty} \chi_{h,N}(\mathbf{k}) = \chi_N(\mathbf{k})$ and we assume $\gamma^N \ll a^{-1}$, that is the lattice cutoff is removed before the fermionic cutoff.

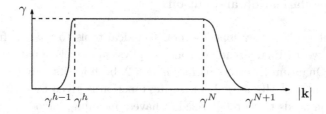

Fig. 3.1  Graphical representation of the cut-off function

-)$A_\mathbf{x} = (A_{0,\mathbf{x}}, A_{1,\mathbf{x}})$ are Euclidean boson fields with periodic boundary conditions and Gaussian measure $P(dA)$ with covariance

$$< A_{\mu,\mathbf{x}} A_{\nu,\mathbf{y}} > = v(\mathbf{x}-\mathbf{y}) \delta_{\mu,\nu} = \delta_{\mu,\nu} \int \frac{d\mathbf{p}}{(2\pi)^2} e^{-i\mathbf{p}(\mathbf{x}-\mathbf{y})} \frac{\chi_{\bar{K}}(\mathbf{p})}{\mathbf{p}^2 + M^2} \quad (3.3)$$

and $\chi_{\bar{K}}(\mathbf{p})$ is a smooth cutoff function selecting momenta $|\mathbf{p}| \leq \gamma^{\bar{K}}$ (we can choose it as identical to the one in the fermionic propagator). At finite $\bar{K}, M$, $||v_{\bar{K}}||_1$ and $|v_{\bar{K}}||_\infty$ are bounded.

The *Schwinger functions* are defined by

$$< \psi_{\mathbf{x}_1}; ...; \psi_{\mathbf{x}_n}; \bar{\psi}_{\mathbf{y}_1}; ...; \bar{\psi}_{\mathbf{y}_n}; A_{\mu_1,\mathbf{z}_1}...; A_{\mu_m,\mathbf{z}_m} >_{h,N} = \quad (3.4)$$

$$\frac{\partial^{2n+m} \mathcal{W}_{h,N}(J^A, \phi)}{\partial \phi_{\mathbf{x}_1}...\partial \phi_{\mathbf{x}_n} \partial \bar{\phi}_{\mathbf{y}_1}...\partial \bar{\phi}_{\mathbf{y}_n} \partial J^A_{\mu_1,\mathbf{z}_1}...\partial J^A_{\mu_m,\mathbf{z}_m}} \bigg|_{J=\phi=0} \quad (3.5)$$

It is easy to see that (see Lemma 3.1 below) that the Schwinger functions are (3.5) are *well defined*, as the presence of the cut-offs $N, \bar{K}$ eliminates the divergences in the Feynman graphs due to the low decay of the propagators at high momenta or the singularities at vanishing momenta. The cut-offs have however no physical meaning and must be removed. This is indeed possible according to the following theorem (originally proved in [27]), proved in the rest of this chapter (for semplicity, only the 2-point Schwinger functions are considered).

**Theorem 3.1.** *For* $|e| \leq C(mM)^{\vartheta}(\|v\| + \|v\|_{\infty})^{-\vartheta}$, *for suitable constants* $C, \vartheta$, *the 2-point Schwinger function is well defined in the limit* $-h, N, L, a^{-1} \to \infty$ *and it is such that*

$$\langle \psi_{\mathbf{k}} \bar{\psi}_{\mathbf{k}} \rangle = \hat{g}(\mathbf{k})[1 + e^2 A(\mathbf{k})] \qquad (3.6)$$

*with* $|A(\mathbf{k})| \leq C$.

The above statement is *non-perturbative*, in the sense that the Schwinger function is expressed by a power series which are convergent when the infrared and ultraviolet cut-offs are removed. The interaction does not modify the large and small distance behaviour of the 2-point Schwinger function, provided that the mass is large enough (the convergence radius shrinks to zero as $m \to 0$). This condition is not technical: the free and interacting 2-point Schwinger function have a similar behaviour *only* if the electron mass is larger than the charge. Note also that the fermionic ultraviolet cut-off can be removed without the introduction of any cut-off dependent counterterm, in agreement with the fact that the model is perturbatively superrinormalizable.

## 3.2 Integration of the bosons

The generating function is an integral over bosons and fermions; we can however integrate over the bosons obtaining a Grassman integral of the type discussed in the previous chapter

$$e^{\mathcal{W}_{h,N}(J^A,\phi)} = \qquad (3.7)$$
$$\int P(d\psi) e^{\frac{1}{2} \int d\mathbf{x} d\mathbf{y} v(\mathbf{x}-\mathbf{y})[e\bar{\psi}_{\mathbf{x}}\gamma^{\mu}\psi_{\mathbf{x}}+J^A_{\mu,\mathbf{x}}][e\bar{\psi}_{\mathbf{y}}\gamma^{\mu}\psi_{\mathbf{y}}+J^A_{\mu,\mathbf{y}}]+\int d\mathbf{x}[\phi_{\mathbf{x}}\bar{\psi}_{\mathbf{x}}+\bar{\phi}_{\mathbf{x}}\psi_{\mathbf{x}}]}$$

As a first application of the determinant bounds seen above, we can show that (3.8) is well defined in the continuum and infinite volume limit $a \to 0, L \to \infty$; we consider for semplicity just the case $\mathcal{W}_{h,N}(0,0) \equiv L^2 E_{h,N}$ in the massless case $m = 0$, but the general case poses no extra difficulties.

**Lemma 3.1.** *For any finite value of* $K, N, h, M$ *and a suitable constant* $C$, *for* $|e| \leq C^{-1}\|v\|^{-1}\gamma^{-2(N-h)}$ *the limit* $L, a^{-1} \to \infty$ *of* $E_{h,N}$ *is analytic in* $e$.

*Proof.* We can write

$$L^2 E_{h,N} = \log \int P(d\psi) e^{\frac{1}{2}\int dxdy v(\mathbf{x}-\mathbf{y})e^2 \bar\psi_\mathbf{x}\gamma^\mu\psi_\mathbf{x}\bar\psi_\mathbf{y}\gamma^\mu\psi_\mathbf{y}} = \sum_{n=0}^{\infty} \frac{1}{n!} \quad (3.8)$$

$$\int \prod_{i=1}^n d\mathbf{x}_i d\mathbf{y}_i \prod_{i=1}^n v(\mathbf{x}_i - \mathbf{y}_i) \mathcal{E}^T(\bar\psi_{\mathbf{x}_1}\gamma^\mu\psi_{\mathbf{x}_1}\bar\psi_{\mathbf{y}_1}\gamma^\mu\psi_{\mathbf{y}_1}; ...; \bar\psi_{\mathbf{x}_n}\gamma^\mu\psi_{\mathbf{x}_n}\bar\psi_{\mathbf{y}_n}\gamma^\mu\psi_{\mathbf{y}_n})$$

We can write the truncated expectations using (2.118); the fermionic propagator can be written in the form (2.65) with

$$A(\mathbf{x}-\mathbf{y}) = \int d\mathbf{k} e^{i\mathbf{k}(\mathbf{x}-\mathbf{y})} \frac{\sqrt{\chi_{h,N}(\mathbf{k})}}{\mathbf{k}^2} \quad (3.9)$$

$$B(\mathbf{x}-\mathbf{y}) = \int d\mathbf{k} e^{i\mathbf{k}(\mathbf{x}-\mathbf{y})} \sqrt{\chi_{h,N}(\mathbf{k})}\, \slashed{k}, \quad (3.10)$$

so that

$$\|A\| \leq C\gamma^{-h} \qquad \|B\| \leq C\gamma^{2N} \quad (3.11)$$

and

$$|E_{h,N}| \leq L^{-2} \sum_{n=0}^{\infty} \frac{1}{n!} \quad (3.12)$$

$$\int \prod_{i=1}^n d\mathbf{x}_i d\mathbf{y}_i \left| \sum_T \left( \prod_{\ell \in T} g^{(h,N)}(\mathbf{r}_\ell) \right) \left( \prod_{\ell \in T^*} e^2 v(\mathbf{r}_\ell) \right) \int dP_T(\mathbf{t}) \det G^T(\mathbf{t}) \right|$$

where $T \cup T^*$ is a set of trees connecting all the points; using that $\int d\mathbf{r} |g^{[h,N]}(\mathbf{r})| \leq C\gamma^{-h}$ we finally obtain, bounding the sum over $T$ by (2.121)

$$|E_{h,N}| \leq \sum_{n=1}^{\infty} e^{2n} \|v\|^n C^n \gamma^{-h(n-1)} \gamma^{(n+1)(2N-h)} \quad (3.13)$$

and the l.h.s. is uniformly convergent if $|e| \leq C^{-1}\|v\|^{-1}\gamma^{-2(N-h)}$. The existence of the limit can be checked by replacing the propagators with their limiting expressions and the finite sums with the corresponding integrals. By considering the difference of the expansions at finite $L, a^{-1}$ and their limits, it is easy to see that the difference can be still represented as a convergent expansion, vanishing as $L, a^{-1} \to \infty$. ∎

Hence the continuum and infinite volume limit can be taken, provided that the coupling is extremely small; the estimated convergence radius of the series shrinks to zero if $h \to -\infty$ or $N \to \infty$. We have to perform the bounds much more carefully if we want to get bound uniform in the values of the fermionic cut-offs.

## 3.3 Propagator decomposition

We can compute $\mathcal{W}_{h,N}(J^A,\phi)$ using the multiscale expansion described in §2.4; in order to do this we write

$$\chi_{h,N}(\mathbf{k}) = [C_{h,N}(\mathbf{k})]^{-1} = \sum_{j=h}^{N} f_j(\mathbf{k}) \qquad (3.14)$$

with

$$f_j(\mathbf{k}) = \chi_0\left(\gamma^{-j}|\mathbf{k}|\right) - \chi_0\left(\gamma^{-j+1}|\mathbf{k}|\right) \qquad (3.15)$$

The functions $f_j(\mathbf{k})$ are compact-support functions with support $\gamma^{j-1} \leq |\mathbf{k}| \leq \gamma^{j+1}$.

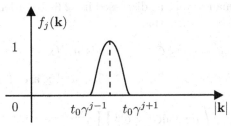

Fig. 3.2  The function $f_j(\mathbf{k})$

By (3.14) we obtain

$$g^{[h,N]}(\mathbf{x}-\mathbf{y}) = \sum_{j=h}^{N} g^{(j)}(\mathbf{x}-\mathbf{y}) \qquad (3.16)$$

with

$$g^{(j)}(\mathbf{x}-\mathbf{y}) = \int \frac{d\mathbf{k}}{(2\pi)^2} \frac{-i\,\slashed{k}+m}{\mathbf{k}^2+m^2} e^{-i\mathbf{k}(\mathbf{x}-\mathbf{y})} f^{(j)}(\mathbf{k}) \qquad (3.17)$$

The single scale propagator verifies the bound given by the following lemma.

**Lemma 3.2.** *Given the positive integers $K, n_0, n_1$ and putting $n = n_0 + n_1$, there exist a constant $C_{K,n}$ such that, if $\mathbf{d}(\mathbf{x}) = (\sin \pi x/L, \sin \pi x_0/L) = (d_L(x), d_L(x_0))$*

$$|g^{(j)}_{\omega,\omega}(\mathbf{x}-\mathbf{y})| \leq C_{K,n} \frac{\gamma^j}{1+(\gamma^j|\mathbf{d}(\mathbf{x}-\mathbf{y})|)^K}$$

$$|g^{(j)}_{\omega,-\omega}(\mathbf{x}-\mathbf{y})| \leq C_{K,n}|\frac{m}{\gamma^j}|\frac{\gamma^j}{1+(\gamma^j|\mathbf{d}(\mathbf{x}-\mathbf{y})|)^K} \qquad (3.18)$$

*Proof* The above bounds follow immediately by integrating by parts: if $N_0, N_1 \geq 0$, such that $K = N_0 + N_1$, and note that

$$d_L(x-y)^{N_1} d_L(x_0-y_0)^{N_0} g^{(h)}_{\omega,\omega'}(\mathbf{x}-\mathbf{y}) = e^{-i\pi(xL^{-1}N_1 + x_0 L^{-1} N_0)}(-i)^{N_0+N_1}$$

$$\frac{1}{L^2} \sum_\mathbf{k} e^{-i\mathbf{k}(\mathbf{x}-\mathbf{y})} \partial_{k'}^{N_1} \partial_{k_0}^{N_0} [f_j(\mathbf{k})[T_j^{-1}(\mathbf{k})]_{\omega,\omega'}] \qquad (3.19)$$

where $\partial_{k'}$ and $\partial_{k_0}$ denote the discrete derivatives. Then one simply uses that, by the compact support properties of the cut-off function, the integral is $O(\gamma^{-(1+n)j})$ and the integration volume is $O(\gamma^{2j})$. ■

We can integrate $\psi^{(N)}, \psi^{(N-1)}, \ldots$ obtaining, as explained in §2.5, a tree expansion for $\mathcal{W}_{h,N}(J^A, \phi)$. Each effective potential $\mathcal{V}^k$ can be written as a sum of Feynnman graphs as discussed in §2.6. In order to bound $\text{Val}(G)$, $G \in \mathcal{G}(\tau), \tau \in \mathcal{T}_{h,N}$, we use that

$$|g^{(j)}(\mathbf{x})| \leq C\gamma^j \qquad \int d\mathbf{x} |g^{(j)}(\mathbf{x})| \leq C\gamma^{-j} \qquad (3.20)$$

Hence, from (2.59), if $G \in \mathcal{G}(\tau), \tau \in \mathcal{T}_{k,N}$ and $\underline{\mathbf{x}}$ are the coordinate of the external lines

$$\int d\underline{\mathbf{x}} |\text{Val} G| \leq C^n \prod_v \gamma^{2h_v(s_v-1)} \gamma^{h_v n_v^0} \qquad (3.21)$$

where $v$ are the vertices of $\tau$, $s_v$ the lines coming out from $v$ and $n_v^0$ the propagator internal to $v$ and not to any inner cluster. We call $n_v^e$ the number of fields external to the cluster $v$ and $m_{4,v}$ the number of end-points internal to $v$. If we define $V_\chi$ the set of vertices such that there is at least a contraction and $v'$ the vertex $\in V_\chi$ preceeding $v$ on the tree, the following relations hold:

$$\sum_{v \in V_\chi(\tau)} (h_v - h)(s_v - 1) = \sum_{v \in V_\chi(\tau)} (h_v - h_{v'})(m_{4,v} - 1) \qquad (3.22)$$

and

$$\sum_{v \in V_\chi(\tau)} (h_v - h) n_v^0 = \sum_{v \in V_\chi(\tau)} (h_v - h_{v'})\left(2m_{4,v} - \frac{n_v^e}{2}\right) \qquad (3.23)$$

Inserting the above two equalities and defining

$$D(P_v) = \frac{n_v^e}{2} - 2, \qquad (3.24)$$

we can rewrite the bound (3.22) as

$$\int d\underline{\mathbf{x}} |\text{Val}(G)| \leq C^n e^{2n} \gamma^{-kD(P_{v_0})} \left(\prod_{v \in V_\chi(\tau)} \gamma^{-(h_v-h_{v'})D(P_v)}\right) \qquad (3.25)$$

if $n$ is the number of the end-points of $\tau$. The quantity $D(P_v)$ is called *dimension* of the vertex $v$. As $D(P_v) \geq -1$ and $|h_v - h_{v'}| \leq N - h$ we get an estimate $O(\gamma^{\vartheta n(N-h)})$, for some finite $\vartheta$, in agreement with the bound (3.13).

However the bound is much better is we restrict to trees with only clusters with six or more external lines; in such a case $D(P_v) \geq 1$ and

$$\sum_{\{h_v\}} \gamma^{-(h_v - h_{v'})D(P_v)} \leq [\sum_{k=0}^{\infty} \gamma^{-nk}]^n \leq C^n \qquad (3.26)$$

that is a bound independent from $N, h$ is found. On the other hand the bound for the cluster with two or more external lines can be improved. Let us consider for instance the following graph and assuming that the three internal lines are at scale $j$, its value is given by

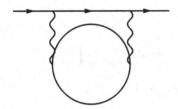

Fig. 3.3  Feynman graph of $\widehat{G}_{a,\omega}^{2,1(2)}$

$$\int d\mathbf{x}_1 d\mathbf{x}_2 d\mathbf{y}_1 d\mathbf{y}_2 v(\mathbf{x}_1 - \mathbf{x}_2) v(\mathbf{y}_1 - \mathbf{y}_2) g^{(j)}(\mathbf{x}_1 - \mathbf{y}_1) g^{(j)}(\mathbf{x}_2 - \mathbf{y}_2) g^{(j)}(\mathbf{x}_2 - \mathbf{y}_2) \qquad (3.27)$$

By integrating over $\mathbf{x}_1 - \mathbf{x}_2, \mathbf{y}_1 - \mathbf{y}_2, \mathbf{x}_1 - \mathbf{y}_1$, we get the bound $\gamma^j = \gamma^k \gamma^{-(k-j)}$ in agreement with the previous estimate. However it is clearly more convenient, for $k \geq 0$, to integrate over a fermionic rather a bosonic line, for instance integrating over $\mathbf{x}_1 - \mathbf{x}_2, \mathbf{x}_2 - \mathbf{y}_2, \mathbf{x}_1 - \mathbf{y}_1$; in this way we we get the much better bound $\gamma^{-j}$.

The above discussion say that our multiscale integration procedure must be somewhat modified in order to take into account the boundedness of the fermionic interaction.

## 3.4 Renormalized expansion

For simplicity we set $h = -\infty$ (keeping the mass finite) and we set

$$e^{W_N(J,\phi)} = \int P(d\psi^{(\leq N)}) e^{\mathcal{V}^{(N)}(\psi^{(\leq N)}) + \mathcal{B}^{(N)}(\psi^{(\leq N)}, J, \phi)} \quad (3.28)$$

and

$$\mathcal{V}^{(N)}(\psi^{(\leq N)}) = -\frac{\lambda}{4} \int dxdy v(\mathbf{x} - \mathbf{y})(\bar\psi_\mathbf{x}^{(\leq N)} \gamma_\mu \psi_\mathbf{x}^{(\leq N)})(\bar\psi_\mathbf{y}^{(\leq N)} \gamma_\mu \psi_\mathbf{y}^{(\leq N)}) =$$

$$-\lambda \int dxdy v(\mathbf{x} - \mathbf{y}) \psi_{\mathbf{x},+}^{(\leq N)+} \psi_{\mathbf{x},+}^{(\leq N)-} \psi_{\mathbf{y},-}^{(\leq N)+} \psi_{\mathbf{y},-}^{(\leq N)-}$$

$$\mathcal{B}^{(N)}(\psi^{(\leq N)}, \phi) = \int d\mathbf{x} J_\mathbf{x} \bar\psi_\mathbf{x}^{(\leq N)} \gamma_\mu \psi_\mathbf{x}^{(\leq N)} + \int d\mathbf{x} [\phi_\mathbf{x} \bar\psi_\mathbf{x}^{(\leq N)} + \bar\phi_\mathbf{x} \psi_\mathbf{x}^{(\leq N)}]$$

and $\lambda = -2e^2$. Using the addition property, we write

$$\int P(d\psi^{(\leq N-1)}) \int P(d\psi^{(N)}) e^{\mathcal{V}^{(N)}(\psi^{(\leq N)}) + \mathcal{B}^{(N)}(\psi^{(\leq N)}, J, \phi)} =$$

$$e^{L^2 E_N} \int P(d\psi^{(\leq N-1)}) e^{\mathcal{V}^{(N-1)}(\psi^{(\leq N-1)}) + \mathcal{B}^{(N-1)}(\psi^{(\leq N-1)}, J, \phi)} \quad (3.29)$$

where

$$\mathcal{V}^{(N-1)}(\psi^{(\leq N-1)}) = \sum_{l,\underline\omega,\underline\varepsilon} \int d\mathbf{x}_1 ... d\mathbf{x}_{2l} W_{2l,\underline\omega,\underline\varepsilon}^{(N-1)} \prod_{i=1}^{2l} \psi_{\mathbf{x}_i,\omega_i}^{(\leq N-1)\varepsilon_i} \quad (3.30)$$

$$\mathcal{B}^{(N-1)}(\psi^{(\leq N-1)}, J, \phi) = \mathcal{B}_1^{(N-1)}(\psi^{(\leq N-1)}, J) +$$
$$\mathcal{B}_2^{(N-1)}(\psi^{(\leq N-1)}, \phi) + \mathcal{B}_3^{(N-1)}(\psi^{(\leq N-1)}, J, \phi)$$

where $\mathcal{B}_1^{(N-1)}$ contains all the terms linear in $J$, $\mathcal{B}_2^{(N-1)}$ all the terms linear in $\phi$ or $\bar\phi$ and $\mathcal{B}_3^{(N-1)}$ is the rest.

We write $\mathcal{V}^{(N-1)}(\psi^{(\leq N-1)})$ in the exponent of r.h.s. of (3.29) as

$$\mathcal{V}^{(N-1)}(\psi^{(\leq N-1)}) = \mathcal{L}\mathcal{V}^{(N-1)}(\psi^{(\leq N-1)}) + \mathcal{R}\mathcal{V}^{(N-1)}(\psi^{(\leq N-1)}) \quad (3.31)$$

where $\mathcal{R} = 1 - \mathcal{L}$ and the $\mathcal{L}$ operation is defined in the following way

$$\mathcal{L} W_{2l}^{(N-1)} = W_{2l}^{(N-1)} \quad \text{if} \quad l = 1, 2$$
$$\mathcal{L} W_{2l}^{(N-1)} = 0 \quad \text{otherwise} \quad (3.32)$$

so that

$$\mathcal{L}\mathcal{V}^{(N-1)}(\psi) = \int \sum_{\underline\varepsilon} [\prod_{i=1}^{4} d\mathbf{x}_i] \lambda_{N-1}(\underline{\mathbf{x}}) \psi_{\mathbf{x}_1,\omega_1}^{\varepsilon_1} \psi_{\mathbf{x}_2,\omega_2}^{\varepsilon_2} \psi_{\mathbf{x}_3,\omega_3}^{\varepsilon_3} \psi_{\mathbf{x}_4,\omega_4}^{\varepsilon_4} +$$

$$\int [\prod_{i=1}^{2} d\mathbf{x}_i] \gamma^{N-1} z_{N-1}(\underline{\mathbf{x}}) \psi_{\mathbf{x}_1,\omega_1}^{+} \psi_{\mathbf{x}_2,\omega_2}^{-} \quad (3.33)$$

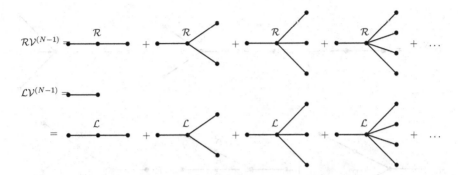

Fig. 3.4 Graphical representation of $\mathcal{R}\mathcal{V}^{(N-1)}$ and $\mathcal{L}\mathcal{V}^{(N-1)}$

Analogously $\mathcal{B}_1^{(N-1)}(\psi^{(\leq N-1)}, J)$ is given by an expression similar to (3.30), sum of momonials with $l$ Grassmann fields and a single $J$ fields, integrated over kernels $W_{2l,1}^{(N-1)}$. We define the $\mathcal{L}$ operation in the following way

$$\mathcal{L}W_{2l,1}^{(N-1)} = W_{2l,1}^{(k)} \quad \text{if} \quad l = 1 \qquad (3.34)$$
$$\mathcal{L}W_{2l,1}^{(N-1)} = 0 \quad \text{otherwise}$$

so that

$$\mathcal{L}\mathcal{B}_1^{(N-1)}(\psi, J) = \int [\prod_{i=1}^{3} d\mathbf{x}_i] Z_{N-1}^{(1)}(\mathbf{x}) \psi_{\mathbf{x}_1,\omega_1}^+ \psi_{\mathbf{x}_2,\omega_2}^- J_{\mathbf{x}_3} \qquad (3.35)$$

Finally $\mathcal{B}_2^{(N-1)}(\psi^{(\leq N-1)}, \phi)$ has the form

$$\mathcal{B}_2^{(N-1)}(\psi, \phi) = \int d\mathbf{x} \left[ \bar{\varphi}_\mathbf{x} \psi_\mathbf{x} + \bar{\psi}_\mathbf{x} \varphi_\mathbf{x} \right] \qquad (3.36)$$
$$+ \int d\mathbf{x} d\mathbf{y} \left[ \bar{\phi}_\mathbf{x} g^{(N)}(\mathbf{x}-\mathbf{y}) \frac{\partial}{\partial \psi_\mathbf{y}} \mathcal{V}^{(N-1)} + \frac{\partial}{\partial \bar{\psi}_\mathbf{y}} \mathcal{V}^{(N-1)} g^{(N)}(\mathbf{x}-\mathbf{y}) \phi_\mathbf{x} \right]$$

and the $\mathcal{L}$ operation is defined decomposing in (3.36) $\mathcal{V}^{(N-1)}$ as $\mathcal{L}\mathcal{V}^{(N-1)} + \mathcal{R}\mathcal{V}^{(N-1)}$. Finally we define $\mathcal{L}\mathcal{B}_3^{(N-1)} = 0$.

Integrating the field $\psi^{(N-1)}$ in (3.29) we get

$$e^{L^2 E_N} \int P(d\psi^{(\leq N-2)}) \int P(d\psi^{(N-1)}) e^{\mathcal{L}\mathcal{V}^{(N-1)}(\psi^{(\leq N-1)}) + \mathcal{R}\mathcal{V}^{(N-1)}(\psi^{(\leq N-1)})}$$
$$e^{\mathcal{L}\mathcal{B}^{(N-1)}(\psi^{\leq N-1}, J, \phi) + \mathcal{R}\mathcal{B}^{(N-1)}(\psi^{\leq N-1}, J, \phi)} =$$
$$e^{L^2 E_{N-1}} \int P(d\psi^{(\leq N-2)}) e^{\mathcal{V}^{(N-2)}(\psi^{(\leq N-2)}) + \mathcal{B}^{(N-2)}(\psi^{(\leq N-2)})} \qquad (3.37)$$

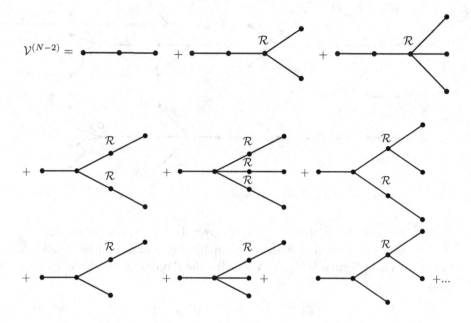

Fig. 3.5  Graphical representation of the effective potential $\mathcal{V}^{(N-2)}$; to the points with scale $N-1$ is associated $\mathcal{L}\mathcal{V}^{(N-1)}$.

As shown in Fig 3.5, $\mathcal{V}^{(N-2)} + \mathcal{B}^{(N-2)}$ can be written as sum of trees, but now the end-points have scale $N$ and $N-1$, and in the latter case $\mathcal{L}\mathcal{V}^{(N-1)} + \mathcal{L}\mathcal{B}^{(N-1)}$ is associated.

The above procedure can be iterated and after the fields $\psi^{(N-2)},...,\psi^{(k+1)}$ are integrated we arrive to expressions similar to (3.36) with $N-1$ replaced by $k$, and in the analogous of (3.36) $g^{(N)}(\mathbf{y}-\mathbf{y})$ is replaced by $\sum_{i=k+1}^{N} g^{(i)}(\mathbf{x}-\mathbf{y})$. The above procedure is iterated up to a scale $K$, defined as the minimum integer such that

$$\|v\| \leq \gamma^{2K} \qquad (3.38)$$

## 3.5  Feynman graph expansion

What we have gained by the introduction of the $\mathcal{L}$ and $\mathcal{R}$ operators in the expansion? $\mathcal{V}^{(k)}$ can be still represented as sums over Feynman graphs, the only difference being that the vertices are not simply associated to $\mathcal{V}^{(N)}$ but also to $\mathcal{LV}^{(j)}$, and to each cluster the $\mathcal{R}$ operation is associated. Proceeding as in §3.3, the bound for a Feynman graph is, for $k \geq K$

$$\int d\underline{x} |\text{Val}(G)| \leq C^n [\sup_{j \geq k}(|\lambda_j| + |z_j|)]^n \gamma^{-kD(P_{v_0})} \left( \prod_{v \in V_\chi}^* \gamma^{-(h_v - h_{v'})D(P_v)} \right) \tag{3.39}$$

where $\prod_v^*$ means that we have the constraint that $n_v^e > 4$ if $h_v \geq K$, from the definition of $\mathcal{R}$ ($\mathcal{R}W = 0$ if $n_v^e = 2, 4$), so that $D(P_v) \geq 1$. This implies a $C^n$-bound for any Feynman graph, contrary to what it was found in §3.3.

Of course this $N$-independent bound is useful only if $\sup_{j \geq k}(|\lambda_j| + |n_j|)$ is indipendent from the ultraviolet cut-off. This fact is not trivial at all, and it is consequence of an improvement of the dimensional bounds due to the boundedness of the bosonic propagator.

In order to prove Theorem 3.1. there are then two technical step to perform; the first is to prove that a bound similar to (3.39) holds for the sum of all Feynman graphs at order $n$ (that is, no factorial arises), and the second is that $\lambda_j$ and $n_j$ are indeed bounded uniformly in $N$.

## 3.6 Convergence of the renormalized expansion

We introduce a *field label* $f$ to distinguish the field variables appearing in the terms of $\mathcal{LV}^{(j)}$ associated with the endpoints. The set of field labels associated with the endpoint $v$ will be called $I_v$. Analogously, if $v$ is not an endpoint, we shall call $I_v$ the set of field labels associated with the endpoints following the vertex $v$; $\mathbf{x}(f)$ will denote the space-time point of the field variable with label $f$. We call *trivial tree* a tree containing only the root and an endpoint.

Let us consider (for definiteness) the case $\phi = 0$ and write

$$\mathcal{V}^{(k)}(\psi^{(\leq k)}, J) + L^2 E_k = \sum_{n=1}^{\infty} \sum_{\tau \in \mathcal{T}_{k,n}} V^{(k)}(\tau) \tag{3.40}$$

where $\mathcal{T}_{k,n}$ is the set of trees defined in the following way.
(1) With each vertex $v$ not end-point an $\mathcal{R}$ operation is associated, up to the first vertex $v_0$ which can have associated either an $\mathcal{R}$ operation or an $\mathcal{L}$ operation.
(2) There are two kind of end-points, *normal* and *special*. With each normal endpoint of scale $h_v$ we associate $\mathcal{LV}^{(h_v - 1)}$ and to each special end-point we

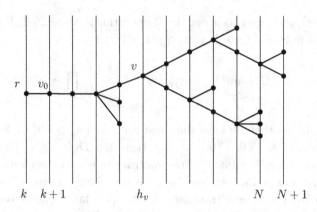

Fig. 3.6  A renormalized tree appearing in the graphic representation of $\mathcal{R}\mathcal{V}^{(k)}$ or $\mathcal{L}\mathcal{V}^{(k)}$

associate $\mathcal{LB}_1^{(h_v-1)}$. Given $v \in \tau$, we shall call $n_v^J$ the number of endpoints of type $J$ following $v$ in the tree.

(3) There is an important constraint on the scale indices of the endpoints. In fact, if $v$ is an endpoint $h_v = h_{v'} + 1$, if $v'$ is the non trivial vertex immediately preceding $v$. This constraint takes into account the fact that at least one of the $\psi$ fields associated with an endpoint normal or of type $J$ has to be contracted in a propagator of scale $h_{v'}$, as a consequence of our definitions.

If $v_0$ is the first vertex of $\tau$ and $\tau_1, .., \tau_s$ ($s = s_{v_0}$) are the subtrees of $\tau$ with root $v_0$, $\bar{V}^{(k)}$ is defined inductively by the relation, $k \leq N - 1$

$$V^{(k)}(\tau) = \frac{(-1)^{s+1}}{s!}\mathcal{E}_{k+1}^T[\bar{V}^{(k+1)}(\tau_1);..;\bar{V}^{(k+1)}(\tau_s)] \qquad (3.41)$$

where $\mathcal{E}_{k+1}^T$ is the *truncated expectation* with integration $P(d\psi^{(k)})$ and $\bar{V}^{(k+1)}(\tau) = \mathcal{R}V^{(k+1)}(\tau)$ if the subtree $\tau_i$ contains more then one endpoint, while if $\tau_i$ contains only one end-point $\bar{V}^{(k+1)}(\tau)$ is equal to one of the following terms

$$\int [\prod_{i=1}^{4} d\mathbf{x}_i]\lambda_{k+1}(\mathbf{x})[\prod_{i=1}^{4} \psi_{\mathbf{x}_i}^{(\leq k+1)\varepsilon_i}] \int [\prod_{i=1}^{2} d\mathbf{x}_i]\gamma^{k+1}z_{k+1}(\mathbf{x})\psi_{\mathbf{x}_1}^{(\leq k+1)+}\psi_{\mathbf{x}_1}^{(\leq k+1)-}$$

$$\int [\prod_{i=1}^{3} d\mathbf{x}_i]Z_k^{(1)}(\mathbf{x})\psi_{\mathbf{x}_1}^{(\leq k+1)+}\psi_{\mathbf{x}_2}^{(\leq k+1)-}J_{\mathbf{x}_3} \qquad (3.42)$$

It is easy to get a simple expression for $V^{(h)}(\tau,\psi^{(\leq k)})$, for any $\tau \in \mathcal{T}_{k,n}$. We associate with any vertex $v$ of the tree a set $P_v$, the *external fermionic fields* of $v$. These subsets must satisfy various constraints. First of all, if $v$ is not an endpoint and $v_1, \ldots, v_{s_v}$ are the vertices immediately following it, then $P_v \subset \cup_i P_{v_i}$. We shall denote $Q_{v_i}$ the intersection of $P_v$ and $P_{v_i}$; this definition implies that $P_v = \cup_i Q_{v_i}$. The subsets $P_{v_i} \setminus Q_{v_i}$, whose union $\mathcal{I}_v$ will be made, by definition, of the *internal fields* of $v$, have to be non empty, if $s_v > 1$.

Given $\tau \in \mathcal{T}_{h,n}$, there are many possible choices of the subsets $P_v$, $v \in \tau$, compatible with all the constraints; we shall denote $\mathcal{P}_\tau$ the family of all these choices and $\mathbf{P}$ the elements of $\mathcal{P}_\tau$. Then we can write

$$V^{(k)}(\tau, \psi^{(\le k)}) = \sum_{\mathbf{P} \in \mathcal{P}_\tau} V^{(k)}(\tau, \mathbf{P}) \tag{3.43}$$

and $V^{(k)}(\tau, \mathbf{P})$ can be represented as

$$V^{(k)}(\tau, \mathbf{P}) = \int d\mathbf{x}_{v_0} \widetilde{\psi}^{(\le k)}(P_{v_0}) \widetilde{J}(\mathbf{z}_{v_0}) K^{(k+1)}_{\tau, \mathbf{P}}(\mathbf{x}_{v_0}) \tag{3.44}$$

where $\mathbf{x}_{v_0}$ are the coordinates of the fermionic or $J$ fields, $\mathbf{z}_{v_0}$ is the subset of the coordinates of the $J$ field, $\widetilde{\psi}^{(\le k)}(P_{v_0}) = \prod_{f \in P_{v_0}} [\psi]^{(\le k) \sigma(f)}_{\mathbf{x}(f)}$, $\widetilde{J}(\mathbf{z}_{v_0})$ is the product of the $J$ field associated with the end-points and $K^{(k+1)}_{\tau, \mathbf{P}}(\mathbf{x}_{v_0}, \mathbf{y}_{v_0})$ defined inductively (recall that $h_{v_0} = h+1$) by the equation, valid for any $v \in \tau$ which is not an endpoint,

$$K^{(h_v)}_{\tau, \mathbf{P}}(\mathbf{x}_v) = \frac{1}{s_v!} \prod_{i=1}^{s_v} [K^{(h_v+1)}_{v_i}(\mathbf{x}_{v_i})] \mathcal{E}^T_{h_v} [\widetilde{\psi}^{(h_v)}(P_{v_1} \setminus Q_{v_1}), \ldots, \widetilde{\psi}^{(h_v)}(P_{v_{s_v}} \setminus Q_{v_{s_v}})] \tag{3.45}$$

If $v$ is an endpoint $K_v^{(h_v)}(\mathbf{x}_v)$ is equal to one of the kernels in (3.42), otherwise if $v$ is not an endpoint $K_v^{(h_v)} = \mathcal{R} K^{(h_v)}_{\tau_i, \mathbf{P}_i}$, where $\tau_1, \ldots, \tau_{s_v}$ are the subtrees of $\tau$ with root $v$, $\mathbf{P}_i = \{P_v, v \in \tau_i\}$.

By using the formula for the truncated expectations (2.118) a more explicit expression for $V^{(k)}(\tau, \mathbf{P})$ is obtained, namely

$$V^{(k)}(\tau, \mathbf{P}) = \sum_{T \in \mathbf{T}} \int d\mathbf{x}_{v_0} W_{\tau, \mathbf{P}, T}(\mathbf{x}_{v_0}) \widetilde{\psi}^{(\le k)}(P_{v_0}) \tag{3.46}$$

where

$$W_{\tau, \mathbf{P}, T,}(\mathbf{x}_{v_0}) = \Big[ \prod_{i=1}^n K^{h_i}_{v_i^*}(\mathbf{x}_{v_i^*}) \Big] \tag{3.47}$$

$$\Big\{ \prod_{v \text{ not e.p.}} \frac{1}{s_v!} \int dP_{T_v}(\mathbf{t}_v) \cdot \det G^{h_v, T_v}(\mathbf{t}_v) \Big[ \prod_{l \in T_v} g^{(h_v)}(\mathbf{x}_l - \mathbf{y}_l) \Big] \Big\}$$

and $v_i^*$ are the end-points, $\mathbf{T}$ is the set of the tree graphs on $\mathbf{x}_{v_0}$, obtained by putting together an anchored tree graph $T_v$ for each non trivial vertex

$v$ and adding lines connecting the space-time points belonging to the set $\mathbf{x}_v$ for each endpoint $v$; $v_1^*, \ldots, v_n^*$ are the endpoints of $\tau$, "e.p." is an abbreviation of "endpoint" and $G^{h_v,T_v}(\mathbf{t}_v)$ is the matrix with elements

$$G^{h_v,T_v}_{ij,i'j'} = t_{v,i,i'} g^{(h_v)}(\mathbf{x}_{ij} - \mathbf{y}_{i'j'}) \qquad (3.48)$$

Note that the definition of $\mathcal{R}$ imposes the constraints on $\mathbf{P}$ that, for each $v$, $|P_v|/2 + n_v^J > 2$.

## 3.7 Determinant bounds

Denoting by $||f|| = \frac{1}{L^2} \int \prod_{i=1}^n d\mathbf{x}_i |f(\mathbf{x}_1,..,\mathbf{x}_n)|$, we will prove the following lemma, if $W^{(k)}_{2l,m}$ are defined as in §3.4.

**Lemma 3.3.** *If* $\sup_{i \geq k}[||\lambda_i||, ||n_i||] \leq \varepsilon_k$ *for* $\varepsilon_k$ *small enough then, for a suitable constant* $C$

$$||W^{(k)}_{2l,m}|| \leq C^{l+m} \varepsilon_k^{\max(l-1,0)} \gamma^{-k(l+m-2)} \qquad (3.49)$$

*Proof* In order to bound $\det G^{h_v,T_v}(\mathbf{t}_v)$ in (3.48) we can use the Gram inequality. It is easy to verify that

$$G^{h_v,T_v}_{ij,i'j'} = t_{i,i'} g^{(h_v)}(\mathbf{x}_{ij} - \mathbf{y}_{i'j'}) = < \mathbf{u}_i \otimes A^{(h_v)}_{\mathbf{x}(f_{ij}^-)}, \mathbf{u}_{i'} \otimes B^{(h_v)}_{\mathbf{x}(f_{i'j'}^+)} > \qquad (3.50)$$

where $\mathbf{u}_i \in \mathbb{R}^s$, $i = 1, \ldots, s$, are the vectors such that $t_{i,i'} = \mathbf{u}_i \cdot \mathbf{u}_{i'}$, and

$$A^{(h_v)}_{\mathbf{x}} = \int d\mathbf{k} e^{i\mathbf{k}(\mathbf{x}-\mathbf{y})} \frac{\sqrt{f^{(h_v)}(\mathbf{k})}}{\mathbf{k}^2 + m^2}$$

$$B^{(h_v)}_{\mathbf{x}} = \int d\mathbf{k} e^{i\mathbf{k}\mathbf{x}} \sqrt{f^{(h_v)}(\mathbf{k})} \left(\slashed{k} + m\right), \qquad (3.51)$$

so that

$$|\det G^{h_v,T_v}(\mathbf{t}_v)| \leq C^{\sum_{i=1}^{s_v}|P_{v_i}|-|P_v|-2(s_v-1)} \gamma^{\frac{h_v}{2}\left(\sum_{i=1}^{s_v}|P_{v_i}|-|P_v|-2(s_v-1)\right)} \qquad (3.52)$$

By using (3.48) and (3.52) we get

$$\int d\mathbf{x}_{v_0} |W_{\tau,\mathbf{P},T}(\mathbf{x}_{v_0})| \leq C^n J_{\tau,\mathbf{P},T} \prod_{v \text{ not e.p.}} \cdot$$

$$\cdot C^{\sum_{i=1}^{s_v}|P_{v_i}|-|P_v|-2(s_v-1)} \gamma^{\frac{h_v}{2}\left(\sum_{i=1}^{s_v}|P_{v_i}|-|P_v|-2(s_v-1)\right)} \qquad (3.53)$$

where

$$J_{\tau,\mathbf{P},T} = \int d\mathbf{x}_{v_0} \left| \left[\prod_{i=1}^n K^{h_i}_{v_i^*}(\mathbf{x}_{v_i^*})\right] \left\{ \prod_{v \text{ not e.p.}} \frac{1}{s_v!} \left[\prod_{l \in T_v} g^{(h_v)}_{\omega_l^-,\omega_l^+}(\mathbf{x}_l - \mathbf{y}_l)\right] \right\} \right| \qquad (3.54)$$

Using that
$$J_{\tau,\mathbf{P},T} \le C^n L^2(\varepsilon_h)^n \prod_{v \text{ not e.p.}} \left[\frac{1}{s_v!} C^{2(s_v-1)} \gamma^{-h_v(s_v-1)}\right] \tag{3.55}$$

and
$$\sum_{\bar{v} \ge v}\{\frac{1}{2}\left(\sum_{i=1}^{s_{\bar{v}}} |P_{\bar{v}_i}| - |P_{\bar{v}}|\right) - 2(s_{\bar{v}} - 1)\} = -\frac{1}{2}|P_v| - n_v^J + 2 \tag{3.56}$$

we find
$$\int d\mathbf{x}_{v_0} |W_{\tau,\mathbf{P},T}(\mathbf{x}_{v_0})| \le \tag{3.57}$$

$$C^n L^2 \varepsilon_h^n \gamma^{-kD(P_{v_0})} \prod_{v \text{ not e.p.}} \left\{\frac{1}{s_v!} C^{\sum_{i=1}^{s_v} |P_{v_i}| - |P_v|} \gamma^{-D(P_v)}\right\} \tag{3.58}$$

with
$$D(P_v) = -2 + \frac{|P_v|}{2} + n_v^J \tag{3.59}$$

In order to sum over $\tau, \mathbf{P}, T$, we first note that the number of addenda in $\sum_{T \in \mathbf{T}}$ is bounded by $\prod_{v \text{ not e.p.}} s_v! \, C^{\sum_{i=1}^{s_v} |P_{v_i}| - |P_v|}$, from the Lemma 2.4.

We define $V_\chi(\tau)$ the set of vertices such that the set of internal fields $\mathcal{I}_v$ is not empty and $v'$ the vertex $\in V_\chi$ preceeding $v$ on the tree. In order to bound the sums over $\tau$ and $\mathbf{P}$ we first use the inequality, if $\alpha > 0$

$$\prod_{v \text{ not e.p.}} \gamma^{-D(P_v)} \le [\prod_{v \in V_\chi} \gamma^{-\alpha(h_v - h_{v'})}][\prod_{v \text{ not e.p.}} \gamma^{-\alpha|P_v|}] \tag{3.60}$$

which follows from the fact that by construction $D(P_v) \ge 1$.

The sum over the trees $\tau \in \mathcal{T}_{k,n}$ can be performed by summing over all the unlabeled trees $\tau^*$ with $n$ end-points and, fixed $\tau^*$, summing over all the possible scale labels of the non trivial vertices of $\tau^*$ (fixing the scale of the root); the value of the scale labels automatically determines the number of trivial vertices between them. Hence we get

$$\sum_{\tau \in \mathcal{T}_{h,n}} \prod_{v \in V_\chi(\tau)} \gamma^{-\alpha(h_v - h_{v'})} \le \sum_{\tau^*} \sum_{\{h_v\}} \gamma^{-\alpha(h_v - h_{v'})} \le C^n \tag{3.61}$$

where $\sum_{\tau^*}$ is bounded using lemma 2.2 and the second sum is bounded by $(\sum_{k=0}^\infty \gamma^{-\alpha k})^n$.

Finally the sum over $\mathbf{P}$ can be bounded by the following combinatorial inequality.

$$\sum_{\mathbf{P}} \left(\prod_{v \text{ not e.p.}} \gamma^{-\alpha|P_v|}\right) \le C^n \tag{3.62}$$

for some constant $C$ depending on $\alpha$. Indeed the sum over $\mathbf{P}$ is such that, if $|P_v| \equiv p_v$ $p_v \leq \sum_{i=1}^{s_v} p_{v_i}$ for all $v$ which are not end-points, hence we get the bound

$$\sum_{\mathbf{P}} \left( \prod_{v \text{ not e.p.}} \gamma^{-\alpha|P_v|} \right) \leq \prod_{v \text{ not e.p.}} \sum_{p_v} \gamma^{-\alpha|P_v|} \frac{(p_{v_1} + \ldots + p_{v_{s_v}})!}{p_v!(p_{v_1} + \ldots + p_{v_{s_v}} - p_v)!} \tag{3.63}$$

We can write the r.h.s. of (3.63) as

$$\prod_v \left[ \sum_{p_v} \gamma^{-\alpha p_v} \frac{(p_{v_1} + \ldots + p_{v_{s_v}})!}{p_v!(p_{v_1} + \ldots + p_{v_{s_v}} - p_v)!} \right] \equiv \prod_v I_v \tag{3.64}$$

which defines the factors $I_v$. In particular we have

$$I_{v_0} = \sum_{p_{v_0}} \gamma^{-\alpha p_{v_0}} \left( \frac{(p_{v_{01}} + \ldots + p_{v_{0s_{v_0}}})!}{p_{v_0}!(p_{v_{01}} + \ldots + p_{v_{0s_{v_0}}} - p_{v_0})!} \right)$$

$$= \left(1 + \gamma^{-\alpha}\right)^{p_{v_{01}} + \ldots + p_{v_{0s_{v_0}}}} = \prod_{j=1}^{s_{v_0}} \left(1 + \gamma^{-\alpha}\right)^{p_{v_{0j}}} \tag{3.65}$$

where $v_{01}, \ldots, v_{0s_{v_0}}$ are the vertices immediately following $v_0$, so that

$$\prod_v \sum_{p_v} \gamma^{-\alpha p_v} \frac{(p_{v_1} + \ldots + p_{v_{s_v}})!}{p_v!(p_{v_1} + \ldots + p_{v_{s_v}} - p_v)!} = \left( \prod_{v > v_0} I_v \right) \prod_{j=1}^{s_{v_0}} \left(1 + \gamma^{-\alpha}\right)^{p_{v_{0j}}} \tag{3.66}$$

If we iterate the procedure we obtain

$$I_{v_0} \prod_{j=1}^{s_{v_0}} I_{v_{0j}} = \prod_{j=1}^{s_{v_0}} \prod_{j'=1}^{s_{v_j}} \left(1 + \gamma^{-\alpha} \left(1 + \gamma^{-\alpha}\right)\right)^{p_{v_{jj'}}} \tag{3.67}$$

where $v_{j1}, \ldots, v_{js_v}$ are the vertices immediately following $v_{0j}$. And so on until we reach all the endpoints of the tree $\tau$. If we denote by $\mathcal{P}$ a path (*i.e.* an oriented connected set of lines) from the root to an endpoint we find

$$\prod_v I_v = \prod_{\mathcal{P}} \left[\left(1 + \gamma^{-\alpha} \left(1 + \gamma^{-\alpha} \left(1 + \gamma^{-\alpha} \left(\ldots\right)\right)\right)\right)\right]^4 \tag{3.68}$$

where we used that the endpoints have at most four external lines and the product is over all the possible paths on $\tau$. Then, if we denote by $\ell(\mathcal{P})$ the "lenght" of the path $\mathcal{P}$, *i.e.* number of vertices along the path $\mathcal{P}$, we have

$$\prod_v I_v = \prod_{\mathcal{P}} \left[ \sum_{k=0}^{\ell(\mathcal{P})} \gamma^{-\alpha k} \right]^4 \leq \left( \frac{\gamma^\alpha}{\gamma^\alpha - 1} \right)^{4n} \equiv C^n \tag{3.69}$$

where $C = \gamma^{4\alpha} \left(\gamma^\alpha - 1\right)^{-4}$. ∎

## 3.8 The short memory property

We have shown in lemma 3.3 that the kernels $W_{2l,m}^{(k)}$ are bounded and smaller than $O(\gamma^{-k(l+m-2)})$, under the assumption that the functions $\lambda_n, n_k, Z_k^{(1)}$ are small.

Let us consider a tree $\tau$ with a non-trivial vertex scale $i$; we can write in (13.13), using that $D(P_v) \geq 1$

$$\int d\mathbf{x}_{v_0} |W_{\tau,\mathbf{P},T}(\mathbf{x}_{v_0})| \leq \qquad (3.70)$$

$$C^n L^2 \varepsilon_h^n \gamma^{(k-i)/2} \gamma^{-hD(P_{v_0})} \prod_{v \text{ not e.p.}} \left\{ \frac{1}{s_v!} C^{\sum_{i=1}^{s_v} |P_{v_i}| - |P_v|} \gamma^{-D(P_v)/2} \right\}$$

and proceeding as above we get for such trees *the same bound times an extra factor* $\gamma^{(k-i)/2}$.

This means that the trees have values smaller and smaller if they have non trivial interations at scales more and more distant from $k$; in other words the final result does not depend from the detail of the ultraviolet region but only on the scales near $k$. This fact motivates the name *short memory property* for this result.

## 3.9 Extraction of loop lines

It remains now to prove that the effective couplings remain small for any $k \geq K$. In order to prove this we note that

$$\lambda_k = W_{4,0}^{(k)}, \quad \gamma^k n_k = W_{2,0}^{(k)}, \quad Z_k^{(1)} = W_{2,0}^{(k)} \qquad (3.71)$$

and the boundedness of the runnning coupling constants is consequence of the following lemma, in which the bounds (3.49) are improved.

**Lemma 3.4.** *For $\lambda$ small enough and $K \leq j \leq N$, for a suitable constants $C_1$*

$$\|W_{2,0}^{(k)}\| \leq C_1 \gamma^k \gamma^{(K-k)} |\lambda| \qquad (3.72)$$

$$\|W_{4,0}^{(k)} - \lambda v\| \leq C_1 \gamma^{(K-k)} \lambda^2 \quad \|W_{2,1}^{(k)} - 1\| \leq C_1 \gamma^{(K-k)}$$

*Proof.* We proceed by induction; we assume that (3.72) holds for $k+1 \leq j \leq N$, so that the bound (3.49) holds with $\varepsilon_k$ replaced by $C|\lambda|$. Note that

$$W_{2,0}^{(k)}(\mathbf{x},\mathbf{y}) = \frac{\partial^2}{\partial \phi_{\mathbf{x},\omega}^+ \partial \phi_{\mathbf{y},\omega}^-} \sum_{n=1}^{\infty} \frac{1}{n!} \mathcal{E}_{k,N}^T (V(\psi + \phi); n) \qquad (3.73)$$

and we can use the following property of truncated expectations

$$\mathcal{E}^T(\widetilde{\psi}(P_1 \cup P_2); \widetilde{\psi}(P_3); ...; \widetilde{\psi}(P_n)) = \quad (3.74)$$

$$\sum_{\substack{K_1,K_2,K_1\cup K_2=0 \\ K_1\cup K_2=(3,...,n)=\{\alpha_i\}_{i=1}^{|K_1|+|K_2|}}} (-1)^\pi \mathcal{E}^T(\widetilde{\psi}(P_1); ...; \widetilde{\psi}(P_{\alpha_{|K_1|}}))$$

$$\mathcal{E}^T(\widetilde{\psi}(P_2); \widetilde{\psi}(P_{\alpha_{|K_1|+1}}); ...; \widetilde{\psi}(P_{\alpha_{|K_1|+|K_2|}})) + \mathcal{E}^T(\widetilde{\psi}(P_1); ...; \widetilde{\psi}(P_n))$$

and $(-1)^\pi$ is the parity of the permutation necessary to bring the Grassmann variables on the r.h.s. of (3.74) to the original order. Note that the number of terms in the sum in the r.h.s. of (3.74) is bounded by $C^n$ for a suitable constant $C$. The above property is quite obvious; the second addend in the l.h.s. in (3.74) represent the graphs in which $\widetilde{\psi}(P_1)$ and $\widetilde{\psi}(P_2)$ are connected, while the first term are the disconnected grahs.

From (3.74) we can write

$$W_{2,0}^{(k)}(\mathbf{x},\mathbf{y}) = \int d\mathbf{y}_1 \lambda v(\mathbf{x}-\mathbf{y}_1) W_{0,1}^{(k)}(\mathbf{y}_1) g^{(k,N)}(\mathbf{x}-\mathbf{y}_2) W_{2,0}^{(k)}(\mathbf{y}_2;\mathbf{y}) +$$

$$\lambda \int d\mathbf{y}_2 v(\mathbf{x}-\mathbf{y}_1) g^{(k,N)}(\mathbf{x}-\mathbf{y}_2) W_{2,1}^{(k)}(\mathbf{y},\mathbf{y}_2;\mathbf{y}_1) + \quad (3.75)$$

$$\lambda \delta(\mathbf{x}-\mathbf{y}) \int d\mathbf{y}_1 v_K(\mathbf{x}-\mathbf{y}_1) W_{0,1}^{(k)}(\mathbf{y}_1)$$

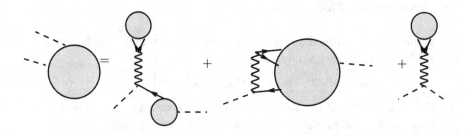

Fig. 3.7 Graphical representation of (3.75); the blobs represent $W_{n,m}^{(h)}$, the paired wiggly lines represent $v$, the paired line $g^{(h,N)}$.

In the massless case $m = 0$ the first and the third addend are vanishing; hence, using that $\|g^{(j)}\|_1 \leq \widetilde{C}\gamma^{-j}$ and $\|v\|_\infty \leq C\gamma^{2K}$, we obtain the following bound

$$\|W_{2,0}^{(k)}\| \leq |\lambda| \cdot \|v\|_\infty \cdot \|W_{2,1}^{(k)}\| \cdot \sum_{j=k}^{N} \|g^{(j)}\|_1 \leq \frac{\widetilde{C}}{1-\gamma^{-1}} C_0 |\lambda| \gamma^k \gamma^{-2(k-K)}$$

$$(3.76)$$

Note that we have a gain $O(\gamma^{-2(k-K)})$, due to the fact that we are integrating over a fermionic instead than over a bosonic line.

Similar arguments can be repeated for $W_{2,0}^{(k)}$, which can be decomposed as in the following picture.

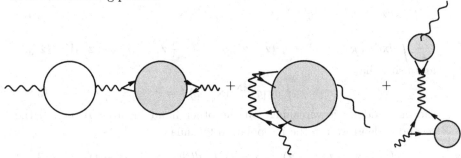

Fig. 3.8 Decomposition of $W_{2,0}^{(k)}$: the blobs represent $W_{n,m}^{(k)}$, the paired wiggly lines represent $v$, the paired line $g^{(k,N)}$

The second term in Fig.3.8 is given by

$$\lambda \int d\mathbf{w}\,d\mathbf{u}'\,d\mathbf{z}'\,d\mathbf{u}\,d\mathbf{w}'\, v(\mathbf{u}' - \mathbf{z}') g^{(k,N)}(\mathbf{w} - \mathbf{u}) g(\mathbf{w} - \mathbf{u}') g^{(k,N)}$$
$$\times (\mathbf{u}' - \mathbf{w}') W_{2,2}^{(k)}(\mathbf{w}', \mathbf{u}; \mathbf{z}, \mathbf{z}') \qquad (3.77)$$

It is convenient to decompose the three propagators into scales, $\sum_{j,i,i'=k}^{N} g^{(j)} g^{(i)} g^{(i')}$ and then, for any realization of $j, i, i'$, to take the $\|\cdot\|_1$ norm on the two propagator on the higher scales, and the $\|\cdot\|_\infty$ norm on the propagator with the lowest one. In this way we can bound (3.77) by:

$$|\lambda| |v|_\infty \cdot \|W_{2;2}^{(k)}\| 3! \sum_{j=k}^{N} \sum_{i=k}^{j} \sum_{i'=k}^{i} \|g^{(j)}\|_1 \|g^{(i)}\|_1 \|g^{(i')}\|_\infty \le$$

$$C_1 |\lambda| \gamma^{2K} \gamma^{-k} \sum_{j=k}^{N} \sum_{i=k}^{j} \sum_{i'=k}^{i} \gamma^{-j} \gamma^{-i} \gamma^{i'} \le C_2 |\lambda| \|v\|_\infty \gamma^{-k} \sum_{j=k}^{N} \gamma^{-j}(j-k) \le$$

$$C_2 |\lambda| \|v\|_\infty \gamma^{-k} \sum_{j=k}^{N} \gamma^{-j} \gamma^{(j-k)/2} \le C_4 |\lambda| \gamma^{-2(k-K)} \qquad (3.78)$$

A similar bound is found for the third term in Fig.3.8; regarding the first term, we can rewrite it as

$$\int dxdz [g^{(k,N)}(\bar{z}-x)]^2 \lambda v_K(x-\bar{z}) W_{0,2}^{(k)}(\bar{z},y) =$$

$$\int dxdz \lambda v_K(\bar{z}-z)[g^{(k,N)}(x-z)]^2 W_{0,2}^{(k)}(\bar{z},y) \qquad (3.79)$$

$$+ \int dxdz [v_K(\bar{z}-x) - v_K(\bar{z}-z)][g^{(k,N)}(x-z)]^2 \lambda v_K(z-z') W_{0,2}^{(k)}(\bar{z},y)$$

and using that

$$\int dx [g^{(k,N)}(x-z)]^2 = 0 \qquad (3.80)$$

the first addend is vanishing; on the other hand in order to bound the second addend we use the intepolation formula

$$v(x'-x) - v(x'-z) = (x-z) \int_0^1 dt \partial v_K(x' + t(x-z)) \qquad (3.81)$$

so that it can be bounded by $C\lambda \gamma^{K-k}$, as by induction $||W_{0,2}^{(k)}|| \leq C|\lambda|$. A similar analysis proves the bound for $W_{4,0}^{k)}$. ∎

Note that in the above lemma two are the crucial points. The first is that, in the graphs which remain connected even cutting a bosonic line, we get a gain $\gamma^{2(K-h)}$ integrating over a bosonic instead than over a fermionic line. Of course there are graphs which are disconnected cutting a bosonic line; for them, the vanishing of the "polarization boubble" (3.80) is what produces the desidered gain.

By using Lemma 3.3 and Lemma 3.4 we the following bounds is derived for $k \geq K$ and $\lambda$ small enough

$$||W_{2l,m}^{(k)}|| \leq \bar{C}|\lambda|\gamma^{-k(l+m-2)} \qquad (3.82)$$

The above results says that the expansion is convergent in the limit $N \to \infty$, provided that the infrared cut-off is greater than the scale $K$, $k \geq K$.

The integration of the scales $\leq K$ can be also performed provided that the fermionic mass is large enough. We call $h^*$ the scale of the fermionic mass, defined as

$$\gamma^{h^*-1} \leq |m| \qquad \gamma^k \geq |m| \quad \text{if} \quad k \geq h^* \qquad (3.83)$$

so that $\gamma^{h^*} = O(|m|)$. It is easy then to check that the propagator of all scales $\leq h^*$ verifies the same bound as a single scale propagator, namely

$$|g^{(\leq h^*)}(x-y)| \leq \gamma^{h^*} \frac{C_P}{1 + (\gamma^{h^*}|x-y|)^P} \qquad (3.84)$$

This implies that the field $\psi^{(\leq h_*)}$ can be integrated without any problem. On the other hand it is clear that in the integration of the scales between $K$ and $h^*$ there is no the gain in lemma 3.4, hence such integration can be done only if $|\lambda| \leq C|m|^{-\alpha}$.

## 3.10 The 2-point Schwinger function

The 2-point Schwinger is still expressed by a sum over trees $\tau \in \mathcal{T}_{h,j}$ similar with the previous ones, with the only difference that there are also two endpoints associated with $\int d\mathbf{k}[\bar{\psi}_\mathbf{k}\phi_\mathbf{k} + \bar{\phi}_\mathbf{k}\psi_\mathbf{k}]$ at a vertex $v$ such that $h_{v'} = h$, and $j$ is the root scale.

We obtain, choosing $|\mathbf{k}| = \gamma^h$

$$\langle \bar{\psi}_\mathbf{k}\psi_\mathbf{k}\rangle = g^{(h)}(\mathbf{k}) + g^{(h)}(\mathbf{k})W_2^{(h)}(\mathbf{k})g^{(h)}(\mathbf{k}) \qquad (3.85)$$

where

$$W_2^{(h)}(\mathbf{k}) = \sum_{\tau \in \mathcal{T}_{h,j}} W_\tau^{(h)} \qquad (3.86)$$

The sum over the scales is done as for the effective potential, with the only difference that instead of fixing the scale of the root, we have now to fix the scale $h$ of the external lines; this has no effect, since we bound the sum over the scales with the sum over the the differences $h_v - h_{v'}$. There is apparently a problem due to the fact that the dimensions can be vanishing; this happens when $|P_v| = 4, n_v^\phi = 2$, that is in the path in the tree connecting the vertex $\bar{v}_0$ with the root $v$ with scale $j_0$. However we can write $\gamma^{j_0} = \gamma^h \gamma^{j_0-h}$ and we can use the factor $\gamma^{j_0-h}$ to make negative the dimension of all vertices. Therefore we get the bound $|W_2^{(h)}(\mathbf{k})| \leq C|\lambda|\gamma^h$ from which Theorem 3.1 follows.

## 3.11 The Yukawa model

Another important example of superrinormalizable QFT in $d = 1 + 1$ is given by the Yukawa model, in which the interaction is

$$\int d\mathbf{x}\phi_\mathbf{x}\bar{\psi}_\mathbf{x}\psi_\mathbf{x} \qquad (3.87)$$

and $\phi_\mathbf{x}$ is a scalar (instead that a vector) massive boson. The construction of Yukawa model with a large fermionic mass was first done integrating out the fermions and reducing to a purely bosonic functional integral, see [3] for references. In [25] it was provided the construction of the Yukawa model in a purely fermionic approach, in a way quite similar to the one followed here for QED2. After the integration of the bosom fields, one obtains a fermionic integral similar to (3.8) with interaction

$$\mathcal{V} = \int d\mathbf{x}d\mathbf{x}v(\mathbf{x}-\mathbf{y})\bar{\psi}_\mathbf{x}\psi_\mathbf{x}\bar{\psi}_\mathbf{y}\psi_\mathbf{y} \qquad (3.88)$$

and the main differences with the previous analysis is that the fermionic boubble is diverging and that the Schwinger functions are Borel summable rather than analytic.

# Chapter 4

# Infrared Problem and Anomalous Behavior

## 4.1 Anomalous dimension

The convergence radius of the expansion seen in the previous chapter is vanishing as $m \to 0$, and convergence is achieved only if the charge is much smaller than the mass. This assumption is not very reasonable physically; in the real word, the electron mass is $O(10^{-20})$ smaller than the electric charge in adimensional units.

This non-uniformity of the expansion seen in chapt. 3 is not simply a technical problem, but it signals that the large distance behavior of the interacting system is qualitatively different with respect to the non-interacting one. This is made clear from the following theorem, proved in the next two chapters.

**Theorem 4.1.** *Given e small enough the 2-point Schwinger function exists in the limit $\lim_{N,-h,a^{-1}\to\infty}$ and it verifys the bound, for any integer $P$ and $|\mathbf{x}-\mathbf{y}| \geq 1$*

$$|\langle \psi_\mathbf{x} \bar\psi_\mathbf{y} \rangle| \leq \frac{1}{1+|\mathbf{x}-\mathbf{y}|^{1+\eta}} \frac{C_P}{1+(m^{1+\eta_\mu}|\mathbf{x}-\mathbf{y}|)^P} \quad (4.1)$$

*with $\eta = a_z e^4 + O(e^6)$, $\eta_\mu = a_\mu e^2 + O(e^4)$, $a_z, a_\mu$ positive constants; for $1 \leq |\mathbf{x}-\mathbf{y}| \leq m^{-1-\eta_\mu}$*

$$|\langle \psi_\mathbf{x} \bar\psi_\mathbf{y} \rangle| = \frac{g(\mathbf{x}-\mathbf{y})}{1+|\mathbf{x}-\mathbf{y}|^\eta}(1 + A(\mathbf{x},\mathbf{y})) \quad (4.2)$$

*with*

$$|A(\mathbf{x},\mathbf{y})| \leq C[e^2 + (m^{1+\eta_\mu}|\mathbf{x}-\mathbf{y}|)^{\frac{1}{2}}] \quad (4.3)$$

*and $C_P, C$ constants.*

*The Fourier transform $\langle \psi_\mathbf{k} \bar\psi_\mathbf{k} \rangle$ in the massless case diverges as $O(|\mathbf{k}|^{-1+\eta})$ as $\mathbf{k} \to 0$, while vanishes as $O(|\mathbf{k}|^{-1})$ as $\mathbf{k} \to \infty$.*

In the free case, one can distinguish two different regimes in the asymptotic behavior of the fermionic propagator, discriminated by an intrinsic length $\xi$ given essentially by the inverse mass $\xi \simeq m^{-1}$; if $1 << |\mathbf{x}| << \xi$, the bounds for the correlation function are the same as in the massless case, namely a $O(|\mathbf{x}-\mathbf{y}|^{-1})$ decay, while if $\xi << |\mathbf{x}|$, there is a faster than any power decay with rate of order $\xi^{-1}$. In the interacting case there are still two different regimes, but the intrinsic length which discriminates between them is $\xi \simeq m^{-1-\eta_\mu}$, where $\eta_\mu$ is a critical index depending on the interaction; moreover in the first regime the exponent of the power law decay is not 1 but $1+\eta$, with $\eta$ a non universal critical index (*anomalus behavior*).

Finally, we see that, while the interaction does not modify the properties for large momenta of the fermionic two-point function, it deeply changes the small momenta behavior, producing a singularity (when $m=0$) with a different exponent with respect to the free case.

A proof of the above theorem will be given in the following two chapters, referring for more details to the original proof in Refs.[28],[29],[30],[31].

## 4.2 Renormalization

If we perform a naive multiscale integration of the fields $\psi^{(K)}, \psi^{(K-1)}, ...$ we arrive to an expansion in terms of Feynman graphs, each one bounded by, if $h \leq K$ and with the same nottaions as in §3.1

$$\int d\underline{\mathbf{x}} |\text{Val}(G)| \leq C^n |\lambda|^n \gamma^{-hD(P_{v_0})} e^{2n} \left( \prod_{v \in V_\chi(\tau)} \gamma^{-(h_v - h_{v'})D(P_v)} \right). \quad (4.4)$$

with $D(P_v) = n_v^e/2 - 2$. Again problems come from the subgraphs with two or four external cases, but we cannot improve the bounds just extracting loop lines as we did for the ultraviolet integration; the factors $\gamma^{K-h}$ one obtains in that way are greater than 1 for $K \geq h$. Such infrared problems are due to the presence of the anomalous behavior; indeed we if try to write the 2-point function (in momentum space) in a power series we get, for small $\mathbf{k}$

$$\langle \psi_\mathbf{k} \bar\psi_\mathbf{k} \rangle \simeq \frac{g_0(\mathbf{k})}{|\mathbf{k}|^\eta} = g_0(\mathbf{k})[1 - \eta \log \mathbf{k} + \eta^2 \log^2 \mathbf{k} + ...] \quad (4.5)$$

so that each term of the expansion is more and more diverging as $\mathbf{k} \to 0$. This suggest that a simple expansion in the coupling cannot work and that a different integration procedure must be found.

Let us consider the graphs with for external lines associated to a cluster. We remember that the non diagonal propagator have an extra factor $\frac{m}{\gamma^j}$, hence if we integrate up to scale $h^*$, they have an extra factor $\gamma^{h^*-j}$ rendering the dimension negative. The dangerous part of the subgraphs with four external lines is then only the one which is independent from $m$; moreover a moment reflection also suggest that the dangerous part is only the one computed at vanishing external momenta. If $\Xi^{(h)}(\mathbf{k}_1, \mathbf{k}_2, \mathbf{k}_3)$ is the value of a graph with four external lines, we can subtract its value at vanishing momenta obtaining so obtaining

$$\Xi^{(h)}(\mathbf{k}_1, \mathbf{k}_2, \mathbf{k}_3) - \Xi^{(h)}(0,0,0) = [\Xi^{(h)}(\mathbf{k}_1, \mathbf{k}_2, \mathbf{k}_3) - \Xi^{(h)}(0, \mathbf{k}_2, \mathbf{k}_3)] +$$
$$[\Xi^{(h)}(0, \mathbf{k}_2, \mathbf{k}_3) - \Xi^{(h)}(0, 0, \mathbf{k}_3)] + [\Xi^{(h)}(0, 0, \mathbf{k}_3) - \Xi^{(h)}(0, 0, 0)] \quad (4.6)$$

Each of such terms can be written by an interpolating formula; for instance

$$\Xi^{(h)}(\mathbf{k}_1, \mathbf{k}_2, \mathbf{k}_3) - \Xi^{(h)}(0, \mathbf{k}_2, \mathbf{k}_3) = \mathbf{k}_1 \cdot \int_0^1 dt \partial_{\mathbf{k}} \Xi^{(h)}(t\mathbf{k}_1, \mathbf{k}_2, \mathbf{k}_3) , \quad (4.7)$$

with $\partial_{\mathbf{k}} = (\partial_k, \partial_{k_0})$. As $\widehat{\Xi}^{(h)}$ is the integral of a product of propagators $g_\ell$ with scales $\geq h_v$, or interactions $v(\mathbf{p})$, the derivative in (4.7) produces an extra dimensional factor $\gamma^{-h_v}$ (if the derivative is applied over $v(\mathbf{p})$, we use that $\partial v = O(\gamma^{-K})$). On the other hand the "zero" $\mathbf{k}_1$ produces an extra factor $\gamma^{h_{v'}}$, as $\mathbf{k}_1$ is the momentum of the external lines of the cluster, so that $|\mathbf{k}_1| \simeq \gamma^{h_{v'}}$. Note that a similar argument cannot apply to $h \geq K$; if the derivatives is applied on $v(\mathbf{p})$ one gets a factor $\gamma^{h_{v'}}$ which produces no gain at large $h_{v'}$. Hence the bound of (4.6) is $O(\gamma^{-(h_v - h_{v'})})$ which means that the subtraction from the graphs of its value at vanishing external momenta makes the dimension equal to 1 instead of 0. Similar considerations could be repeated for the clusters with two external lines and motivate the integration procedure for the scales $\leq K$ we will explain below.

## 4.3 Modification of the fermionic interaction

After the integration of the fields $\psi^{(N)}, \psi^{(N-1)}, ..., \psi^{(K)}$ we obtain an expression of the form

$$\int P(d\psi^{(\leq K)}) e^{\mathcal{V}^{(K)} + \mathcal{B}^{(K)}} \quad (4.8)$$

We integrate the remaining fields $\psi^{(K)}, \psi^{(K-1)}, ..., \psi^{(h)}$ according to the following iterative procedure.

Assume that we have integrated up to scale $j$ and we have obtained the following expression

$$e^{\mathcal{W}(\phi,J)} = e^{-L^2 E_j} \int P_{\widetilde{Z}_j}(d\psi^{(\leq j)}) e^{-\mathcal{V}^{(j)}(\sqrt{Z_j}\psi^{(\leq j)}) + \mathcal{B}^{(j)}(\sqrt{Z_j}\psi^{(\leq j)}, \phi, J)} \quad (4.9)$$

where $Z_K = 1, \mu_K = m$ and $P_{\widetilde{Z}_j}(d\psi^{(\leq j)})$ is

$$P_{\widetilde{Z}_j}(d\psi^{(\leq j)}) = \prod_{k:C_j(k)>0} \prod_{\omega,\omega'=\pm 1} \frac{d\widehat{\psi}_{k,\omega}^{(\leq j))+} d\widehat{\psi}_{k,\omega}^{(\leq j)-}}{\mathcal{N}_j(k)}$$

$$\exp\left\{-\frac{1}{L^2} \sum_k C_j(k) \widetilde{Z}_j(k) \sum_{\omega \pm 1} \widehat{\psi}_\omega^{(\leq j)+} T_{\omega,\omega'}^{(j)}(k) \widehat{\psi}_{k,\omega,\omega'}^{(\leq j)-}\right\} \quad (4.10)$$

with

$$T^j(k) = \begin{pmatrix} D_+(k) & -\mu_j(k) \\ -\mu_j(k) & D_-(k) \end{pmatrix}$$

and $D_\omega(k) = -ik_0 + \omega k_1$.

The *effective potential on scale* $K$, $\mathcal{V}^{(K)}(\psi)$, is a sum of monomial of Grassmanian variables multiplied by suitable kernels. *i.e.* it is of the form

$$\mathcal{V}^{(j)}(\psi) = \sum_{n=1}^{\infty} \frac{1}{L^{4n}} \sum_{\substack{k_1,...,k_{2n} \\ \omega_1,...,\omega_{2n}}} \prod_{i=1}^{2n} \widehat{\psi}_{k_i,\omega_i}^{\sigma_i} \widehat{W}_{2n,\underline{\omega}}^{(j)}(k_1,...,k_{2n-1}) \delta\left(\sum_{i=1}^{2n} \sigma_i k_i\right)$$

$$(4.11)$$

where $\sigma_i = +$ for $i = 1,\ldots,n$, $\sigma_i = -$ for $i = n+1,\ldots,2n$ and $\underline{\omega} = (\omega_1,\ldots,\omega_{2n})$;

The *effective source term at scale* $j$, $\mathcal{B}^{(j)}(\sqrt{Z_j}\psi, \phi, J)$, is a sum of monomials of Grassmanian variables and $\phi^\pm, J$ field, with at least one $\phi^\pm$ or one $J$ field; we shall write it in the form

$$\mathcal{B}^{(j)}(\sqrt{Z_j}\psi, \phi, J) = \mathcal{B}_\phi^{(j)}(\sqrt{Z_j}\psi) + \mathcal{B}_J^{(j)}(\sqrt{Z_j}\psi) + \mathcal{W}_R^{(j)}(\sqrt{Z_j}\psi, \phi, J),$$

$$(4.12)$$

where $\mathcal{B}_\phi^{(j)}(\psi)$ and $\mathcal{B}_J^{(j)}(\psi)$ denote the sums over the terms containing only one $\phi$ or $J$ field, respectively; of course also $\mathcal{B}^{(j)}(\sqrt{Z_j}\psi, \phi, J)$ can be written as sum over monomials of $\psi, \phi, J$ mutiplied by kernels $\widehat{W}_{2n,n_\phi,n_J\underline{\omega}}^{(j)}$.

The integration procedure is similar to the one described in the previous chapter, but a different definition of the $\mathcal{L}, \mathcal{R}$ operators is necessary, acording to the previous considerations. We introduce fist an operator $\mathcal{P}_j$, $j=0,1$, acting on the kernels $W^{(j)}$ in the following way

$$\mathcal{P}_0 W^{(j)} = W^{(j)}|_{\mu_0,...\mu_j=0} \quad \mathcal{P}_1 W^{(j)} = \sum_{k \geq j} \mu_k(k) \frac{\partial W^{(j)}}{\partial \mu_k}|_{\mu_0,...\mu_j=0} \quad (4.13)$$

that is extracting from $\mathcal{W}^{(j)}$ the part independent or linear in the mass.

We introduce also the operators $\mathcal{L}_0, \mathcal{L}_1$ acting on the kernels $W^{(j)}$ in the following way (in the limit $L \to \infty$ for simplicity):

1) If $n = 1$,
$$\mathcal{L}_0 \widehat{W}^{(j)}_{2,\omega}(\mathbf{k}) = \widehat{W}^{(j)}_{2,\omega}(0)$$
$$\mathcal{L}_1 \widehat{W}^{(j)}_{2,\omega}(\mathbf{k}) = k_0 \partial_{k_0} \widehat{W}^{(j)}_{2,\omega}(0) + k \partial_k \widehat{W}^{(j)}_{2,\omega}(0) \qquad (4.14)$$

2) If $n = 2$, $\mathcal{L}_1 \widehat{W}_{4,\omega} = 0$ and
$$\mathcal{L}_0 \widehat{W}^{(j)}_{4,\omega}(\mathbf{k}_1, \mathbf{k}_2, \mathbf{k}_3) = \widehat{W}^{(j)}_{4,\omega}(0, 0, 0) . \qquad (4.15)$$

3) If $n > 2$, $\mathcal{L}_0 \widehat{W}_{2n,\omega} = \mathcal{L}_1 \widehat{W}_{2n,\omega} = 0$.

Given $\mathcal{L}_k, \mathcal{P}_k$, $k = 0, 1$ as above, we define the action of $\mathcal{L}$ on the kernels $\widehat{W}_{2n,\omega}$ as follows.

1) If $n = 1$, then
$$\mathcal{L}\widehat{W}^{(j)}_{2,\omega,\omega'} = (\mathcal{L}_0 + \mathcal{L}_1)\mathcal{P}_0 \widehat{W}^{(j)}_{2,\omega,\omega'} + \mathcal{L}_0 \mathcal{P}_1 \widehat{W}^{(j)}_{2,\omega,\omega'} \qquad (4.16)$$

2) If $n = 2$, then $\mathcal{L}\widehat{W}^{(j)}_{4,\omega} = \mathcal{L}_0 \mathcal{P}_0 \widehat{W}^{(j)}_{4,\omega}$.

3) If $n > 2$, then $\mathcal{L}\widehat{W}^{(j)}_{2n,\omega} = 0$.

Note that $\mathcal{L}_0 \widehat{W}_{2,\omega,\omega} = 0$. Furthermore, since the only way of constructing a graph contributing to $\widehat{W}_{2,\omega,-\omega}$ is to use at least one antidiagonal propagator, it also holds $\mathcal{P}_0 \widehat{W}^{(j)}_{2,\omega,-\omega} = 0$. Therefore
$$\mathcal{L}\widehat{W}^{(j)}_{2,\omega,\omega} = \mathcal{L}_1 \mathcal{P}_0 \widehat{W}^{(j)}_{2,\omega,\omega} \qquad \mathcal{L}\widehat{W}^{(j)}_{2,\omega,-\omega} = \mathcal{L}_0 \mathcal{P}_1 \widehat{W}^{(j)}_{2,\omega,-\omega} \qquad (4.17)$$

By the above definitions it follows
$$\mathcal{L}^2 \mathcal{V}^{(j)} = \mathcal{L}\mathcal{V}^{(j)} \qquad \mathcal{R}^2 \mathcal{V}^{(j)} = \mathcal{R}\mathcal{V}^{(j)} \qquad (4.18)$$

The effect of $\mathcal{L}$ on $\mathcal{V}^{(j)}$ is such that
$$\mathcal{L}\mathcal{V}^{(j)}(\psi^{(\leq j)}) = z_j F^{(\leq j)}_\zeta + s_j F^{(\leq j)}_\sigma + l_j F^{(\leq j)}_\lambda , \qquad (4.19)$$

where $z_j$, $a_j$ and $l_j$ are real numbers and
$$F^{(\leq j)}_\zeta = \frac{1}{L^2} \sum_\omega \sum_{\mathbf{k}: C_j^{-1}(\mathbf{k}) > 0} D_\omega(\mathbf{k}) \widehat{\psi}^{(\leq j)+}_{\mathbf{k},\omega} \widehat{\psi}^{(\leq j)-}_{\mathbf{k},\omega}$$
$$F^{(\leq j)}_\sigma = \frac{1}{L^2} \sum_\omega \sum_{\mathbf{k}: C_j^{-1}(\mathbf{k}) > 0} \widehat{\psi}^{(\leq j)+}_{\mathbf{k},\omega} \widehat{\psi}^{(\leq j)-}_{\mathbf{k},-\omega} \qquad (4.20)$$
$$F^{(\leq j)}_\lambda = \frac{1}{L^8} \sum_{\mathbf{k}_1,\ldots,\mathbf{k}_4: C_j^{-1}(\mathbf{k}_i) > 0} \widehat{\psi}^{(\leq j)+}_{\mathbf{k}_1,+} \widehat{\psi}^{(\leq j)-}_{\mathbf{k}_2,+} \widehat{\psi}^{(\leq j)+}_{\mathbf{k}_3,-} \widehat{\psi}^{(\leq j)-}_{\mathbf{k}_4,-} \delta(\mathbf{k}_1 - \mathbf{k}_2 + \mathbf{k}_3 - \mathbf{k}_4)$$

Note that there are no terms $D_\omega(\mathbf{k})\psi^+_\omega\psi^-_{-\omega}$ or $\psi^+_\omega\psi^-_\omega$, since they are proportional to $\mathcal{L}_1\widehat{W}_{2,\omega,-\omega}=0$ and to $\mathcal{L}_0\widehat{W}_{2,\omega,\omega}=0$; note also that the graph contributing to $\psi^+_\omega\psi^-_{-\omega}$ contains an odd number of non diagonal propagators.

Analogously we write $\mathcal{B}^{(j)}=\mathcal{L}\mathcal{B}^{(j)}+\mathcal{R}\mathcal{B}^{(j)}$, $\mathcal{R}=1-\mathcal{L}$, according to the following definition. First of all, we put $\mathcal{L}W_R^{(j)}=0$. Let us consider now $\mathcal{B}_J^{(j)}(\sqrt{Z_j}\psi)$. It is easy to see that the field $J$ is equivalent, from the point of view of dimensional considerations, to two $\psi$ fields. Hence, the only terms which need to be renormalized are those of second order in $\psi$, which are indeed marginal. We shall use for them the definition

$$\mathcal{B}_J^{(j,2)}(\sqrt{Z_j}\psi)=\sum_{\omega,\tilde\omega}\int\frac{d\mathbf{p}}{(2\pi)^2}\frac{d\mathbf{k}}{(2\pi)^2}\widehat{B}_{\omega,\tilde\omega}(\mathbf{p},\mathbf{k})\widehat{J}(\mathbf{p})(\sqrt{Z_j}\widehat\psi^+_{\mathbf{p}+\mathbf{k},\tilde\omega})(\sqrt{Z_j}\widehat\psi^-_{\mathbf{k},\tilde\omega}) \quad (4.21)$$

We regularize $\mathcal{B}_J^{(j,2)}(\sqrt{Z_j}\psi)$, in analogy to what we did for the effective potential, by decomposing it as the sum of $\mathcal{L}\mathcal{B}_J^{(j,2)}(\sqrt{Z_j}\psi)$ and $\mathcal{R}\mathcal{B}_J^{(j,2)}(\sqrt{Z_j}\psi)$, where $\mathcal{L}$ is defined through its action on $\widehat{B}_\omega(\mathbf{p},\mathbf{k})$ in the following way:

$$\mathcal{L}\widehat{B}_{\omega,\tilde\omega}(\mathbf{p},\mathbf{k})=\mathcal{L}_0 P_0 \widehat{B}_{\omega,\tilde\omega}(0,\bar{\mathbf{k}}) , \quad (4.22)$$

note that, since the graph expansion of $\widehat{B}_{\omega,-\omega}$ requires at least one antidiagonal propagator, $\mathcal{L}_0 P_0 \widehat{B}_{\omega,-\omega}=0$ so that

$$\mathcal{L}\mathcal{B}_J^{(j,2)}(\sqrt{Z_j}\psi)=\sum_\omega \frac{Z_j^{(2)}}{Z_j}\int d\mathbf{x}\, J_{\mathbf{x},\omega}\left(\sqrt{Z_j}\psi^+_{\mathbf{x},\omega}\right)\left(\sqrt{Z_j}\psi^-_{\mathbf{x},\omega}\right) \quad (4.23)$$

which defines the renormalization constant $Z_j^{(2)}$.

Finally we have to define $\mathcal{L}$ for $\mathcal{B}_\phi^{(j)}(\sqrt{Z_j}\psi)$; we want to show that, by a suitable choice of the localization procedure, if $j\leq N-1$, it can be written in the form

$$\mathcal{B}_\phi^{(j)}(\sqrt{Z_j}\psi)=\sum_\omega\sum_{i=j+1}^N\int d\mathbf{x}d\mathbf{y}$$

$$\left[\phi^+_{\mathbf{x},\omega}g_\omega^{Q,(i)}(\mathbf{x}-\mathbf{y})\frac{\partial}{\partial\psi^+_{\mathbf{y},\omega}}\mathcal{V}^{(j)}(\sqrt{Z_j}\psi)+\frac{\partial}{\partial\psi^-_{\mathbf{y},\omega}}\mathcal{V}^{(j)}(\sqrt{Z_j}\psi)g_\omega^{Q,(i)}(\mathbf{y}-\mathbf{x})\phi^-_{\mathbf{x},\omega}\right]$$

$$+\sum_\omega\int\frac{d\mathbf{k}}{(2\pi)^2}\left[\widehat\psi^{(\leq j)+}_{\mathbf{k},\omega}\widehat{Q}^{(j+1)}_\omega(\mathbf{k})\widehat\phi^-_{\mathbf{k},\omega}+\widehat\phi^+_{\mathbf{k},\omega}\widehat{Q}^{(j+1)}_\omega(\mathbf{k})\widehat\psi^{(\leq j)-}_{\mathbf{k},\omega}\right] \quad (4.24)$$

where $\widehat{g}_\omega^{Q,(i)}(\mathbf{k})=\widehat{g}_\omega^{(i)}(\mathbf{k})\widehat{Q}_\omega^{(i)}(\mathbf{k})$ and $\widehat{Q}_\omega^{(j)}(\mathbf{k})$ is defined inductively by the relations

$$\widehat{Q}_\omega^{(j)}(\mathbf{k})=\widehat{Q}_\omega^{(j+1)}(\mathbf{k})-z_j Z_j D_\omega(\mathbf{k})\sum_{i=j+1}^N \widehat{g}_\omega^{Q,(i)}(\mathbf{k}) , \quad \widehat{Q}_\omega^{(0)}(\mathbf{k})=1 \quad (4.25)$$

The $\mathcal{L}$ operation for $\mathcal{B}_\phi^{(j)}$ is defined by decomposing $\mathcal{V}^{(j)}$ in the r.h.s. of (4.24) as $\mathcal{L}\mathcal{V}^{(j)} + \mathcal{R}\mathcal{V}^{(j)}$, $\mathcal{L}\mathcal{V}^{(j)}$ being defined by (4.19).

We write
$$\mathcal{V}^{(j)} = \mathcal{L}\mathcal{V}^{(j)} + \mathcal{R}\mathcal{V}^{(j)} \qquad \mathcal{B}^{(j)} = \mathcal{L}\mathcal{B}^{(j)} + \mathcal{R}\mathcal{B}^{(j)} \qquad (4.26)$$

and we note that the quadratic part in $\mathcal{L}\mathcal{V}^{(j)}$ has exactly the same form than the one appearing in the fermionic integration; they describe a renormalization of the mass and of the fermionic wave function normalization. By using the addition property explained above, we can add the quadratic part of the r.h.s. of (4.19) to $P_{\widetilde{Z}_j,C_j}(d\psi^{(\leq j)})$, by adding to it part of the r.h.s. of (4.19) and using the addition property. We get that (4.9) can be written as

$$e^{-L^2 t_j} \int P_{\widetilde{Z}_{j-1}}(d\psi^{(\leq j)}) e^{-l_j F_\lambda(\sqrt{Z_j}\psi^{(\leq j)}) - \mathcal{R}\mathcal{V}(\sqrt{Z_j}\psi^{(\leq j)}) + \widetilde{\mathcal{B}}^{(j)}(\sqrt{Z_j}\psi^{(\leq j)})}$$
(4.27)

where
$$\widetilde{Z}_{j-1}(\mathbf{k}) = Z_j(\mathbf{k})[1 + C_j^{-1}(\mathbf{k}) z_j], \quad \mu_{j-1}(\mathbf{k}) = \mu_j + C_j^{-1}(\mathbf{k}) s_j] \qquad (4.28)$$

and the factor $\exp(-L^2 t_j)$ in (4.27) takes into account the different normalization of the two measures. Moreover

$$\widetilde{\mathcal{B}}^{(j)}(\sqrt{Z_j}\psi^{[h,j]}) = \widetilde{\mathcal{B}}_\phi^{(j)}(\sqrt{Z_j}\psi^{(\leq j)}) + \mathcal{B}_J^{(j)}(\sqrt{Z_j}\psi^{(\leq j)}) + W_R^{(j)}, \qquad (4.29)$$

where $\widetilde{\mathcal{B}}_\phi^{(j)}$ is obtained from $\mathcal{B}_\phi^{(j)}$ by inserting (4.26) in the second line of (4.24) and by absorbing the terms proportional to $z_j$ in the terms in the third line of (4.24).

The r.h.s of (4.27) can be written as

$$e^{-L^2 t_j} \int P_{\widetilde{Z}_{j-1},C_{j-1}}(d\psi^{(\leq j-1)}) \int P_{Z_{j-1},\widetilde{f}_j^{-1}}(d\psi^{(j)}) \qquad (4.30)$$
$$e^{-\widetilde{\mathcal{V}}^{(j)}\left(\sqrt{Z_j}[\psi^{(\leq j-1)} + \psi^{(j)}]\right) + \widetilde{\mathcal{B}}^{(j)}\left(\sqrt{Z_j}[\psi^{(\leq j-1)} + \psi^{(j)}]\right)}$$

where $P_{Z_{j-1},\widetilde{f}_j^{-1}}(d\psi^{(j)})$ is the integration with propagator $\widehat{g}_\omega^{(j)}(\mathbf{k}) = \frac{1}{Z_{j-1}} \frac{\widetilde{f}_j(\mathbf{k})}{D_\omega(\mathbf{k})}$ with $\widetilde{f}_j(\mathbf{k}) = f_j(\mathbf{k}) Z_{j-1} [\widetilde{Z}_{j-1}(\mathbf{k})]^{-1}$.

The propagator $\widehat{g}_\omega^{Q,(i)}(\mathbf{k})$ is equivalent to $\widehat{g}_\omega^{(i)}(\mathbf{k})$, as concerns the dimensional bounds.

We now *rescale* the field so that

$$\widetilde{\mathcal{V}}^{(j)}(\sqrt{Z_j}\psi^{(\leq j)}) = \widehat{\mathcal{V}}^{(j)}(\sqrt{Z_{j-1}}\psi^{(\leq j)}), \qquad (4.31)$$
$$\widetilde{\mathcal{B}}^{(j)}(\sqrt{Z_j}\psi^{(\leq j)}) = \widehat{\mathcal{B}}^{(j)}(\sqrt{Z_{j-1}}\psi^{(\leq j)})$$

it follows that
$$\mathcal{L}\widehat{\mathcal{V}}^{(j)}(\psi^{(\leq j)}) = \lambda_j \int d\mathbf{x} \psi_{\mathbf{x},+}^{(\leq j)+} \psi_{\mathbf{x},+}^{(\leq j)-} \psi_{\mathbf{x},-}^{(\leq j)+} \psi_{\mathbf{x},-}^{(\leq j)-} \quad (4.32)$$
where $\lambda_j = (Z_j Z_{j-1}^{-1})^2 l_j$. If we now define
$$e^{-\mathcal{V}^{(j-1)}\sqrt{Z_j}(\psi^{(\leq j-1)}) + \mathcal{B}^{(j-1)}(\sqrt{Z_j}\psi^{(\leq j-1)}) - L^2 E_j} =$$
$$= \int P_{Z_{j-1},\tilde{f}_j^{-1}}(d\psi^{(j)})\, e^{-\widehat{\mathcal{V}}^{(j)}\left(\sqrt{Z_j}[\psi^{(\leq j-1)} + \psi^{(j)}]\right) + \widehat{\mathcal{B}}^{(j)}\left(\sqrt{Z_j}[\psi^{(\leq j-1)} + \psi^{(j)}]\right)} \quad (4.33)$$
it is easy to see that $\mathcal{V}^{(j-1)}$ and $\mathcal{B}^{(j-1)}$ are of the same form of $\mathcal{V}^{(j)}$ and $\mathcal{B}^{(j)}$ and that the procedure can be iterated up to a scale $h^*$ defined as the minimal scale such that
$$\gamma^k \leq |\mu_k| \quad \text{for} \quad k \geq h^*; \quad \gamma^{h^*-1} \geq |\mu_{h^*-1}| \quad (4.34)$$
As in the previous chapter, all the scales $\leq h^*$ can be integrated is a single step.

The iterative procedure described above essentially means that at momentum scale $\gamma^j$ one has an "effective theory" describing interacting fermions with wave function renormalization $Z_j$, current renormalization $Z_j^{(2)}$, effective mass $\mu_j$ and effective coupling $\lambda_j$.

The dependence of such parameters on the scale $j$ can be deduced from a set of recursive *flow equations*; indeed the definition of the $\mathcal{L}$ operation, implies that the running coupling constants at scale $j$ can expressed in terms in terms of $\lambda_{j'}$, $j' \geq j$ and obey to equations of the form
$$\lambda_{j-1} = [\frac{Z_j}{Z_{j-1}}]^2[\lambda_j + \bar{\beta}_{j,\lambda}(\lambda_{j+1},\ldots,\lambda_K)]$$
$$\frac{Z_{j-1}}{Z_j} = 1 + \bar{\beta}_{j,z}(\lambda_{j+1},\ldots,\lambda_K)$$
$$\frac{Z_{j-1}^{(2)}}{Z_j^{(2)}} = 1 + \bar{\beta}_{j,2}(\lambda_{j+1},\ldots,\lambda_K)$$
$$\frac{\mu_{j-1}}{\mu_j} = \frac{Z_j}{Z_{j-1}}[1 + \bar{\beta}_{j,\mu}(\lambda_{j+1},\ldots,\lambda_K)] \quad (4.35)$$
which can be equivalently written as
$$\lambda_{j-1} = \lambda_j + \beta_{j,\lambda}(\lambda_{j+1},\ldots,\lambda_K)$$
$$\frac{Z_{j-1}}{Z_j} = 1 + \beta_{j,z}(\lambda_{j+1},\ldots,\lambda_K)$$
$$\frac{\xi_{j-1}}{\xi_j} = 1 + \beta_{j,\xi}(\lambda_{j+1},\ldots,\lambda_K)$$
$$\frac{\mu_{j-1}}{\mu_j} = 1 + \beta_{j,\mu}(\lambda_{j+1},\ldots,\lambda_K) \quad (4.36)$$

where $\xi_j = \frac{Z_{j-1}^{(2)}}{Z_j}$. The r.h.s. of (4.36) is called *Beta function*. Note that by construction the functions $\beta_{j,\lambda}, \beta_{j,\mu}, \beta_{j,z}$ are independent from the mass $\mu$; moreover the running coupling constants in the massless and in the massive theory (for $j \geq h^*$) are the same.

## 4.4 Bounds for the renormalized expansion

The bounds for the renormalized expansion are done similarly to the ones in the previous chapter, the only difference being that the $\mathcal{R}$ operation has to be taken into account. As the bounds are done in the coordinate space, we have to express the $\mathcal{R}$-operation in the $x$-space representation.

If $2n = 4$

$$\mathcal{L}_0 \int d\underline{x} W(\underline{x}) \prod_{i=1}^{4} \psi_{\mathbf{x}_i,\omega_i}^{(\leq h)\sigma_i} = \int d\underline{x} W(\underline{x}) \prod_{i=1}^{4} \psi_{\mathbf{x}_4,\omega_i}^{(\leq h)\sigma_i}, \qquad (4.37)$$

where $\underline{x} = (\mathbf{x}_1, \ldots, \mathbf{x}_4)$, $W(\underline{x}) = W_{4,\underline{\sigma},\underline{\omega}}^{(h)}(\mathbf{x}_1, \mathbf{x}_2, \mathbf{x}_3, \mathbf{x}_4)$; moreover

$$\mathcal{R}_0 \int d\underline{x} W(\underline{x}) \prod_{i=1}^{4} \psi_{\mathbf{x}_i,\omega_i}^{(\leq h)\sigma_i} = \int d\underline{x} W(\underline{x}) \left[ \prod_{i=1}^{4} \psi_{\mathbf{x}_i,\omega_i}^{(\leq h)\sigma_i} - \prod_{i=1}^{4} \psi_{\mathbf{x}_4,\omega_i}^{(\leq h)\sigma_i} \right] \qquad (4.38)$$

The term in square brackets in the above equation can be written as

$$\psi_{\mathbf{x}_1,\omega_1}^{(\leq h)\sigma_1} \psi_{\mathbf{x}_2,\omega_2}^{(\leq h)\sigma_2} D_{\mathbf{x}_3,\mathbf{x}_4,\omega_3}^{(\leq h)\sigma_3} \psi_{\mathbf{x}_4,\omega_4}^{(\leq h)\sigma_4} + \qquad (4.39)$$
$$+ \psi_{\mathbf{x}_1,\omega_1}^{(\leq h)\sigma_1} D_{\mathbf{x}_2,\mathbf{x}_4,\omega_2}^{(\leq h)\sigma_2} \psi_{\mathbf{x}_4,\omega_3}^{(\leq h)\sigma_3} \psi_{\mathbf{x}_4,\omega_4}^{(\leq h)\sigma_4} + D_{\mathbf{x}_1,\mathbf{x}_4,\omega_1}^{(\leq h)\sigma_1} \psi_{\mathbf{x}_4,\omega_2}^{(\leq h)\sigma_2} \psi_{\mathbf{x}_4,\omega_3}^{(\leq h)\sigma_3} \psi_{\mathbf{x}_4,\omega_4}^{(\leq h)\sigma_4}$$

where

$$D_{\mathbf{y},\mathbf{x},\omega}^{(\leq h)\sigma} = \psi_{\mathbf{y},\omega}^{(\leq h)\sigma} - \psi_{\mathbf{x},\omega}^{(\leq h)\sigma} \qquad (4.40)$$

This field can be rewritten as

$$\psi_{\mathbf{y},\omega}^{(\leq h)\sigma} - \psi_{\mathbf{x},\omega}^{(\leq h)\sigma} = (\mathbf{y} - \mathbf{x}) \cdot \int_0^1 dt\, \partial \psi_{\xi(t),\omega}^{(\leq h)\sigma}, \quad \xi(t) = \mathbf{x} + t(\bar{\mathbf{y}} - \mathbf{x}) \quad (4.41)$$

where $\partial = (\partial_1, \partial_0)$ is the gradient. We can see the effect of the $\mathcal{R}$ operation in this case as replacing a $\psi$ fields with a $D$-field, which can be written as in (4.41); with respect to previous bounds, one has an extra derivative, producing a $\gamma^{h'}$, and an extra zero producing a $\gamma^{-h}$. Moreover, the coordinate of the field is also modified by the $\mathcal{R}_0$ operation, which produces an "interpolated point" $\xi(t) = \mathbf{x} + t(\bar{\mathbf{y}} - \mathbf{x})$.

Instead as acting on the fields, we can equivalently see the $\mathcal{R}_0$ operation as acting on the kernels, in the following way

$$\mathcal{R}_0 \int d\underline{\mathbf{x}} \prod_{i=1}^{4} \psi_{\mathbf{x}_i} W(\underline{\mathbf{x}}) = \int d\underline{\mathbf{x}} \prod_{i=1}^{4} \psi_{\mathbf{x}_i} \left[ W(\underline{\mathbf{x}}) - \delta(\mathbf{x}_3 - \mathbf{x}_4) \int d\mathbf{y}_3 W(\mathbf{x}_1, \mathbf{x}_2, \mathbf{y}_3, \mathbf{x}_4) \right]$$
$$+ \int d\underline{\mathbf{x}} \prod_{i=1}^{4} \psi_{\mathbf{x}_i} \delta(\mathbf{x}_3 - \mathbf{x}_4) \int d\mathbf{y}_3 \Big[ W(\mathbf{x}_1, \mathbf{x}_2, \mathbf{y}_3, \mathbf{x}_4) -$$
$$- \delta(\mathbf{x}_2 - \mathbf{x}_4) \int d\mathbf{y}_2 W(\mathbf{x}_1, \mathbf{y}_2, \mathbf{y}_3, \mathbf{x}_4) \Big] + \int d\underline{\mathbf{x}} \prod_{i=1}^{4} \psi_{\mathbf{x}_i} \delta(\mathbf{x}_2 - \mathbf{x}_4) \delta(\mathbf{x}_3 - \mathbf{x}_4)$$
$$\int d\mathbf{y}_2 \int d\mathbf{y}_3 \Big[ W(\mathbf{x}_1, \mathbf{y}_2, \mathbf{y}_3, \mathbf{x}_4) - \delta(\mathbf{x}_1 - \mathbf{x}_4) \int d\mathbf{y}_1 W(\mathbf{y}_1, \mathbf{y}_2, \mathbf{y}_3, \mathbf{x}_4) \Big]$$
(4.42)

In the new representation, the action of $\mathcal{R}_0$ is seen as the decomposition of the original term in the sum of three terms with a different kernel, containing suitable delta functions; note also that, for $n \geq 1$

$$\int d\underline{\mathbf{x}} \prod_{i=1}^{4} \psi_{\mathbf{x}_i} (\mathbf{x}_1 - \mathbf{x}_2)^n \mathcal{R} W(\underline{\mathbf{x}}) = \int d\underline{\mathbf{x}} \prod_{i=1}^{4} \psi_{\mathbf{x}_i} (\mathbf{x}_1 - \mathbf{x}_2)^n W(\underline{\mathbf{x}}) \quad (4.43)$$

Similar considerations can be repeated in all cases in which $\mathcal{L}_0$ is non trivial; the conclusion is that the $\mathcal{R}_0$ operation can be seem either as an operation acting on the kernels that an operation acting on the fields.

Let us represent first the $\mathcal{R}$ operation as acting on kernels. Proceeding as in the previous chapter and considering the $\phi = J = 0$ case for definiteness we get

$$\mathcal{V}^{(h)}(\sqrt{Z_h}\psi^{(\leq h)}) + L^2 \widetilde{E}_{h+1} = \sum_{n=1}^{\infty} \sum_{\tau \in \mathcal{T}_{h,n}} V^{(h)}(\tau, \sqrt{Z_h}\psi^{(\leq h)}) \quad (4.44)$$

where $\mathcal{T}_{h,n}$ are a set of trees defined as in the previous chapter with the following modification;

1) the scales are from $K+1$ to $h$ (the scale of the root).

2) To each endpoint of scale $h_v \leq K$ is associates $\mathcal{L}\widehat{\mathcal{V}}^{(h_v-1)}$ given by (4.32); to the endpoint with scale $h_v = K+1$ is associated $\mathcal{L}\widehat{\mathcal{V}}^{(K)}$ or one of the terms in $\mathcal{R}\mathcal{V}^{(K)}$.

3) If $v$ is an endpoint to which is associated $\mathcal{L}\widehat{\mathcal{V}}^{(h_v-1)}(\psi^{(\leq h_v-1)}])$, there is the constraint that $h_v = h_{v'} + 1$ is $v'$ is the nontrivial vertex immediately preceding $v$.

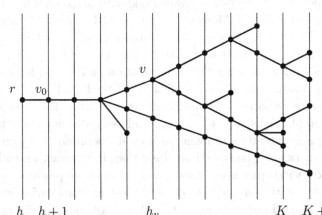

Fig. 4.1  A renormalized tree appearing in the graphic representation of $\mathcal{RV}^{(h)}$ or $\mathcal{LV}^{(h)}$

If $v_0$ is the first vertex of $\tau$ and $\tau_1, .., \tau_s$ ($s = s_{v_0}$) are the subtrees of $\tau$ with root $v_0$, $V^{(h)}(\tau, \sqrt{Z_h}\psi^{(\leq h)})$ is defined inductively by the relation

$$V^{(h)}(\tau, \sqrt{Z_h}\psi^{(\leq h)}) = \frac{(-1)^{s+1}}{s!}\mathcal{E}^T_{h+1}[\bar{V}^{(h+1)}(\tau_1, \sqrt{Z_h}\psi^{(\leq h+1)}); ..; \bar{V}^{(h+1)}(\tau_s, \sqrt{Z_h}\psi^{(\leq h+1)})] \quad (4.45)$$

and $\bar{V}^{(h+1)}(\tau_i, \sqrt{Z_h}\psi^{(\leq h+1)})$

a) is equal to $\mathcal{R}\hat{V}^{(h+1)}(\tau_i, \sqrt{Z_h}\psi^{(\leq h+1)})$ if the subtree $\tau_i$ is not trivial;

b) if $\tau_i$ is trivial it is equal $\mathcal{LV}^{(h+1)}$ if $h < K-1$ or, if $h = K-1$, to $\mathcal{LV}^{(K)}$ or to one of the terms in $\mathcal{RV}^{(K)}$.

Given $\tau \in \mathcal{T}_{h,n}$, there are many possible choices of the subsets $P_v$, $v \in \tau$, compatible with all the constraints; we shall denote $\mathcal{P}_\tau$ the family of all these choices and $\mathbf{P}$ the elements of $\mathcal{P}_\tau$. Then we can write

$$V^{(h)}(\tau, \sqrt{Z_h}\psi^{(\leq h)}) = \sum_{\mathbf{P}\in\mathcal{P}_\tau} V^{(h)}(\tau, \mathbf{P}) \quad (4.46)$$

$V^{(h)}(\tau, \mathbf{P})$ can be represented as

$$V^{(h)}(\tau, \mathbf{P}) = \sqrt{Z_h}^{|P_{v_0}|} \int d\mathbf{x}_{v_0} \widetilde{\psi}^{(\leq h)}(P_{v_0}) K^{(h+1)}_{\tau,\mathbf{P}}(\mathbf{x}_{v_0}) \quad (4.47)$$

with $K^{(h+1)}_{\tau,\mathbf{P}}(\mathbf{x}_{v_0})$ defined inductively (recall that $h_{v_0} = h+1$) by the equation, valid for any $v \in \tau$ which is not an endpoint,

$$K^{(h_v)}_{\tau,\mathbf{P}}(\mathbf{x}_v) = \frac{1}{s_v!}\left(\frac{Z_{h_v}}{Z_{h_v-1}}\right)^{\frac{|P_v|}{2}} \prod_{i=1}^{s_v}[K^{(h_v+1)}_{v_i}(\mathbf{x}_{v_i})]$$
$$\cdot \widetilde{\mathcal{E}}^T_{h_v}[\widetilde{\psi}^{(h_v)}(P_{v_1}\backslash Q_{v_1}), \ldots, \widetilde{\psi}^{(h_v)}(P_{v_{s_v}}\backslash Q_{v_{s_v}})] \quad (4.48)$$

where $\widetilde{\mathcal{E}}_h^T$ denotes the truncated expectation with propagator $g^{(h)}$ (without the scaling factor $Z_{h-1}$). Moreover, if $v$ is an endpoint to which is associated $\mathcal{L}\mathcal{V}^{(h_v+1)}$ then $K_v^{(h_v)}(\mathbf{x}_v) = \lambda_{h_v-1}$, otherwise $K_v^{(h_v)} = \mathcal{R} K_{\tau_i, \mathbf{P}_i}^{(h_v)}$, where $\tau_1, \ldots, \tau_{s_v}$ are the subtrees of $\tau$ with root $v$, $\mathbf{P}_i = \{P_v, v \in \tau_i\}$ and the action of $\mathcal{R}$ is defined using the representation (4.42) (and its analogues) of the $\mathcal{R}_0$ operation, seen as a modification of the kernel.

In order to perform the bounds, we have to write the $\mathcal{R}_0$ operation, instead as in (4.42), as in (4.40). If such replacement is done in a too rough way, there is apparently a problem; as each $\mathcal{R}$ operation (4.40) produces a "zero" factor $(\mathbf{x} - \mathbf{y})$, one could be afraid that, in trees with $n$ end-points, a factor $(\mathbf{x} - \mathbf{y})^n$ is produced, whose integration would produce a constant $C_n = O(n!^\alpha)$, with some $\alpha > 1$ ($C_n$ is the constant appearing in Lemma 3.2 with $K = n$). In order to check that this is not the case, one can follow an iterative procedure starting from the clusters with higher scale on which the $\mathcal{R}$ operation acts non trivially; starting from (4.48), we write the $\mathcal{R}$ operation as in (4.40) and we decompose the "zero" along the propagators belonging to the anchored tree $T_v$ of the truncated expectation or over $\mathcal{R} K_{\tau_i, \mathbf{P}_i}^{(h_v)}$. A priori such zero could compulate with the zero eventually coming from $\mathcal{R} K_{\tau_i, \mathbf{P}_i}^{(h_v)}$; however, from (4.43), this is not the case. The final result (see §3 of Ref.[28] for a detailed derivation) can be written in the following way, if $\alpha$ is an index labelling the different terms obtained by the $\mathcal{R}$ operation:

$$V^{(h)}(\tau, \mathbf{P}) = \sqrt{Z_h}^{|P_{v_0}|} \tag{4.49}$$

$$\sum_{T \in \mathbf{T}} \sum_{\alpha \in A_T} \int d\mathbf{x}_{v_0} W_{\tau, \mathbf{P}, T, \alpha}(\mathbf{x}_{v_0}) \Big\{ \prod_{f \in P_{v_0}} [\partial^{q_\alpha(f)} \psi]_{\mathbf{x}_\alpha(f), \omega(f)}^{(\leq h) \sigma(f)} \Big\}$$

where

$$W_{\tau, \mathbf{P}, T, \alpha}(\mathbf{x}_{v_0}) = \Big[ \prod_{v \text{ not e.p.}} \left( Z_{h_v} / Z_{h_v-1} \right)^{|P_v|/2} \Big] \tag{4.50}$$

$$\Big[ \prod_{i=1}^n (\mathbf{x}_i - \mathbf{y}_i)^{b_\alpha(v_i^*)} K_{v_i^*}^{h_i}(\mathbf{x}_{v_i^*}) \Big] \Big\{ \prod_{v \text{ not e.p.}} \frac{1}{s_v!} \int dP_{T_v}(\mathbf{t}_v)$$

$$\det G_\alpha^{h_v, T_v}(\mathbf{t}_v) \Big[ \prod_{l \in T_v} \partial^{q_\alpha(f_l^-)} \partial^{q_\alpha(f_l^+)} [(\mathbf{x}_l - \mathbf{y}_l)^{b_\alpha(l)} g^{(h_v)}(\mathbf{x}_l - \mathbf{y}_l)] \Big] \Big\}$$

$\mathbf{T}$ is the set of the tree graphs on $\mathbf{x}_{v_0}$, obtained by putting together an anchored tree graph $T_v$ for each non trivial vertex $v$; $A_T$ is a set of indices which allows to distinguish the different terms produced by the non trivial

$\mathcal{R}$ operations and the iterative decomposition of the zeros; $v_1^*, \ldots, v_n^*$ are the endpoints of $\tau$, $f_l^-$ and $f_l^+$ are the labels of the two fields forming the line $l$.

The propagators can be written more explicitly as

$$\partial^{q_\alpha(f_l^-)} \partial^{q_\alpha(f_l^+)} [(\mathbf{x}_l - \mathbf{y}_l)^{b_\alpha(l)} g^{(h_v)}(\mathbf{x}_l - \mathbf{y}_l)] = \qquad (4.51)$$

$$= \int_0^1 dt_l \int_0^1 ds_l \partial^{q_\alpha(f_l^-)} \partial^{q_\alpha(f_l^+)} [(\mathbf{x}_l'(t_l) - \mathbf{y}_l'(s_l))^{b_\alpha(l)} g^{(h_v)}(\mathbf{x}_l'(t_l) - \mathbf{y}_l'(s_l))]$$

Finally $G_\alpha^{h_v,T_v}(\mathbf{t}_v)$ is obtained from the matrix $G^{h_v,T_v}(\mathbf{t}_v)$, associated with the vertex $v$ and $T_v$ by substituting $G_{ij,i'j'}^{h_v,T_v} = t_{v,i,i'} g_{\omega_l^-,\omega_l^+}^{(h_v)}(\mathbf{x}_{ij} - \mathbf{y}_{i'j'})$ with

$$G_{\alpha,ij,i'j'}^{h_v,T_v} = t_{v,i,i'} \partial^{q_\alpha(f_{ij}^-)} \partial^{q_\alpha(f_{ij}^+)} g_{\omega_l^-,\omega_l^+}^{(h_v)}(\mathbf{x}_{ij} - \mathbf{y}_{i'j'}) \qquad (4.52)$$

It would be very difficult to give a precise description of the various contributions to the sum over $A_T$, but fortunately we only need to know some very general properties, which easily follows from the discussion in the previous sections.

1) There is a constant $C$ such that, $\forall T \in \mathbf{T}$, $|A_T| \leq C^n$.
2) $b = 0, 1, 2, 3$ and $q = 0, 1, 2, 3$.
3) For any $\alpha \in A_T$, the following inequality is satisfied

$$\left[\prod_f \gamma^{h_\alpha(f) q_\alpha(f)}\right] \left[\prod_l \gamma^{-h_\alpha(l) b_\alpha(l)}\right] \leq C^n \prod_{v \text{ not e.p.}} \gamma^{-z(P_v)}, \qquad (4.53)$$

where $z(P_v) = 1$ if $|P_v| = 4$ or $|P_v| = 2$ and $\sum_{f \in P_v} \omega(f) \neq 0$, $z(P_v) = 2$ if $|P_v| = 2$ and $\sum_{f \in P_v} \omega(f) = 0$ and $z(P_v) = 0$ otherwise.

Another important difference with respect to the estimates in §3.7 is that the coordinates in the propagators in $T$ are "interpolated points", as a consequence of the action of the $\mathcal{R}$ operation; nevertheless, fixed a point $\bar{\mathbf{x}}$

$$\int d(\mathbf{x}_{v_0} \backslash \bar{\mathbf{x}}) \prod_v \prod_{l \in T_v} \frac{1}{1 + [\gamma^{h_v} |\mathbf{x}_l'(t_l) - \mathbf{y}_l'(s_l)|]^M} \leq \prod_v C \gamma^{-h_v(s_v - 1)} \qquad (4.54)$$

Let us call $\tilde{T} = \cup_v \tilde{T}_v$, where $\tilde{T}_v$ is the set of lines connecting $\mathbf{x}_l'(t_l)$ with $\mathbf{y}_l'(s_l)$, for any $l \in T_v$. $\tilde{T}$ is not a tree in general; however, for any $v$, $\tilde{T}_v$ is still an anchored tree graph between the clusters of points $\mathbf{x}^{(i)}$, $i = 1, \ldots, s_v$. Hence, the proof of (4.54) becomes trivial thanks to the following lemma.

**Lemma 4.1.** *For a given $\tilde{T}$*

$$d(\mathbf{x}_{v_0} \backslash \bar{\mathbf{x}}) = \prod_{l \in \tilde{T}} d\mathbf{r}_l \qquad (4.55)$$

*where $\mathbf{r}_l = \mathbf{x}_l'(t_l) - \mathbf{y}_l'(s_l)$.*

*Proof.* Let us consider first a vertex $v$ with $|T_v| > 0$, which is maximal with respect to the tree order; hence either $v$ is a non local endpoint with $h_v = 2$ or it is a non trivial vertex with no vertex $v'$ with $|T_{v'}| > 0$ following it. In this case $\tilde{T}_v = T_v$, that is no line depends on the interpolation parameters, and $\tilde{T}_v$ is a tree on the set $\mathbf{x}_v$, so that we get immediately the identity

$$d\mathbf{x}_v = d\bar{\mathbf{x}}^{(v)} \prod_{l \in \tilde{T}_v} d\mathbf{r}_l , \qquad (4.56)$$

where $\bar{\mathbf{x}}^{(v)}$ is an arbitrary point of $\mathbf{x}_v$. If we use (4.56) for the family $S_0$ of all maximal vertices with $|T_v| > 0$, we get

$$d\mathbf{x}_{v_0} = \prod_{v \in S_0} \left[ d\bar{\mathbf{x}}^{(v)} \prod_{l \in \tilde{T}_v} d\mathbf{r}_l \right] \qquad (4.57)$$

Let us now consider a line $\bar{l} \in \tilde{T}$, which connects two clusters of points $\mathbf{x}_{v_1}$ and $\mathbf{x}_{v_2}$, with $v_i \in S_0$, $i = 1, 2$. Note that

$$\mathbf{r}_{\bar{l}} = \mathbf{x}'_{\bar{l}}(t_{\bar{l}}) - \mathbf{y}'_l(s_{\bar{l}}) = t_{\bar{l}}\mathbf{x}_{\bar{l}} + (1 - t_{\bar{l}})\bar{\mathbf{x}}_{\bar{l}} - \mathbf{y}'_l(s_{\bar{l}}) \qquad (4.58)$$

implying that

$$\bar{\mathbf{x}}^{(v_1)} = \mathbf{r}_{\bar{l}} + \bar{\mathbf{x}}^{(v_1)} - \mathbf{r}_{\bar{l}} = \mathbf{r}_{\bar{l}} + t_{\bar{l}}(\bar{\mathbf{x}}^{(v_1)} - \mathbf{x}_{\bar{l}}) + (1 - t_{\bar{l}})(\bar{\mathbf{x}}^{(v_1)} - \bar{\mathbf{x}}_{\bar{l}}) + \mathbf{y}'_{\bar{l}}(s_{\bar{l}}) . \quad (4.59)$$

Since $\mathbf{y}'_{\bar{l}}(s_{\bar{l}})$ depends only on the variables $\mathbf{x}_{v_2}$ and $(\bar{\mathbf{x}}^{(v_1)} - \mathbf{x}_{\bar{l}})$ and $(\bar{\mathbf{x}}^{(v_1)} - \bar{\mathbf{x}}_{\bar{l}})$ both depend only on $\{\mathbf{r}_l, l \in \tilde{T}_{v_1}\}$, we get

$$\prod_{i=1}^{2} \left[ d\bar{\mathbf{x}}^{(v_i)} \prod_{l \in \tilde{T}_{v_i}} d\mathbf{r}_l \right] = d\mathbf{r}_{\bar{l}} d\bar{\mathbf{x}}^{(v_2)} \prod_{i=1}^{2} \prod_{l \in \tilde{T}_{v_i}} d\mathbf{r}_l \qquad (4.60)$$

Vy iterating this procedure we get (4.55). ∎

By using the Gram inequality for $\det G_\alpha^{h_v, T_v}$ and by (4.53),(4.54) we obtain, calling $\sup_{k \geq h} |\lambda_k| = \bar{\lambda}_h$

$$\int d\mathbf{x}_{v_0} |W_{\tau,\mathbf{P},T,\alpha}(\mathbf{x}_{v_0})| \leq C^n L^2 \bar{\lambda}_h^n \gamma^{-hD_k(P_{v_0})} \prod_{v \in V_2} \frac{|\mu_{h_v}|}{\gamma^{h_v}}$$

$$\prod_{v \text{ not e.p.}} \left\{ \frac{1}{s_v!} C^{\sum_{i=1}^{s_v} |P_{v_i}| - |P_v|} \left( Z_{h_v} / Z_{h_v - 1} \right)^{|P_v|/2} \gamma^{-[-2 + \frac{|P_v|}{2} + z(P_v)]} \right\}$$

where $V_2$ is the set of vertices, which are not endpoints, with at least a non diagonal contraction.

Moreover, if $v \in V_2$,

$$\frac{|\mu_{h_v}|}{\gamma^{h_v}} = \frac{|\mu_h|}{\gamma^h} \frac{|\mu_{h_v}|}{|\mu_h|} \gamma^{h - h_v} \leq \frac{|\mu_h|}{\gamma^h} \gamma^{(h - h_v)(1 - c_1 \bar{\lambda}_h)} \leq C \gamma^{(h - h_v)(1/2)} , \qquad (4.61)$$

It follows that

$$\prod_{v\in V_2} \frac{|\mu_{h_v}|}{\gamma^{h_v}} \le C^n \prod_{v \text{ not e.p.}} \gamma^{-\frac{1}{2}\tilde{z}(P_v)}, \qquad (4.62)$$

where $\tilde{z}(P_v) = 1$ if $|P_v| = 2$ and $\sum_{f\in P_v} \omega(f) \ne 0$, and 0 otherwise.

$$\int d\mathbf{x}_{v_0} |W_{\tau,\mathbf{P},T,\alpha}(\mathbf{x}_{v_0})| \le C^n L^2 \bar{\lambda}_h^n \gamma^{-hD_k(P_{v_0})} \qquad (4.63)$$

$$\prod_{v \text{ not e.p.}} \left\{ \frac{1}{s_v!} C^{\sum_{i=1}^{s_v} |P_{v_i}| - |P_v|} \left(Z_{h_v}/Z_{h_v-1}\right)^{|P_v|/2} \gamma^{-[-2+\frac{|P_v|}{2}+z(P_v)+\frac{\tilde{z}(P_v)}{2}]} \right\}$$

with

$$-2 + \frac{|P_v|}{2} + z(P_v) + \frac{\tilde{z}(P_v)}{2} \ge \frac{1}{2}, \quad \forall v \text{ not e.p.}. \qquad (4.64)$$

Again, as in chapter.3, to each vertex $v$ is associated a negative dimension; we can proceed as in §3.7 and (4.64) immediately implies the following lemma.

**Lemma 4.2.** *If*

$$\bar{\lambda}_h \le \varepsilon \qquad \left|\frac{Z_h}{Z_{h-1}}\right| \le 1 + \varepsilon \qquad (4.65)$$

*for $\varepsilon$ small enough, then*

$$\|W_{2n,m}^j(\mathbf{k})\| \le C\varepsilon^{n+m-1} \gamma^{-j(n+m-2)} \qquad (4.66)$$

When $J \ne 0, \phi \ne 0$ the bound (4.64) is replaced by

$$\int d\mathbf{\underline{x}} |S_{2m^\phi, n^J, \tau, T, \underline{\omega}}(\mathbf{\underline{x}})| \le L^2 (C\varepsilon_{j_0})^n \gamma^{-j_0(-2+m^\phi+n^J)} \qquad (4.67)$$

$$\prod_{i=1}^{2m^\phi} \frac{\gamma^{-h_i}}{(Z_{h_i})^{1/2}} \prod_{r=1}^{n^J} \frac{Z_{\bar{h}_r}^{(2)}}{Z_{\bar{h}_r}} \prod_{v \text{ not e.p}} \left(\frac{Z_{h_v}}{Z_{h_v-1}}\right)^{|P_v|/2} \gamma^{-D(P_v)}$$

where $h_i$ is the scale of the propagator linking the $i$-th endpoint of type $\phi$ to the tree, $\bar{h}_r$ is the scale of the $r$-th endpoint of type $J$, $n_v^\phi$, $n_v^J$ the number of enpoints of type $\phi, J$ followng $v$ and

$$D(P_v) = -2 + |P_v|/2 + n_v^J + \tilde{z}(P_v) \qquad (4.68)$$

with $\tilde{z}(P_v) = z(P_v)$ if $n_v^\phi \le 1, n_v^J = 0$; $\tilde{z}(P_v) = 3/4$ if $n_v^\phi = 0$, $n_v^J = 1$, $|P_v| = 2$; $\tilde{z}(P_v) = 0$ otherwise; moreover $z(P_v) = \frac{3}{4}$ if $|P_v| = 4$, $n_v^\phi = 0$; $z(P_v) = 2$ if $|P_v| = 2$, $\omega_1 = \omega_2$, $n_v^\phi = 0$ and $z(P_v) = \frac{3}{2}$ if $|P_v| = 2$, $\omega_1 = -\omega_2$ and $n_v^\phi = 0$.

## 4.5 The beta function at lowest orders

Lemma 4.2 says that the kernels of the effective potential are bounded *provided that* the effective couplings $\lambda_j$ remain small for any $j$. The dependence of the effective parameters from the scale can be deduced from the recursive equations (4.36). Let us write explicitly the value of the contribution of the lowest order terms

$$\lambda_{j-1} = [\frac{1}{1 + b_z^1 \lambda_j^2 + ...}]^2 [\lambda_j + b_\lambda^{(2)} \lambda_j^2 + b_\lambda^{(3)} \lambda_j^3 + ...] \qquad (4.69)$$

$$\frac{Z_{j-1}}{Z_j} = 1 + b_z^{(2)} \lambda_j^2 + ...$$

and the coefficients $b_\lambda^{(i)}, b_z^{(i)}$ are $O(1)$ constants.

The difference with respect to the situation described in the previous chapter is clear; in that case the coefficients of the beta function were $O(\gamma^{K-h})$, for $K \geq h$; this ensures that the running couplings remain close to their initial value. In the present case, on the contrary, the coefficients are $O(1)$ and their explicit value determines the dependence from the scale of the running coupling constants.

Suppose for instance that in (4.69) $b_\lambda^{(2)} < 0$; one finds that $\lambda_j \to 0$ as $j \to -\infty$. This would mean that the effective interaction becomes weaker and weaker, and the series expansion would be convergent if $\lambda$ is chosen small enough. Model in which the beta function is negative are called *asymtotically free*.

If $b_\lambda^{(2)} > 0$, on the contrary, one would obtain that the effective coupling becomes larger and larger, and at a certain point it exits from the convergence radius of the series, which cannot provide then a way of computing the Schwinger functions, unless the mass is taken large enough.

Remarkably, no one of such two possibilities is effectively realized in the case we are considering, as it turns out that

$$b_\lambda^{(2)} = 0 \qquad (4.70)$$

Indeed $b_\lambda^{(2)}$ is given by the graphs in fig.4.2, whose values are

$$\int d\mathbf{x} g_\omega^{(h)}(\mathbf{x}) g_\omega^{[K,h]}(\mathbf{x}) + \int d\mathbf{x} g_\omega^{(h)}(-\mathbf{x}) g_\omega^{[K,h]}(\mathbf{x}) = 0 \qquad (4.71)$$

From the vanishing of $b_\lambda^{(2)}$ we cannot of course deduce anything about the flow of $\lambda_j$, as it will be complely determined by the next non vanishing order. The third order is less easy to compute; according to (4.69), it is

Fig. 4.2  $\beta_{j,\lambda}$ at second order

given by $\beta_\lambda^{(3)} - 2\beta_z^{(2)}$, where $\beta_\lambda^{(2)}$ and $\beta_z^{(3)}$ are given by the graphs in Fig.4.3 and 4.4. The computation is of course much more involved in this case but the final result is again

$$2b_z^{(2)} - 2b_\lambda^{(3)} = 0 \qquad (4.72)$$

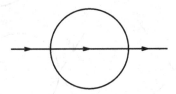

Fig. 4.3  $\beta_{j,z}$ at second order

What we have achieved by such computations? Not so much; as we cannot exclude (unless we perform an explicit computations) that the fourth order is non-vanishing, all we have obtained is that $|\beta_{j,\lambda}| \leq C\lambda^4$ from which

$$|\lambda_j - \lambda_K| \leq \sum_{k=j}^{K} |\beta_\lambda^j| \leq C|j - K|\lambda^4 \qquad (4.73)$$

This means that the effective coupling remains close to its initial value only for scales $|j - K| \leq O(\lambda^{-3})$ so that the infrared cut-off cannot be removed. Of course it is natural to conjecture that the beta function is vanishing to all orders; this is indeed just what happens, but it is impossible to verify this property by an explicit computation.

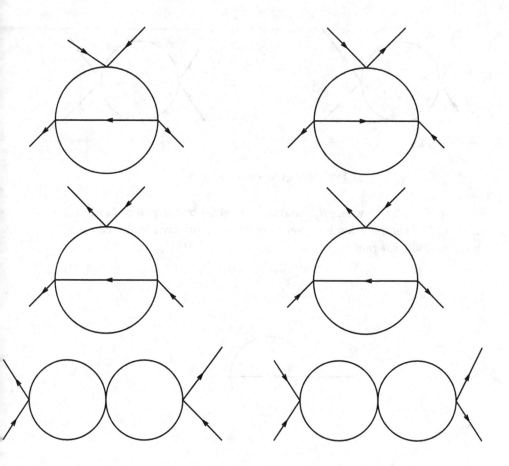

Fig. 4.4  $\beta_{j,\lambda}$ at third order; clusters and $\mathcal{R}$ operations are not represented

Another cancellation of the same type happens in the flow equation for $Z_h^{(1)}/Z_h$. Indeed we can write

$$\frac{Z_{j-1}^{(2)}}{Z_j^{(2)}} = 1 + b_2^{(2)} \lambda_j^2 + ... \qquad (4.74)$$

and $b_2^{(2)}$ is given by the graphs in Fig.4.5; by an explicit computation one can verify that its value essentially coincides with $b_z^{(2)}$; this means that the beta function for $Z_h^{(2)}/Z_h$ at the second order is vanishing.

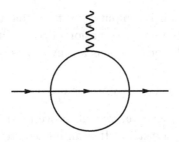

Fig. 4.5 $\bar\beta_{j,2}$ at second order

## 4.6 Boundedness of the flow

We have seen that the beta function for the effective coupling can be written as sum over trees

$$\beta_{h,\lambda}(\lambda_h,\ldots,\lambda_K,\lambda) = \sum_{n=1}^{\infty} \sum_{\tau \in \mathcal{T}_{h,n}} \beta_{\lambda,\tau}(\lambda_h,\ldots,\lambda_K,\lambda) \qquad (4.75)$$

We call $\bar{\mathcal{T}}_{h,n}$ the subset of $\mathcal{T}_{h,n}$ of the trees with end-points associated to $\mathcal{LV}^{(h_v)}$, and we define

$$\beta_{h,\lambda,L}(\lambda_h,\ldots,\lambda_K) = \sum_{n=1}^{\infty} \sum_{\tau \in \bar{\mathcal{T}}_{h,n}} \beta_{\lambda,\tau}(\lambda_h,\ldots,\lambda_K) \qquad (4.76)$$

We write

$$\beta_{h,\lambda}(\lambda_h,\ldots,\lambda_K,\lambda) = \beta_{h,\lambda,L}(\lambda_h,\ldots,\lambda_K) + r_{h,\lambda}(\lambda_h,\ldots,\lambda_K,\lambda) \qquad (4.77)$$

and

$$|r_{h,\lambda}(\lambda_h,\ldots,\lambda_K,\lambda)| \le C(\bar\lambda_h)^2 \gamma^{-\vartheta(K-h)} \qquad (4.78)$$

Indeed the trees contributing to $r_{h,\lambda}$ have surely an end-point at scale $K+1$ to which is associated $\mathcal{RV}^{(K)}$; the dimensions of the trivial vertices between such end-point and the first non trivial vertex, with scale $j$, are all negative, so that by the short memory property we get (4.78).

We can decompose the flow equation for the effective coupling $\lambda_h$ as

$$\lambda_{h-1} = \lambda_h + \beta_{h,\lambda,L}(\lambda_h,\ldots,\lambda_h) + \sum_{k=h+1}^{K} D_{h,k} + r_{h,\lambda}, \qquad (4.79)$$

where

$$D_{h,k} = \beta_{h,\lambda,L}(\lambda_h,\ldots,\lambda_h,\lambda_k,\lambda_{k+1},\ldots) - \beta_{h,\lambda,L}(\lambda_h,\ldots,\lambda_h,\lambda_h,\lambda_{k+1},\ldots). \qquad (4.80)$$

As we have seen, an explicit computation shows that there are cancellations at the first orders of the series expansion of $\beta_{h,\lambda,L}$; it is natural to conjecture that such cancellations happens at any orders so that , for a suitable constant $\vartheta$

$$|\beta_{j,\lambda,L}(\lambda_j,\ldots,\lambda_j)| \leq C\lambda_j^2 \gamma^{-\vartheta(K-j)} \tag{4.81}$$

Indeed in the following chapter we will prove (4.81), as a consequence of certain approximate symmetries. It is an immediate consequence of (4.81) that there exist positive constants $c_1$ such that

$$|\lambda_h - \lambda_K| \leq c_1|\lambda|^2 \tag{4.82}$$

The above equation says that the effective coupling remain close to its initial value, for any scale $h$; by choosing $\lambda$ small enough, convergence of the series expansion follows.

The proof of (6.24) is done by induction; assume that, for any $k \geq h$

$$|\lambda_k - \lambda_{k-1}| \leq c|\lambda|\gamma^{-\vartheta(K-k)} \tag{4.83}$$

Of course $D_{h,k}$ admits a tree expansion similar to that of $\beta_\lambda^h(\lambda_h,\ldots,\lambda_1)$, with the property that all trees giving a non zero contribution must have an endpoint of scale $h$, associated with a difference $\lambda_k - \lambda_h$. Hence, by the short memory property

$$|D_{h,k}| \leq C|\bar\lambda_h|\gamma^{-2\vartheta(k-h)}|a_k - a_h| \ . \tag{4.84}$$

hence

$$|\lambda_{h-1} - \lambda_h| \leq C\lambda^2\gamma^{-\vartheta(K-h)} + C\lambda^2 \sum_{k=h+1}^{K} \gamma^{-2\vartheta(k-h)}\gamma^{-\vartheta(K-h')} \tag{4.85}$$

which immediately implies (4.83) with $k \to h-1$.

Similar consideration can be done for the flow of $\frac{Z_h^{(2)}}{Z_h}$; if we define

$$\beta_{h,\xi,L}(\lambda_h,\ldots,\lambda_K) = \sum_{n=1}^{\infty} \sum_{\tau\in\bar T_{h,n}} \beta_{\xi,\tau}^h(\lambda_h,\ldots,\lambda_K) \tag{4.86}$$

we will prove in the next chapter that $\beta_{h,\xi,L}$ verifies a bound like (4.81) and as a consequence

$$|\frac{Z_h^{(2)}}{Z_h} - 1| \leq C\lambda^2 \tag{4.87}$$

Regarding the flow of the effective renormalization $Z_h$, let us define

$$\eta = \log_\gamma(1+z_{-\infty}) \ , \tag{4.88}$$

and $\eta = a\lambda^2 + O(\lambda^3)$. The previous analysis says that that there exists a positive $\vartheta < 1$, such that

$$|z_h - z_{h+1}| \leq C\lambda^2 \gamma^{-\vartheta(K-h)} . \tag{4.89}$$

We can write

$$\log_\gamma Z_h = \sum_{h'=h+1}^{K} \log_\gamma [1 + z_{-\infty} + (z_{h'} - z_\infty)] = -\eta h + \sum_{h'=h+1}^{K} r_{h'} . \tag{4.90}$$

and

$$|r_h| \leq C \sum_{h'=-\infty}^{h-1} |z_{h'} - z_{h'+1}| \leq C\lambda_1^2 \gamma^{-\vartheta(K-h)} \tag{4.91}$$

Hence

$$Z_h = c_h \gamma^{-\eta h} \tag{4.92}$$

with $|c_h - 1| \leq C\lambda^2$. From (4.92) we see that $Z_h$ is exponentially diverging as $h \to -\infty$; from (4.87) also $Z_h^{(2)}$ is exponentially diverging, but theit ratio remain close to 1 for any $h$.

Finally in the massive case by similar considerations it follows that

$$\mu_h = \widehat{c}_h \gamma^{-\eta_\mu h} \tag{4.93}$$

with $\eta_\mu = c\lambda + O(\lambda^2)$ and $c_h = 1 + O(\lambda)$. By (4.34)

$$\gamma^{h^*} = \mu^{\frac{1}{1-\eta_\mu}} (1 + O(\lambda)) \tag{4.94}$$

## 4.7 The 2-point Schwinger function

In order to conclude the proof of Theorem 4.1 we have to analyze the 2-point Schwinger function $\langle \psi_\mathbf{x} \bar{\psi}_\mathbf{y} \rangle$, which can be written as sum over tress defined in §4.6. For semplicity we just consider the contributions from trees with scales $\leq K - 1$, with only end-point associated to local terms, which are the dominant ones at large distances (the other are negligible for the short memory property).

Let us call $h_\mathbf{x}, h_\mathbf{y}$ the scale of the two special end-points $v_\mathbf{x}, v_\mathbf{y}$ and $h_{\mathbf{x},\mathbf{y}}$ the scale of the higher vertex $\bar{v}_0$ such that two special end-points follows $v$. We can write

$$\langle \psi_\mathbf{x} \bar{\psi}_\mathbf{y} \rangle = \sum_{k=h^*}^{K} \frac{1}{Z_k} [g^{(k)}(\mathbf{x},\mathbf{y}) + S^{(k)}(\mathbf{x},\mathbf{y})] \tag{4.95}$$

with $S^{(h)}(\mathbf{x},\mathbf{y})$ given by the sum over all trees such that $h_{\mathbf{x},\mathbf{y}} = k$. It is an easy conequence of (4.68) that

$$|S^{(k)}(\mathbf{x},\mathbf{y})| \leq C_P |\lambda| \frac{\gamma^k}{1+(\gamma^k|\mathbf{x}-\mathbf{y}|)^P} \qquad (4.96)$$

In fact, given the tree graph $T$ on $x_{v_0}$, let us call $T_{\mathbf{x},\mathbf{y}}$ its subtree connecting the points of $\mathbf{x}_{\bar{v}_0}$, and $T_{\mathbf{x},\mathbf{y}} = \cup_{v \geq \bar{v}_0} T_v$. We want to bound $|\mathbf{x}-\mathbf{y}|$ in terms of the distances between the points connected by the lines $l \in T_{\mathbf{x},\mathbf{y}}$.

Let us call $\bar{v}^{(i)}$, $i = 1,\ldots,s_{\bar{v}_0}$ the non trivial vertices or endpoints following $\bar{v}_0$. The definition of $\bar{v}_0$ implies that $s_{\bar{v}_0} > 1$ and that $\mathbf{x}$ and $\mathbf{y}$ belong to two different sets $\mathbf{x}_{\bar{v}^{(i)}}$; note also that $T_{\bar{v}_0}$ is an anchored tree graph between the sets of points $\mathbf{x}_{\bar{v}^{(i)}}$. Hence there is an integer $r$, a family $l_1,\ldots,l_r$ of lines belonging to $\widetilde{T}_{\bar{v}_0}$ and a family $v^{(1)},\ldots,v^{(r+1)}$ of vertices to be chosen among $\bar{v}^{(1)},\ldots,\bar{v}^{(s_{\bar{v}_0})}$, such that $1 \leq r \leq s_{\bar{v}_0} - 1$ and

$$|\mathbf{x}-\mathbf{y}| \leq \sum_{j=1}^{r} |\mathbf{x}_{l_j} - \mathbf{y}_{l_j}| + \sum_{j=1}^{r+1} |\mathbf{x}^{(j)} - \mathbf{y}^{(j)}| \leq \sum_{l \in \widetilde{T}_{\bar{v}_0}} |\mathbf{x}_l - \mathbf{y}_l| + \sum_{j=1}^{r+1} |\mathbf{x}^{(j)} - \mathbf{y}^{(j)}| \qquad (4.97)$$

where $\mathbf{x}^{(1)} = \mathbf{x}$, $\mathbf{y}^{(r+1)} = \mathbf{y}$. Iterating we get the bound

$$|\mathbf{x}-\mathbf{y}| \leq \sum_{l \in T_{\mathbf{x},\mathbf{y}}} |\mathbf{x}-\mathbf{y}|. \qquad (4.98)$$

Since there are at most $2n+1$ lines in $T$, there exists at least one line $l \in T_{\mathbf{x},\mathbf{y}}$, such that

$$|\mathbf{x}_l - \mathbf{y}_l| \geq \frac{|\mathbf{x}-\mathbf{y}|}{2n+1}. \qquad (4.99)$$

If $\mathbf{x}_l, \mathbf{y}_l$ is such couple of points belonging to a fermionic propagator with scale $h_l$, using that $\gamma^{h_l} \geq \gamma^k$

$$|g^{(h_l)}(\mathbf{x}_l - \mathbf{y})| \leq \frac{1}{1+[\gamma^{h_l}|\mathbf{x}_l - \mathbf{y}_l|]^3} \frac{C_P(2n+1)^P}{1+[\gamma^k|\mathbf{x}-\mathbf{y}|]^P} \qquad (4.100)$$

This means that, for $|\mathbf{x}-\mathbf{y}| \geq \gamma^{-h^*}$, we can extract from the propagators and the interactions a factor $\frac{C_P(2n+1)^N}{1+[\gamma^k|\mathbf{x}-\mathbf{y}|]^P}$.

The sum over the scales is done as for the effective potential, with the only difference that instead of fixing the scale of the root, we have now to fix the scale of $v_{\mathbf{x},\mathbf{y}}$; this has no effect, since we bound the sum over the scales with the sum over the the differences $h_v - h_{v'}$. There is apparently a problem due to the fact that the dimensions can be vanishing; this happens when $|P_v| = 4, n_v^\phi = 2$, that is in the path in the tree connecting the vertex

$\bar{v}_0$ with the root $v$ with scale $j_0$. However we can write $\gamma^{j_0} = \gamma^{h_{\mathbf{x},\mathbf{y}}} \gamma^{j_0 - h_{\mathbf{x},\mathbf{y}}}$ and we can use the factor $\gamma^{j_0 - h_{\mathbf{x},\mathbf{y}}}$ to make the dimension of all vertices $> 0$; this implies (4.96). We obtain

$$\sum_{h=h^*}^{K} \frac{1}{Z_h} |S^{(h)}(\mathbf{x},\mathbf{y})| \le$$

$$C_{N,n} \sum_{h=h^*}^{K} \frac{\gamma^{(1+\eta)h}}{[1+(\gamma^h|\mathbf{x}-\mathbf{y}|)^P]} \le \frac{C_P}{|\mathbf{d}(\mathbf{x})|^{1+\eta}} H_{N,1+\eta}(|\mathbf{d}(\mathbf{x})|) \quad (4.101)$$

where

$$H_{N,\alpha}(r) = \sum_{h=h^*}^{0} \frac{(\gamma^h r)^\alpha}{1+(\gamma^h r)^P} \cdot \quad (4.102)$$

and, for $P \ge 2$

$$H_{P,\alpha}(r) \le \frac{C_{N,\alpha}}{1+(\Delta r)^{P-\alpha}}, \quad \Delta = \gamma^{h^*}. \quad (4.103)$$

# Chapter 5

# Ward Identities and Vanishing of the Beta Function

## 5.1 Schwinger functions and running couplings

We analyze in the chapter the properties of the beta function $\beta_{j,\lambda,L}, \beta_{j,\xi,L}$ defined in (4.76),(4.86).

**Theorem 5.1.** *For $\lambda_j$ small enough, the functions $\beta_{j,\lambda,L}, \beta_{j,\xi,L}$ are analytic for $\lambda_j$ small enough and verify the following bounds, for suitable positive constants $C, \vartheta > 0$*

$$|\beta_{j,\lambda,L}(\lambda_j,\ldots,\lambda_j)| \leq C\lambda_j^2 \gamma^{-\vartheta(K-j)} \qquad |\beta_{j,\xi,L}(\lambda_j,\ldots,\lambda_j)| \leq \lambda_j^2 \gamma^{-\vartheta(K-j)} \tag{5.1}$$

The above property, usually called *vanishing of the beta function*, is the key result for proving the anomalous infrared behaviour of QED2, as we have seen in the previous chapter. Moreover, we will see in the rest of this book that the inequalities (5.1) are central for the understanding of a number of models; it will turn out that in several cases their beta function can be written as a sum of two terms, one which is universal and given by $\beta_{h,\lambda,L}, \beta_{h,\xi,L}$ and a rest which is model-dependent, and that the resulting flow is bounded one that one knows (5.1).

To simplify some technical aspects of the analysis, we consider the functional integral (3.1) with the cut-off function $[C_{h,N}^\varepsilon(\mathbf{k})]^{-1}$ replacing $[C_{h,N}(\mathbf{k})]^{-1}$, where $[C_{h,N}^\varepsilon(\mathbf{k})]^{-1}$ is a cut-off function selecting momenta $\gamma^h \leq |\mathbf{k}| \leq \gamma^N$, essentially equivalent to $[C_{h,N}(\mathbf{k})]^{-1} = \sum_{k=h}^{N} f_k(\mathbf{k})$ as far as the scaling properties are considered; the main difference is that the support of $[C_{h,N}^\varepsilon(\mathbf{k})]^{-1}$ is the set of all the space-time momenta $\mathcal{D}$ while the support of $[C_{h,N}(\mathbf{k})]^{-1}$ is the set $\widetilde{\mathcal{D}}$ containing the $\mathbf{k} \in \mathcal{D}$ such that $\gamma^{h-1} \leq |\mathbf{k}| \leq \gamma^{N+1}$; in the limit $\varepsilon \to 0$ $[C_{h,N}^\varepsilon(\mathbf{k})]^{-1}$ reduces to $[C_{h,N}(\mathbf{k})]^{-1}$.

# NON-PERTURBATIVE RENORMALIZATION

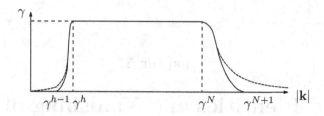

Fig. 5.1  Graphical representation of the cut-off function

More exactly we can choose

$$C_{h,N}(\mathbf{k})^{-1} = \sum_{k=h}^{N} f_k^\varepsilon(\mathbf{k}) \tag{5.2}$$

where $f_k^\varepsilon(\mathbf{k}) = f_k(\mathbf{k})$ for $h+1 \leq k \leq N-1$, and $f_N^\varepsilon(\mathbf{k})$ with support $|\mathbf{k}| \geq \gamma^N$ and $f_k^\varepsilon(\mathbf{k})$ with support in $|\mathbf{k}| \leq \gamma^h$ and with the same scaling properties as $f_N(\mathbf{k}), f_h(\mathbf{k})$.

If we call $\lambda_j^{(h)}, Z_j^{(h)}, Z_j^{(2)(h)}$, $0 \geq j \geq h$, the coupling constants and the renormalizations of the model with infrared cut-off $\gamma^h$, and $\lambda_j, Z_j, Z_j^{(2)}$ the same quantities in the limit $h \to -\infty$, it holds that, by the compact support properties of the decomposition, if $\sup|\lambda_h| \leq \varepsilon_0$

$$\lambda_j = \lambda_j^{(h)} \quad , \quad Z_j = Z_j^{(h)} \quad , \quad Z_j^{(2)} = Z_j^{(h)(2)} \quad , \quad j = 0, \ldots, h+1 \tag{5.3}$$

and

$$\lambda_h = \lambda_h^{(h)}(1+O(\varepsilon_0^2)) \quad , \quad Z_h = Z_h^{(h)}(1+O(\varepsilon_0^2)) \quad , \quad Z_h^{(2)} = Z_h^{(2,h)}(1+O(\varepsilon_0)) \tag{5.4}$$

This means that the coupling constants and the renormalizations are essentially insensitive to presence of the infrared cut-off. Note also that $\lambda_j^{(h)}, Z_j^{(h)}, Z_j^{(2)(h)}$ are very weakly depending from $\varepsilon$, and at the end of the analysis the limit $\varepsilon \to 0$ can be taken without problems.

Moreover by the analysis in the previous chapter it follows that the Fourier transform of the Schwinger functions computed at the cut-off scale are given by

$$\langle \rho_{\omega,2\bar{\mathbf{k}}}; \psi^+_{\omega',\bar{\mathbf{k}}} \psi^-_{\omega',-\bar{\mathbf{k}}} \rangle_{h,N} \equiv G^{(2,1)} = -\frac{Z_h^{(2)}}{Z_h^2 D_\omega(\bar{\mathbf{k}})^2}[1+O(\bar{\lambda}_h^2)] \tag{5.5}$$

$$\langle \psi^+_{+,\bar{\mathbf{k}}} \psi^-_{+,-\bar{\mathbf{k}}}; \psi^+_{-,\bar{\mathbf{k}}} \psi^-_{-,-\bar{\mathbf{k}}} \rangle_{h,N} \equiv G^{(4)} = Z_h^{-2}|\bar{\mathbf{k}}|^{-4}[-\lambda_h + O(\bar{\lambda}_h^2)]$$

$$\langle \psi^+_{\omega,\bar{\mathbf{k}}} \psi^-_{\omega,\bar{\mathbf{k}}} \rangle_{h,N} \equiv G^{(2)} = \frac{1}{Z_h D_\omega(\bar{\mathbf{k}})}[1+O(\bar{\lambda}_h^2)]$$

where $|\bar{\mathbf{k}}| = \gamma^h$, $\rho_{\omega,\mathbf{k}}$ is the Fourier transform of $\psi^+_{\omega,\mathbf{x}}\psi^-_{\omega,\mathbf{x}}$. (5.5) suggests that relations between the running coupling constants at a certain scale $h$ can be obtained from relations between the Schwinger functions computed at infrared cut-off scale.

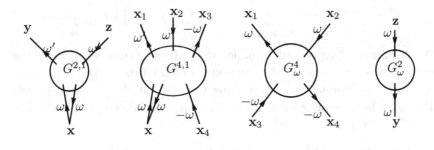

Fig. 5.2    Graphical representation of the Schwinger functions

## 5.2   Ward identities in presence of cut-offs

As discussed in chapt.1, non trivial relations between the Schwinger functions can obtained by performing the local trasformation in (3.1)

$$\psi^{\pm}_{\mathbf{x},\bar{\omega}} \to e^{\pm i\alpha_{\mathbf{x}}}\psi^{\pm}_{\mathbf{x},\bar{\omega}} \tag{5.6}$$

The generating function can be written as, if $\int d\mathbf{x} = aa_0 \sum_{\mathbf{x}\in\Lambda}$

$$e^{\mathcal{W}} = \int P(d\psi^{[h,N]})e^{-\int d\mathbf{x}\psi^+_{\bar{\omega},\mathbf{x}}(e^{i\alpha_{\mathbf{x}}}D^{[h,N]}e^{-i\alpha_{\mathbf{x}}}-D^{[h,N]})\psi^-_{\bar{\omega},\mathbf{x}}} \tag{5.7}$$

$$e^{-\mathcal{V}^{(N)}+\int d\mathbf{x}[\psi^+_{-\bar{\omega},\mathbf{x}}\phi^-_{-\bar{\omega},\mathbf{x}}+\phi^+_{-\bar{\omega},\mathbf{x}}\psi^-_{-\bar{\omega},\mathbf{x}}+e^{i\alpha_{\mathbf{x}}}\psi^+_{\bar{\omega},\mathbf{x}}\phi^-_{\bar{\omega},\mathbf{x}}+e^{-i\alpha_{\mathbf{x}}}\phi^+_{\bar{\omega},\mathbf{x}}\psi^-_{\bar{\omega},\mathbf{x}}]} \tag{5.8}$$

where

$$D^{[h,N]}\psi^{\pm}_{\omega,\mathbf{x}} = \int d\mathbf{k} e^{\pm i\mathbf{k}\mathbf{x}}C^{\varepsilon}_{h,N}(\mathbf{k})(-ik_0+\omega k)\psi^{\pm}_{\omega,\mathbf{k}} \tag{5.9}$$

In deriving the above expression we have used that that the Jabobian corresponding to the transformation (5.6) is 1; this is particularly simple to check if the support of the cut-off function is in all $\mathcal{D}$ (this is the reason why we are considering here the cut-off function $[C_{h,N}^{\varepsilon}(\mathbf{k})]^{-1}$ instead of $[C_{h,N}(\mathbf{k})]^{-1}$).

In fact

$$\int \mathcal{D}\psi [\prod_{\mathbf{x}\in A^+} \bar{\psi}_{\mathbf{x}}][\prod_{\mathbf{y}\in A^-} \psi_{\mathbf{y}}] = \frac{1}{L^{4|A^+|}} \sum_{\mathbf{k}_1 \ldots \mathbf{k}_{|A^+|}} \qquad (5.10)$$

$$e^{i\sum_{i=1}^{|A^+|} \mathbf{k}_i \mathbf{x}_i} \frac{1}{L^{4|A^-|}} \sum_{\mathbf{p}_1 \ldots \mathbf{p}_{|A^-|}} e^{-i\sum_{i=1}^{|A^-|} \mathbf{p}_i \mathbf{y}_i} \int \mathcal{D}\psi [\prod_{i=1}^{|A^+|} \bar{\psi}_{\mathbf{k}_i}][\prod_{i=1}^{|A^-|} \psi_{\mathbf{p}_i}]$$

By definition of Grassman variables the only nonvanishing term in the sum in the r.h.s. correspond to an addend with all the $\mathbf{k}_i, \mathbf{p}_i$ are distinct; moreover one can have a nonvanishing term only if $\{\mathbf{k}_i\}_{i=1}^{A^+} = \{\mathbf{p}_i\}_{i=1}^{A^-} = \mathcal{D}$, and this is possible only if the $\mathbf{x}_i, \mathbf{y}_i$ are all different and such that $\{\mathbf{x}_i\}_{i=1}^{A^+} = \{\mathbf{y}_i\}_{i=1}^{A^-} = \Lambda_a$, so that

$$\int \mathcal{D}\psi [\prod_{\mathbf{x}\in A^+} \bar{\psi}_{\mathbf{x}}][\prod_{\mathbf{y}\in A^-} \psi_{\mathbf{y}}] = \int \mathcal{D}\psi [\prod_{\mathbf{x}\in A^+} e^{i\alpha_{\mathbf{x}}} \bar{\psi}_{\mathbf{x}}][\prod_{\mathbf{x}\in A^-} e^{-i\alpha_{\mathbf{x}}} \psi_{\mathbf{x}}] \qquad (5.11)$$

We write the exponent of (5.7) as

$$\int d\mathbf{x} [\psi_{\mathbf{x},\omega}^+ D^{[h,N]}(\alpha_{\mathbf{x}} \psi_{\omega,\mathbf{x}}^-) - \alpha_{\mathbf{x}} \psi_{\mathbf{x},\omega}^+ D^{[h,N]} \psi_{\omega,\mathbf{x}}^-] = \int d\mathbf{x} \alpha_{\mathbf{x}} \qquad (5.12)$$

$$[(D^{[h,N]}\psi_{\mathbf{x},\omega}^+)\psi_{\omega,\mathbf{x}}^- + \psi_{\mathbf{x},\omega}^+(D^{[h,N]}\psi_{\omega,\mathbf{x}}^-)] = \int d\mathbf{x} \alpha_{\mathbf{x}} [D(\psi_{\mathbf{x},\omega}^+ \psi_{\omega,\mathbf{x}}^-) + \alpha_{\mathbf{x}} \delta T_{\mathbf{x}}]$$

where

$$\delta T_{\mathbf{x}} = \frac{1}{L^4} \sum_{\mathbf{k}^+ \neq \mathbf{k}^-} e^{i(\mathbf{k}^+ - \mathbf{k}^-)\mathbf{x}} C(\mathbf{k}^+, \mathbf{k}^-) \psi_{\mathbf{k}^+,\omega}^+ \psi_{\mathbf{k}^-,\omega}^+ \qquad (5.13)$$

$$C_{\omega}(\mathbf{k}^+, \mathbf{k}^-) = (C_{h,N}^{\varepsilon}(\mathbf{k}^-) - 1)D_{\omega}(\mathbf{k}^-) - (C_{h,N}^{\varepsilon}(\mathbf{k}^+) - 1)D_{\omega}(\mathbf{k}^+)$$

and $D(\psi_{\mathbf{x},\omega}^+ \psi_{\omega,\mathbf{x}}^-) = \frac{1}{L^2} \sum_{\mathbf{p}} e^{i\mathbf{p}\mathbf{x}} D_{\omega} \rho_{\mathbf{p},\omega}$. Differentiating with respect to $\alpha_{\mathbf{x}}$ both sides of (5.7), and setting $\alpha_{\mathbf{x}} = 0$ we find

$$0 = \frac{1}{Z(\phi)} \int P(d\psi) [D^{[h,N]}(\psi_{\bar{\omega},\mathbf{x}}^+ \psi_{\bar{\omega},\mathbf{x}}^-) + \qquad (5.14)$$

$$\delta T_{\bar{\omega},\mathbf{x}} - \phi_{\bar{\omega},\mathbf{x}}^+ \psi_{\bar{\omega},\mathbf{x}}^- + \psi_{\bar{\omega},\mathbf{x}}^+ \phi_{\bar{\omega},\mathbf{x}}^-] e^{-V + \int d\mathbf{x} \sum_{\omega}(\psi_{\bar{\omega},\mathbf{x}}^+ \psi_{\bar{\omega},\mathbf{x}}^- + \psi_{\bar{\omega},\mathbf{x}}^+ \psi_{\bar{\omega},\mathbf{x}}^-)}$$

and finally making a derivative with respect to $\phi_{\mathbf{x}}, \phi_{\mathbf{y}}$ and passing to Fourier trasform

$$D_{\omega}(\mathbf{p}) \langle \hat{\rho}_{\mathbf{p},\omega} \widehat{\psi}_{\mathbf{k}-\mathbf{p},\omega'}^- \widehat{\psi}_{\mathbf{k},\omega'}^+ \rangle_{h,N} =$$

$$\delta_{\omega,\omega'} \left[ \langle \widehat{\psi}_{\mathbf{k}-\mathbf{p},\omega'}^- \widehat{\psi}_{\mathbf{k}-\mathbf{p},\omega'}^+ \rangle_{h,N} - \langle \widehat{\psi}_{\mathbf{k},\omega'}^- \widehat{\psi}_{\mathbf{k},\omega'}^+ \rangle_{h,N} \right] + \Delta_{\omega',\omega}^{1;2}(\mathbf{p};\mathbf{k}) \qquad (5.15)$$

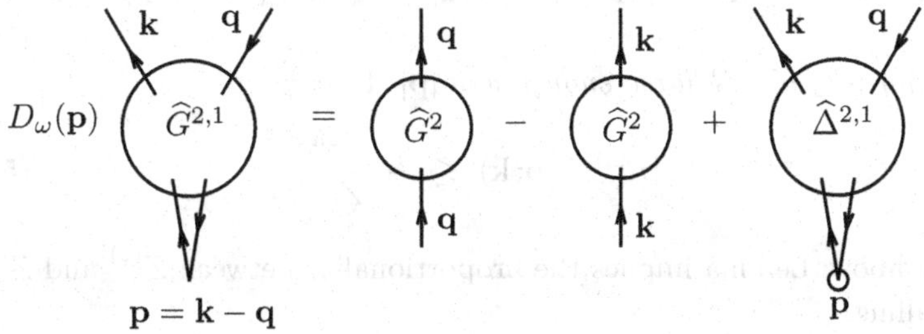

$$D_\omega(\mathbf{p}) \; \widehat{G}^{2,1} \;=\; \widehat{G}^2 \;-\; \widehat{G}^2 \;+\; \widehat{\Delta}^{2,1}$$

Fig. 5.3  Graphical representation of the Ward identity (5.15); the small circle in $\widehat{\Delta}^{2,1}$ represents the function $C_+$ in (5.14)

with

$$\widehat{\Delta}^{2,1}_{\omega,\omega'}(\mathbf{p},\mathbf{k}) = \frac{1}{L^2}\sum_{\mathbf{k}'} C_+(\mathbf{k}',\mathbf{k}'-\mathbf{p}) <\widehat{\psi}^+_{\mathbf{k}',\omega}\widehat{\psi}^-_{\mathbf{k}'-\mathbf{p},\omega};\widehat{\psi}^-_{\mathbf{k},\omega'}\widehat{\psi}^+_{\mathbf{k}-\mathbf{p},\omega'}>_{h,N}$$

(5.16)

If we compare (5.15) with the formal Ward Identity (1.9), obtained from the functional integral without cut-offs, we see that they differ for the presence of $\widehat{\Delta}^{2,1}_{\omega,\omega'}$ in (5.15). One can indeed hope that such term, produced by breaking of the phase symmetry due to the cut-offs, is negligible, at least at momenta far from the cut-offs; this is in some sense reasonable (there is no such a term if we formally replace the cut-off function with the identity) and it would imply that $Z_h$ and $Z_h^{(2)}$ are proportional, by using (5.5), in apparent agreement with the perturbative computations in §4.5.

## 5.3  The correction identity

Things are however more subtle; $\widehat{\Delta}^{2,1}_{\omega,\omega'}$ is *not negligible* at all, but still the proportionality between $Z_h$ and $Z_h^{(2)}$ can be derived by (5.15); this is consequence of the following lemma.

**Lemma 5.1.** *Assuming* $\bar\lambda_h \leq \varepsilon$, *in* (5.15) $\Delta^{2,1}_{\omega,\omega'}$ *can be written as*

$$\Delta^{2,1}_{\omega,\omega'}(\mathbf{p};\mathbf{k}) = \frac{\lambda}{4\pi}v(\mathbf{p})D_{-\omega}(\mathbf{p})\langle\widehat\rho_{\mathbf{p},-\omega}\widehat\psi^-_{\mathbf{k}-\mathbf{p},\omega'}\widehat\psi^+_{\mathbf{k},\omega'}\rangle_{h,N} + D_\omega(\mathbf{p})R^{2,1,N}_{\omega,\omega'}(\mathbf{p};\mathbf{k}) \qquad (5.17)$$

*where, for* $h \leq K$, $N$ *large enough and* $|\mathbf{p}|, |\mathbf{k}| = \gamma^h$

$$|R^{2,1,N}(\mathbf{p};\mathbf{k})| \leq C\varepsilon\frac{\gamma^{-2h}}{Z_h} \qquad (5.18)$$

The above Lemma implies the proportionality between $Z_h^{(2)}$ and $Z_h$; in fact, calling

$$a(\mathbf{p}) = \left[1 - \frac{\lambda}{4\pi}v(\mathbf{p})\right]^{-1} \qquad \bar a = \left[1 + \frac{\lambda}{4\pi}v(\mathbf{p})\right]^{-1} \qquad (5.19)$$

it is easy to see that

$$D_{\omega'}(\mathbf{q})\langle\rho_{\mathbf{q},\omega'}\widehat\psi^-_{\mathbf{k}-\mathbf{q},\omega}\widehat\psi^+_{\mathbf{k},\omega}\rangle = \frac{a(\mathbf{p}) + \omega\omega'\bar a(\mathbf{p})}{2}\left[\langle\widehat\psi^-_{\mathbf{k}-\mathbf{p},\omega}\widehat\psi^+_{\mathbf{k}-\mathbf{p},\omega}\rangle - \langle\widehat\psi^-_{\mathbf{k},\omega}\widehat\psi^+_{\mathbf{k},\omega}\rangle\right]$$
$$+ \sum_{\varepsilon=\pm}D_{\varepsilon\omega'}(\mathbf{p})\frac{a(\mathbf{p}) + \bar a(\mathbf{p})\varepsilon\omega\omega'}{2}R^{1;2}_{\varepsilon\omega',\omega}(\mathbf{p};\mathbf{k}) \qquad (5.20)$$

Hence by (5.5) and that the running coupling constant are independent from the cut-off scale, we find for any $h \leq K$

$$\left|\frac{Z_h^{(2)}}{Z_h} - 1\right| \leq C\bar\lambda_h^2 \qquad (5.21)$$

Hence $Z_h^{(2)}$ and $Z_h^{(2)}$ are both diverging in the limit $h \to -\infty$, but their ratio remains finite.

*Proof.* It is convenient to write $R^{2,1,N}_{\omega,\omega'}(\mathbf{p};\mathbf{k})$ in the following way

$$R^{2,1,N}_{\omega,\omega'}(\mathbf{p};\mathbf{k}) = \frac{\partial^3}{\partial J_{\mathbf{p},\omega}\partial\phi_{\mathbf{k},\omega'}\partial\phi_{\mathbf{k}-\mathbf{p},\omega'}}\mathcal{W}_\Delta(J,\widehat J,\phi) \qquad (5.22)$$

where

$$e^{\mathcal{W}_\Delta(J,\widehat J,\phi)} = \int P(d\psi)e^{-V(\psi) + \sum_\omega \int d\mathbf{z}[\psi^+_{\omega,\mathbf{z}}\phi^-_{\omega,\mathbf{z}} + \phi^+_{\omega,\mathbf{z}}\psi^-_{\omega,\mathbf{z}}] + T_0(\widehat J,\psi) - T_-(\widehat J,\psi)} \qquad (5.23)$$

with

$$T_0(\psi,J) = \sum_{\omega,\sigma}\int\frac{d^2\mathbf{p}}{(2\pi)^2}\frac{d^2\mathbf{k}}{(2\pi)^2}\bar\chi(\mathbf{p})v(\mathbf{p})C_\omega(\mathbf{k},\mathbf{k}+\mathbf{p})J_{\mathbf{p},\omega}\widehat\psi^+_{\mathbf{k},\omega}\widehat\psi^-_{\mathbf{k}+\mathbf{p},\omega} \qquad (5.24)$$

$$T_-(\psi,J) = \sum_\omega \int \frac{d^2\mathbf{p}}{(2\pi)^2} \frac{d^2\mathbf{k}}{(2\pi)^2} \bar{\chi}(\mathbf{p}) \frac{\lambda}{4\pi} v(\mathbf{p}) J_{\mathbf{p},\omega} D_{-\omega}(\mathbf{p}) \widehat{\psi}^+_{\mathbf{k},-\omega} \widehat{\psi}^-_{\mathbf{k}+\mathbf{p},-\omega}$$
(5.25)

where $\bar{\chi}(\mathbf{p})$ is a smooth compact support function $= 0$ for $|\mathbf{p}| \geq 2\gamma^{N+1}$ and $= 1$ for $|\mathbf{p}| \geq 2\gamma^N$ (such function takes into account that we are only interrested in $|\mathbf{p}| \leq 2\gamma^N$, the only possible momenta in the limit $\varepsilon \to 0$ for the compact support properties of the cut-off function).

The functional integral (5.22) (5.23) is quite similar to (3.28) considered in chapt.3: the only difference is that the local term

$$\int \frac{d\mathbf{p}}{(2\pi)^2} \frac{d\mathbf{k}'}{(2\pi)^2} \bar{J}_{\mathbf{p}} \psi^+_{\omega,\mathbf{k}'} \psi_{\omega,\mathbf{k}'-\mathbf{p}}$$
(5.26)

is replaced by the two integrals $T_0 - T_1$.

The properties of $R^{2,1,N}_{\omega,\omega'}$ depend crucially from some peculiar features of $T_0$; indeed when either the two fields in $T_0$ are contracted one gets the function

$$\Delta^{i,j}(\mathbf{k}^+,\mathbf{k}^-) = g^{(i)}(\mathbf{k}^+) C_\omega(\mathbf{k}^+,\mathbf{k}^-) g^{(j)}(\mathbf{k}^-) =$$
(5.27)

$$\bar{\chi}(\mathbf{p}) \frac{f_i(\mathbf{k}^+)}{D_\omega(\mathbf{k}^+)} [\frac{f_j(\mathbf{k}^-)}{\chi_N(\mathbf{k}^-)} - f_j(\mathbf{k}^-)] - \frac{f_j(\mathbf{k}^-)}{D_\omega(\mathbf{k}^-)} [\frac{f_i(\mathbf{k}^+)}{\chi_N(\mathbf{k}^+)} - f_i(\mathbf{k}^+)]$$
(5.28)

and it holds that

$$\Delta^{h,k}(\mathbf{k}^+,\mathbf{k}^-) = 0 \qquad h<i, h<N$$
(5.29)

The above equations says that *at least one of the two fields in $T_0$ has to be at the scale of the ultraviolet or infrared cut-off*.

On the other hand, when it is non vanishing $\Delta^{i,j}$ is essentially equivalent to the product of two propagators at scales $i$ and $j$. Assume $i \geq j$; if $h < i < N$ ($Z_N = 1$ and $Z_i = 1$ for $i \geq K$)

$$\Delta^{N,i}(\mathbf{k}^+,\mathbf{k}^-) = -\frac{1}{Z_i} \bar{\chi}(\mathbf{k}^+ - \mathbf{k}^-) \frac{f_j(\mathbf{k}^-) u_N(\mathbf{k}^+)}{D_\omega(\mathbf{k}_-)} =$$
(5.30)

$$-\chi(\mathbf{p}) \frac{f_j(\mathbf{k}^-)}{i \not{\mathbf{k}}_-} p_i \int_0^1 dt \partial_i u_N(\mathbf{k}_+ + t\mathbf{p}) = \mathbf{p}_i S^{N,j}_i(\mathbf{k}_+,\mathbf{k}_-)$$
(5.31)

where $u_N(\mathbf{k}) = 0$ for $|\mathbf{k}| \leq \gamma^N$ and $u_N(\mathbf{k}) = 1 - f_N(\mathbf{k})$ for $|\mathbf{k}| \geq \gamma^N$; we have used that $f_j(\mathbf{k}^-) u_N(\mathbf{k}^+) = f_j(\mathbf{k}^-)[u_N(\mathbf{k}^+) - u_N(\mathbf{k}_-)]$. The Fourier trasform admits the bound

$$|S^{N,i}(\mathbf{z}-\mathbf{x},\mathbf{z}-\mathbf{y})| \leq C_n \frac{1}{Z_i} \frac{\gamma^N}{1+[\gamma^N|\mathbf{z}-\mathbf{x}|]^n} \frac{\gamma^i}{1+[\gamma^i|\mathbf{z}-\mathbf{y}|]^n}$$
(5.32)

The bound follows integrating by parts, noting that volume factors are $\gamma^{2j}$ (for the support of $f_j$) and $\gamma^{2N}$ (for the support of $\bar{\chi}(\mathbf{p})u_N(\mathbf{k}^+)$), and each derivative with respect to $\mathbf{k}_-$ gives and extra $\gamma^{-j}$ while each derivative with respect to $\mathbf{k}_+$ gives an extra $\gamma^{-N}$. As similar bound is obtained when $i = N$.

Finally for $j = h, i < N - 1$

$$\Delta_\omega^{(i,h)}(\mathbf{k}^+,\mathbf{k}^-) = \frac{1}{\widetilde{Z}_{h-1}(\mathbf{k}^-)Z_{i-1}}\frac{\widetilde{f}_i(\mathbf{k}^+)u_h(\mathbf{k}^-)}{D_\omega(\mathbf{k}^+)}, j = k < i \leq N-1 \quad (5.33)$$

from which

$$|S^{i,h}(\mathbf{z}-\mathbf{x},\mathbf{z}-\mathbf{y})| \leq C_M \frac{1}{Z_{i-1}}\gamma^{h-i}\frac{\gamma^h}{1+[\gamma^h|\mathbf{z}-\mathbf{x}|]^M}\frac{\gamma^i}{1+[\gamma^i|\mathbf{z}-\mathbf{y}|]^M} \quad (5.34)$$

Note in the above bound we have an extra factor $\gamma^{h-i}$ and a $Z_h$ missing.

We can integrate (5.23) following an iterative procedure similar to the one used in chapt. 3, for the scales $N, N-1, .., K$; considering first the case $\phi = 0$, we get a sequence of effective potentials

$$\widetilde{\mathcal{V}}^{(h)}(\psi,\bar{J}) = \sum_{\substack{n,m \\ n+m \geq 0}} G_{2n,m}^{(h)}(\mathbf{z};\mathbf{x},\mathbf{y})\prod_{i=1}^m J_{\mathbf{z}_i}\prod_{i=1}^n \psi_{\mathbf{x}_i,\omega_i}^{+(\leq h)}\prod_{i=1}^n \psi_{\mathbf{x}_i,\omega_i}^{-(\leq h)} \quad (5.35)$$

we have just to say how to define the $\mathcal{L}$ operation for the terms linear in $J$ (the terms $J$-independent are identical as the ones in chapt. 3).

Let us consider in more detail the terms $G_{2,1}^{(h)}$ whose dimension, according to power counting, is vanishing. We can distinguish the case in which only one or both the two fields in $T_0 - T_1$ are contracted; in the first case one can check that the dimension is indeed negative so that one defines $\mathcal{L} = 0$ for such terms. In fact, when such terms are obtained contracting $T_0$ the improvment in the dimension is due to the extra factor $\gamma^{h-j}$ in (5.34); when $T_-$ is contracted, the gain is due to the fact that we can replace the contracted propagator $g^i(\mathbf{k})$ with $g^i(\mathbf{k}) - g^i(0)$, where $\mathbf{k}$ is an external momentum.

On the other hand, we define $\mathcal{L} = 1$ if no fields in $T_0 - T_-$ is external, so that the effective potential is given by, $k \geq K$

$$\mathcal{L}\bar{\mathcal{V}}^{(h)}(\psi,J) = \int \frac{d\mathbf{p}}{(2\pi)^2}\frac{d\mathbf{k}'}{(2\pi)^2}[\nu_{+,h}(\mathbf{k},\mathbf{p})\widehat{J}_\mathbf{p}\psi_{\mathbf{k},\omega}^+\psi_{\mathbf{k}+\mathbf{p},\omega}^-$$
$$+\nu_{-,h}(\mathbf{k},\mathbf{p})\widehat{J}_\mathbf{p}\psi_{\mathbf{k},-\omega}^+\psi_{\mathbf{k}+\mathbf{p},-\omega}^-] \quad (5.36)$$

The above integration procedure can be iterated with no important differences up to scale $K$. By proceeding as in chapt 3, and assuming that the running coupling $\nu_{k,\mathbf{p}}$ verigy the following bound, for $N \geq k \geq K$ and $0 < \vartheta < 1$ is a constant

$$|\nu_{k,\mathbf{p}}| \leq C\bar{\lambda}_h \gamma^{-\vartheta(N-k)} \tag{5.37}$$

then, for $k \geq K$ and $n \geq 2$

$$\|G_{n,1}^{(k)}\| \leq C\bar{\lambda}_h^{\max(0,\frac{n}{2}-1)} \gamma^{k(1-\frac{n}{2})} \gamma^{\vartheta(k-N)} \tag{5.38}$$

The above bound is identical to the one in Lemma 3.3, up to the extra terms $\gamma^{\vartheta(k-N)}$; indeed in the trees contributing to $G_{n,1}^{(k)}$ there is surely an end-point associated to $\nu_i$, hence by the short memory property and (5.37) we get the extra factor $\gamma^{\vartheta(k-N)}$.

It remains to prove (5.37), and in order to do this we perform an analysis very similar to the one in Lemma 3.4. We write

$$\nu_h = G_{a,2,1}^{(h)} + G_{b,2,1}^{(h)} \tag{5.39}$$

where

$$G_{a,2,1}^{(h)} = \frac{1}{2} \frac{\partial}{\partial \psi_\mathbf{x}^{(\leq h)}} \frac{\partial}{\partial \bar\psi_\mathbf{y}^{(\leq h)}} \sum_{k=1}^{\infty} \frac{1}{(k-1)!} \tag{5.40}$$

$$\mathcal{E}_{h+1,N}^T([\int \frac{d\mathbf{p}}{(2\pi)^2} \frac{d\mathbf{k}'}{(2\pi)^2} C_\omega(\mathbf{k}', \mathbf{k}' - \mathbf{p})\psi_{\omega,\mathbf{k}'}^+ \psi_{\omega,\mathbf{k}'-\mathbf{p}}]\mathcal{V}^{(N)}...\mathcal{V}^{(N)})\Big|_{\psi^{(\leq h)}=0}$$

$$G_{a,2,1}^{(h)} = \frac{1}{2} \frac{\partial}{\partial \psi_\mathbf{x}^{(\leq h)}} \frac{\partial}{\partial \bar\psi_\mathbf{y}^{(\leq h)}} \sum_{k=1}^{\infty} \frac{1}{(k-1)!}$$

$$\mathcal{E}_{h+1,N}^T(\int \frac{d\mathbf{p}}{(2\pi)^2} \frac{\lambda_N}{4\pi} D_{-\omega}(\mathbf{p})\widehat{v}_K(\mathbf{p})\psi_{-\omega,\mathbf{k}'}^+ \psi_{-\omega,\mathbf{k}'-\mathbf{p}}\mathcal{V}^{(N)}...\mathcal{V}^{(N)})\Big|_{\psi^{(\leq h)}=0}$$

We can decompose $G_{a,2,1}^{(h)}$ as explained in Fig. 5.5, which is the analogous

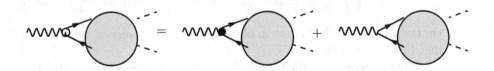

Fig. 5.4 Graphical representation of (5.39)

of the decomposition in Fig. 3.8. Regarding the second term, given by

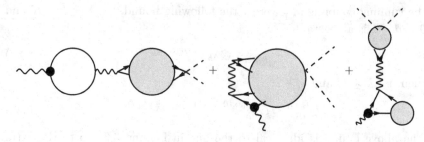

Fig. 5.5 Decomposition of $G^h_{a,2,1}$

$$\sum_{i,j=h}^{N} \int du du' dw dw' \mathbf{S}^{(i,j)}(\mathbf{z};\mathbf{u},\mathbf{w})g(\mathbf{u}-\mathbf{u}')v(\mathbf{u}-\mathbf{w}')H^{(h)}_{1,4}(\mathbf{w}';\mathbf{u}',\mathbf{w},\mathbf{x},\mathbf{y})$$

(5.41)

we can proceed exactly as for (3.78), the main difference being that, from (5.28), either $i$ or $j$ has to be $N$, so that (5.41) is bounded by

$$C_1|\lambda|\gamma^{2K}\gamma^{-h}\sum_{i=h}^{N}\sum_{i'=h}^{i}\gamma^{-N}\gamma^{-i}\gamma^{i'} \leq$$

(5.42)

$$C_2|\lambda|\gamma^{2K}\gamma^{-h-N}(N-h) \leq C_3|\lambda|\gamma^{-2(h-K)}\gamma^{-(1/2)(N-h)}$$

A similar bound holds for the third term in Fig. 5.5; note that such terms obeys to similar bounds as their analogous in Fig 3.8, up to an extra $\gamma^{-(1/2)(N-h)}$.

The main difference with the analysis in §3.9 is in the first term of Fig 5.5; while the local part of the "boubble" in Fig 3.8 was vanishing, here the value of the boubble is given by

$$\nu = \lambda \int \frac{d\mathbf{k}}{(2\pi)^2} \frac{C_{\omega,N}(\mathbf{k},\mathbf{k}-\mathbf{p})}{D_{-\omega}(\mathbf{p})} g^{(\leq N)}_{\omega}(\mathbf{k}) g^{(\leq N)}_{\omega}(\mathbf{k}-\mathbf{p})|_{\mathbf{p}=0}$$

$$= -\frac{\lambda}{4}\int \frac{d\mathbf{k}}{(2\pi)^2} \frac{k_0}{|\mathbf{k}|}\chi'_0(|\mathbf{k}|)D^{-1}_{\omega}(\mathbf{k}) = \frac{\lambda}{\pi}\int_0^{\infty} d\rho \chi'_0(\rho) = \frac{\lambda}{4\pi}$$

(5.43)

On the other hand the first term of Fig 5.5 is *exactly canceled* by $G^{(h)}_{b,2,1}$, so that (5.37) is proved.

In the integration of the fields $\psi^{(K)}, \psi^{(K-1)}, ..$ we proceed as in chapt 4; the $\mathcal{L}$ operation for the terms $J\psi^+\psi^-$ is defined as in §5.3 and the corresponding couplings are $O(\gamma^{C|\lambda h|}\gamma^{\vartheta(-\vartheta(N-K))})$, for $h \leq K$. In the expansion for $R^{1,2}$ we have contributions from trees in which there is $T_0$ with one field contracted at scale $j$ and the other one contracted at scale $h$; they verify the bound $\lambda \frac{\gamma^{-2h}}{Z_h^2}$, where we have used that there is no contribution of $O(1)$

and the fact that in (5.33) there is a $Z_h$ missing. On the other hand the contribution from trees in which in which there is a $\nu_j$ have a dimensional bound $\frac{\gamma^{-2h}}{Z_k^2}$ times a factor $O(\gamma^{C|\lambda j|}\gamma^{\vartheta(-\vartheta(N-K))})$ (coming from $\nu_j$); hence for $N$ large enough the lemma is proved. ∎

## 5.4 The Schwinger-Dyson equation

In order to prove the boundedness of the effective coupling, we write a *Schwinger-Dyson* equation for the 4-point Schwinger function. By the property of the truncated expectation we can write

$$\langle \psi^-_{\mathbf{x}_1,+}\psi^+_{\mathbf{x}_2,+};\psi^-_{\mathbf{x}_3,-}\psi^+_{\mathbf{x}_4,-}\rangle_{h,N} = -\lambda \int d\mathbf{z}_1 d\mathbf{z}_2 v(\mathbf{z}_1-\mathbf{z}_2)[ \qquad (5.44)$$

$$g^{(h,N)}_-(\mathbf{z}_1-\mathbf{x}_4)\langle\psi^-_{\mathbf{x}_1,+};\psi^+_{\mathbf{x}_2,+}\psi^+_{\mathbf{z}_2,+}\psi^-_{\mathbf{z}_2,+}\rangle_{h,N}\langle\psi^-_{\mathbf{x}_3,-};\psi^+_{\mathbf{z}_1,-}\rangle_{h,N}$$

$$+g^{(h,N)}_-(\mathbf{z}_1-\mathbf{x}_4)\langle\psi^+_{\mathbf{z}_2,+}\psi^-_{\mathbf{z}_2,+};\psi^-_{\mathbf{x}_1,+};\psi^+_{\mathbf{x}_2,+};\psi^-_{\mathbf{x}_3,-};\psi^+_{\mathbf{z}_1,-}\rangle_{h,N}$$

$$g^{(h,N)}_-(\mathbf{z}_1-\mathbf{x}_4)\langle\psi^+_{\mathbf{z}_2,+}\psi^-_{\mathbf{z}_2,+}\rangle_{h,N}\langle\psi^-_{\mathbf{x}_1,+};\psi^+_{\mathbf{x}_2,+};\psi^-_{\mathbf{x}_3,-};\psi^+_{\mathbf{z}_1,-}\rangle_{h,N}] (5.45)$$

and the last term is vanishing as $\langle\psi^+_{\mathbf{z}_2,+}\psi^-_{\mathbf{z}_2,+}\rangle_{h,N} = 0$. In Fourier transform the Schwinger-Dyson equation can be written as, if $\rho_{\omega,\mathbf{z}} = \psi^+_{\omega,\mathbf{x}}\psi^-_{\omega,\mathbf{x}}$

$$-\langle\psi^-_{\mathbf{k}_1,+}\psi^+_{\mathbf{k}_2,+};\psi^-_{\mathbf{k}_3,-}\psi^+_{\mathbf{k}_4,-}\rangle_{h,N} = \lambda \hat{g}^{(h,N)}_-(\mathbf{k}_4)\Big[v(\mathbf{k}_1-\mathbf{k}_2)$$

$$\langle\psi^-_{\mathbf{k}_1,+};\psi^+_{\mathbf{k}_2,+};\rho_{\mathbf{k}_1-\mathbf{k}_2,+}\rangle_{h,N}\langle\psi^-_{\mathbf{k}_3,-}\psi^+_{\mathbf{k}_3,-}\rangle_{h,N}$$

$$+\frac{1}{L^2}\sum_{\mathbf{p}}\lambda_N\hat{v}(\mathbf{p})\langle\rho_{\mathbf{p},+};\psi^-_{\mathbf{k}_1,+};\psi^+_{\mathbf{k}_2,+};\psi^-_{\mathbf{k}_3,-};\psi^+_{\mathbf{k}_4-\mathbf{p},-}\rangle_{h,N}\Big] \qquad (5.46)$$

see Fig. 5.6.

We want to prove that (5.46) implies

$$\lambda_h = \lambda \hat{v}(2\gamma^h) + O(\bar{\lambda}_h^2) \qquad (5.47)$$

which means that the effective coupling stays close to $\lambda$ for any $h \leq K$.

If we compute (5.46) for momenta at the cut-off scale, we get, by (5.5)

$$Z_h^{-2}|\bar{\mathbf{k}}|^{-4}[\lambda_h + O(\bar{\lambda}_h^2)] = \lambda g_-(\bar{\mathbf{k}})\Big[v_K(2\gamma^h)\frac{1}{Z_h D_\omega(\bar{\mathbf{k}})^2}\frac{1}{Z_h D_-(\bar{\mathbf{k}})^2}[1+O(\bar{\lambda}_h^2)]$$

$$+\frac{1}{L^2}\sum_{\mathbf{p}}\hat{v}(\mathbf{p})\langle\rho_{\mathbf{p},+};\psi^-_{\mathbf{k}_1,+};\psi^+_{\mathbf{k}_2,+};\psi^-_{\mathbf{k}_3,-};\psi^+_{\mathbf{k}_4-\mathbf{p},-}\rangle_{h,N}\Big] \qquad (5.48)$$

so that (5.47) is proved provided that we show that the last term in the r.h.s. of (5.48) is $O(\bar{\lambda}_h^2\gamma^{-4h}Z_h^{-2})$. If the sum over **p** is restricted over

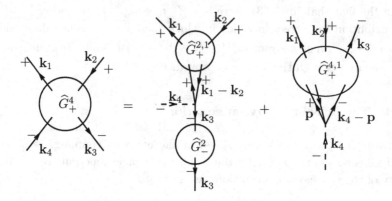

Fig. 5.6 Graphical representation of the Dyson equation (5.46): the dotted line represents the "bare" propagator

$|\mathbf{p}| \leq M\gamma^h$, $M$ a constant, the sum is bounded by, if $h_\mathbf{p}$ is the scale of $\mathbf{p}$,

$$Z_h^{-2} \sum_{h_\mathbf{p}=-\infty}^{h} \widehat{v}(\gamma^{h_\mathbf{p}})\gamma^{-5h}\gamma^{h_\mathbf{p}}\gamma^{-(h-h_\mathbf{p})/2} \qquad (5.49)$$

where a factor $\gamma^{(h-h_\mathbf{p})/2}$ is consumed for making negative the dimension of the cluster with a $J$ and a $\phi$ external lines; hence such contribution is $O(\bar{\lambda}_h^2 \gamma^{-4h} Z_h^{-2})$.

When the sum over $\mathbf{p}$ is over is over $|\mathbf{p}| \geq M\gamma^h$ simple dimensional arguments are not enough and we need Ward Identities. By proceeding as in §5.1 we can derive the following identities

$$D_+(\mathbf{p})\langle \rho_{\mathbf{p},+}; \psi^-_{\mathbf{k}_1,+}; \psi^+_{\mathbf{k}_2,+}; \psi^-_{\mathbf{k}_3,-}; \psi^+_{\mathbf{k}_4-\mathbf{p},-}\rangle_{h,N} = \qquad (5.50)$$
$$\langle \psi^-_{\mathbf{k}_1-\mathbf{p},+}\psi^+_{\mathbf{k}_2,+}; \psi^-_{\mathbf{k}_3,-}\psi^+_{\mathbf{k}_4-\mathbf{p},-}\rangle_{h,N} - \langle \psi^-_{\mathbf{k}_1,+}\psi^+_{\mathbf{k}_2-\mathbf{p},+}; \psi^-_{\mathbf{k}_3,-}\psi^+_{\mathbf{k}_4-\mathbf{p},-}\rangle_{h,N}$$
$$+\frac{v(\mathbf{p})\lambda}{4\pi}D_-(\mathbf{p})\langle \rho_{\mathbf{p},-}; \psi^-_{\mathbf{k}_1,+}; \psi^+_{\mathbf{k}_2,+}; \psi^-_{\mathbf{k}_3,-}; \psi^+_{\mathbf{k}_4-\mathbf{p},-}\rangle_{h,N} + R^{4,1,N}_+(\mathbf{p};\underline{\mathbf{k}})$$

and

$$D_-(\mathbf{p})\langle \rho_{\mathbf{p},-}; \psi^-_{\mathbf{k}_1,+}; \psi^+_{\mathbf{k}_2,+}; \psi^-_{\mathbf{k}_3,-}; \psi^+_{\mathbf{k}_4-\mathbf{p},-}\rangle_{h,N} = \qquad (5.51)$$
$$\langle \psi^-_{\mathbf{k}_1,+}\psi^+_{\mathbf{k}_2,+}; \psi^-_{\mathbf{k}_3-\mathbf{p},-}\psi^+_{\mathbf{k}_4-\mathbf{p},-}\rangle_{h,N} - \langle \psi^-_{\mathbf{k}_1,+}\psi^+_{\mathbf{k}_2,+}; \psi^-_{\mathbf{k}_3,-}\psi^+_{\mathbf{k}_4,-}\rangle_{h,N}$$
$$+\frac{v(\mathbf{p})\lambda}{4\pi}D_+(\mathbf{p})\langle \rho_{\mathbf{p},+}; \psi^-_{\mathbf{k}_1,+}; \psi^+_{\mathbf{k}_2,+}; \psi^-_{\mathbf{k}_3,-}; \psi^+_{\mathbf{k}_4-\mathbf{p},-}\rangle_{h,N} + R^{4,1,N}_-(\mathbf{p};\underline{\mathbf{k}})$$

so that we obtain, if $\chi_M(\mathbf{p})$ is a smooth function with support in $|\mathbf{p}| > M\gamma^h$

$$\frac{1}{L^2}\sum_{\mathbf{p}}\chi_M(\mathbf{p})\widehat{v}(\mathbf{p})\langle\rho_{\mathbf{p},+};\psi^-_{\mathbf{k}_1,+};\psi^+_{\mathbf{k}_2,+};\psi^-_{\mathbf{k}_3,-};\psi^+_{\mathbf{k}_4-\mathbf{p},-}\rangle_{h,N} =$$

$$\frac{1}{L^2}\sum_{\mathbf{p}}\chi_M(\mathbf{p})\frac{\widehat{v}(\mathbf{p})}{D_+(\mathbf{p})}\Big\{\frac{a(\mathbf{p})+\bar{a}(\mathbf{p})}{2}\big[\langle\psi^-_{\mathbf{k}_1-\mathbf{p},+}\psi^+_{\mathbf{k}_2,+};\psi^-_{\mathbf{k}_3,-}\psi^+_{\mathbf{k}_4-\mathbf{p},-}\rangle_{h,N} -$$

$$\langle\psi^-_{\mathbf{k}_1,+}\psi^+_{\mathbf{k}_2-\mathbf{p},+};\psi^-_{\mathbf{k}_3,-}\psi^+_{\mathbf{k}_4-\mathbf{p},-}\rangle_{h,N}\big] + \frac{a(\mathbf{p})-\bar{a}(\mathbf{p})}{2}$$

$$\big[\langle\psi^-_{\mathbf{k}_1,+}\psi^+_{\mathbf{k}_2,+};\psi^-_{\mathbf{k}_3-\mathbf{p},-}\psi^+_{\mathbf{k}_4-\mathbf{p},-}\rangle_{h,N} - \langle\psi^-_{\mathbf{k}_1,+}\psi^+_{\mathbf{k}_2,+};\psi^-_{\mathbf{k}_3,-}\psi^+_{\mathbf{k}_4,-}\rangle_{h,N}\big] +$$

$$\frac{a(\mathbf{p})+\bar{a}(\mathbf{p})}{2}R^{4,1,N}_+(\mathbf{p};\underline{\mathbf{k}}) + \frac{a(\mathbf{p})-\bar{a}(\mathbf{p})}{2}R^{4,1,N}_-(\mathbf{p};\underline{\mathbf{k}})\Big\} \qquad (5.52)$$

We have now to bound all the sums in the r.h.s. of (5.52). Note first that, by parity

$$\frac{1}{L^2}\sum_{\mathbf{p}}\chi_M(\mathbf{p})\frac{\widehat{v}(\mathbf{p})}{D_+(\mathbf{p})}\langle\psi^-_{\mathbf{k}_1,+}\psi^+_{\mathbf{k}_2,+};\psi^-_{\mathbf{k}_3,-}\psi^+_{\mathbf{k}_4,-}\rangle_{h,N} = 0 \qquad (5.53)$$

Moreover the first term in the r.h.s. of (5.52) verifies the bound

$$\frac{1}{L^2}\sum_{\mathbf{p}}\chi_M(\mathbf{p})\Big|\frac{\widehat{v}(\mathbf{p})}{D_+(\mathbf{p})}\frac{a(\mathbf{p})+\bar{a}(\mathbf{p})}{2}\langle\psi^-_{\mathbf{k}_1-\mathbf{p},+}\psi^+_{\mathbf{k}_2,+};\psi^-_{\mathbf{k}_3,-}\psi^+_{\mathbf{k}_4-\mathbf{p},-}\rangle_{h,N}\Big|$$

$$\leq C\bar\lambda_h\frac{\gamma^{-3h}}{Z_h^2} \qquad (5.54)$$

and a similar bound is true for the second and third term. Indeed $\langle\psi^-_{\mathbf{k}_1-\mathbf{p},+}\psi^+_{\mathbf{k}_2,+};\psi^-_{\mathbf{k}_3,-}\psi^+_{\mathbf{k}_4-\mathbf{p},-}\rangle_{h,N}$ is given by a sum of trees, in which each vertex has negative dimension except the ones in the path from the first vertex at which $n_\phi = 2$ to the root; as such vertex has scale $h_\mathbf{p}$ (the scale of $\mathbf{p}$), if we multiply times a factor $\gamma^{\frac{(h-h_\mathbf{p})}{2}}$ we see that all vertices has negative dimension and we can sum over the scales, obtaining a bound $\frac{1}{Z_h^2}\gamma^{-2h-2h_\mathbf{p}}\gamma^{\frac{(-h+h_\mathbf{p})}{2}}\bar\lambda_h$; hence the l.h.s. of (5.54) is bounded by

$$\frac{1}{Z_h^2}\sum_{h_\mathbf{p}}\chi_M(\mathbf{p})v(2\gamma^{h_\mathbf{p}})\gamma^{-2h-h_\mathbf{p}}\gamma^{\frac{(-h+h_\mathbf{p})}{2}}\bar\lambda_h = \frac{1}{Z_h^2}\sum_{h_\mathbf{p}}v(2\gamma^{h_\mathbf{p}})\gamma^{-3h}\gamma^{\frac{(h-h_\mathbf{p})}{2}}\bar\lambda_h$$

$$(5.55)$$

which gives the desidered bound.

In order to prove (5.47) we have still to show that, for $\varepsilon = \pm$

$$\frac{1}{L^2}\sum_{\mathbf{p}}\chi_M(\mathbf{p})\Big|\frac{\widehat{v}(\mathbf{p})}{D_+(\mathbf{p})}g_-(\bar{\mathbf{k}})\frac{a(\mathbf{p})-\varepsilon\bar{a}(\mathbf{p})}{2}R^{4,1,N}_\varepsilon(\mathbf{p};\underline{\mathbf{k}})\Big| \leq C\bar\lambda_h\frac{\gamma^{-4h}}{Z_h^2}$$

$$(5.56)$$

which will be proven in the next section.

## 5.5 Analysis of the cut-off corrections

In order to bound the l.h.s. of (5.56) it is convenient to write it as

$$g_-^{(h,N)}(\mathbf{k}_4) \sum_{\varepsilon=\pm} \frac{a(\mathbf{p}) + \varepsilon \bar{a}(\mathbf{p})}{2} \frac{\partial^4}{\partial h_{\mathbf{k}_4,-} \partial \phi^+_{\mathbf{k}_3,-} \partial \phi^+_{\mathbf{k}_1,+} \partial \phi^-_{\mathbf{k}_2,+}} \mathcal{W}_{\varepsilon,N}|_{h=\phi=0} \tag{5.57}$$

where, if $\varepsilon = \pm$

$$e^{\mathcal{W}_{\varepsilon,N}(h,\phi)} = \int P(d\psi^{(h,N)}) e^{\mathcal{V}+T_0(\psi^{(\leq N)},h)+T_1(\psi^{(\leq N)},h)+\int \frac{d\mathbf{k}}{(2\pi)^2} \phi^{+(\leq N)}_{\mathbf{k},\omega} \psi_{\mathbf{k},\omega}} \tag{5.58}$$

and

$$T_0 = \int \frac{d\mathbf{p}}{(2\pi)^2} \frac{d\mathbf{k}'}{(2\pi)^2} \widehat{v}(\mathbf{p}) h_{\mathbf{k}_4,-} \psi_{\mathbf{k}_4-\mathbf{p},-} \frac{C_\varepsilon(\mathbf{k}', \mathbf{k}'-\mathbf{p})}{D_+(\mathbf{p})} \psi^+_{\varepsilon,\mathbf{k}'} \psi_{\varepsilon,\mathbf{k}'-\mathbf{p}} \tag{5.59}$$

$$T_1 = \frac{\lambda}{4\pi} \int \frac{d\mathbf{p}}{(2\pi)^2} \frac{d\mathbf{k}'}{(2\pi)^2} \widehat{v}(\mathbf{p}) h_{\mathbf{k}_4,-} \psi_{\mathbf{k}_4-\mathbf{p},-} \frac{D_{-\varepsilon}(\mathbf{p})}{D_+(\mathbf{p})} \psi^+_{-\varepsilon,\mathbf{k}'} \psi_{-\varepsilon,\mathbf{k}'-\mathbf{p}}$$

Note that $\partial^4 \mathcal{W}_{\varepsilon,\omega,N}$ in (5.57) is very similar to the 4-point Schwinger function; the difference is that there is a new interaction $T_0 + T_1$ and that the external propagator is necessarily connected to this interaction.

Again the integration of the ultraviolet scales is done as before, and after the integration of $N, N-1, ..h+1$ the exponent in the functional integration is

$$\widehat{\mathcal{V}}^{(h)}(\psi, h) = \mathcal{V}^{(h)}(\psi) + \tag{5.60}$$

$$\sum_{\substack{n,m \\ n+m>0, m>0}} D^{(h)}_{m,2n-m}(\underline{\mathbf{x}}, \underline{\mathbf{y}}, \underline{\mathbf{z}}) [\prod_{i=1}^{n} \psi^{-(\leq h)}_{\mathbf{x}_i}][\prod_{i=1}^{n-m} \psi^{+(\leq h)}_{\mathbf{y}_i, \omega_i}][\prod_{i=1}^{m} h_{\mathbf{z}_i, \omega_i}]$$

We proceed as before analyzing in more detail the kernels $D^{(h)}_{1,1}$ and $D^{(h)}_{1,3}$ which have non negative dimension, in order to improve their bounds.

There are several possible contributions. In the truncated expectations contributing to $D^{(h)}_{1,1}$ and $D^{(h)}_{1,3}$ there is necessarily a $T_0$ or a $T_1$; we decompose $D^{(h)}_{1,2n-1}$ as

$$D^{(h)}_{1,2n-1} = D^{\alpha(h)}_{1,2n-1} + D^{\beta(h)}_{1,2n-1} \tag{5.61}$$

where in $D^{\alpha(h)}_{1,2n-1}$ are the terms such that the field $\psi_{\mathbf{k}_4-\mathbf{p},\omega}$ appearing in (5.59) is an external field, while in $D^{\beta(h)}_{1,2n-1}$ the field $\psi_{\mathbf{k}_4-\mathbf{p},\omega}$ is contracted. One immediately recognizes that

$$D^{\alpha(h)}_{1,3} = G^{(h)}_{2,1} \tag{5.62}$$

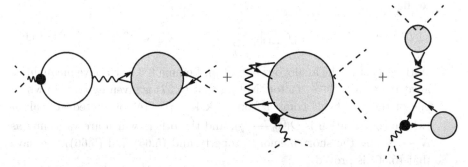

Fig. 5.7 Graphical representation of the contributiond to $D_{1,3}^{\alpha(h)}$ from the first of (5.59)

so that, from (5.38), for $h \geq K$

$$\|D_{1,3}^{\alpha(h)}\| \leq C|\lambda|\gamma^{\vartheta(h-N)} \tag{5.63}$$

On the other hand $D_{n,1}^{\beta(h)}$, $n = 1, 3$, has the structure shown in the following picture.

Fig. 5.8 Graphical representation of $D_{3,1}^{\beta(h)}$ and $D_{1,1}^{\beta(h)}$

We can write
$$D_{1,3}^{\beta(h)}(\mathbf{x}_1, \mathbf{x}_2, \mathbf{x}_3, \mathbf{x}_4) = \int \lambda v(\mathbf{x}_1 - \mathbf{z}_1) g^{(h,N)}(\mathbf{x}_1 - \mathbf{z}_2) G_{4,1}^{(h)}(\mathbf{x}_1, \mathbf{x}_2, \mathbf{x}_3, \mathbf{z}_2; \mathbf{z}_1)$$
(5.64)

and by (5.38), for $N \geq h \geq K$

$$\|D_{1,3}^{\beta(h)}\| \leq C_1 |\lambda| \|v\|_\infty \|G_{4,1}^{(h)}\| \cdot \sum_{j=h}^{N} \|g^{(j)}\|_1 \leq C_2 |\lambda| \gamma^{2K} \gamma^{-2h} \gamma^{\vartheta(h-N)} \tag{5.65}$$

We can proceed in the same way for $D_{1,1}^{\beta(h)}$ writing

$$D_{1,1}^{\beta(h)}(\mathbf{x}_1, \mathbf{x}_2) = \int \lambda v_K(\mathbf{x}_1 - \mathbf{z}_1) g^{(h,N)}(\mathbf{x}_1 - \mathbf{z}_2) G_{4,1}^{(h)}(\mathbf{x}_2, \mathbf{z}_2; \mathbf{z}_1) \tag{5.66}$$

so that

$$\|D_{1,1}^{\beta(h)}\| \leq C_1 |\lambda| \|v\|_\infty \|G_{1,1}^{(h)}\| \cdot \sum_{j=h}^{N} \|g^{(j)}\|_1 \leq C_2 |\lambda| \gamma^{2K} \gamma^{-h} \gamma^{\vartheta(h-N)} \quad (5.67)$$

With respect to the analogous bounds in Lemma 3.3, the above inequalities have an extra $\gamma^{\vartheta(h-N)}$ factor. Noting that (5.57) is given by trees in which one of the two fields coming from $C(\mathbf{k}, \mathbf{k} - \mathbf{p})$ are contracted at scale $h$ (whose contribution is $O(\bar{\lambda}_h \frac{\gamma^{-4h}}{Z_h})$), and the others which are vanishing as $N \to \infty$ (by the short memory property and (5.65) and (5.66)), we have that (5.56) is proved.

## 5.6 Vanishing of Beta function

The above results implies that the infrared cut-off can be removed.

**Lemma 5.2.** *The model (3.1) is well defined in the massless limit. In fact there are constants $\varepsilon_1$ and $c_2$ such that $|\lambda| \leq \varepsilon_1$ implies $\bar{\lambda}_j \leq c_2 \varepsilon_1$, for any $j < 0$.*

*Proof.* The proof is done by contradiction. Assume that there exists a $h \leq 0$ such that

$$\bar{\lambda}_{h+1} \leq c_2 \varepsilon_1 < |\bar{\lambda}_h| \leq 2 c_2 \varepsilon_1 \leq \varepsilon_0 , \quad (5.68)$$

where $\varepsilon_0$ is the same as in Theorem 4.1. We show that this is not possible, if $\varepsilon_1, c_2$ are suitably chosen.

Let us consider the model with cutoff $\gamma^h$. In the previous sections we have proved that

$$|\lambda_h - \lambda| \leq c_3 \bar{\lambda}_{h+1}^2 . \quad (5.69)$$

However $\lambda_j$ is the same, for any $j \geq h$, in the model with or without cutoff; in fact $g^{(j)}(\mathbf{k})$ does not depend on $h$ for $j > h$, and $\lambda_h$ only depends on the propagators $g^{(j)}(\mathbf{k})$ with $j > h$ by definition. Hence we get the bound $|\lambda_h| \leq \varepsilon_1 + c_3 c_2^2 \varepsilon_1^2$, which is in contradiction with (5.68) if, for instance, $c_2 = 2$ and $\varepsilon_1 \leq 1/(4c_3)$. ∎

The proof of Theorem 5.1 is a corollary of the above results. We fix $K = 0$ for definiteness (the general case can be found by simply replacing $j$ with $j - K$). The Beta function is an analytic function of $\lambda_j$ and it can be written as

$$\beta_{j,\lambda,L}(\lambda_j, ..., \lambda_j) = \sum_n b_{j,n} (\lambda_j)^n \quad (5.70)$$

Note that $b_{j,n} = b_n + O(\gamma^{\vartheta j})$ and we prove that for any $n$
$$b_n = 0 \tag{5.71}$$
The proof is by contradiction. Assume that, for some $\bar{n}$
$$\beta_j(\lambda_j,...,\lambda_j) = b_{j,n}(\lambda_j)^{\bar{n}} + O(\lambda_j)^{\bar{n}+1}, \tag{5.72}$$
with $b_{\bar{n}}$ a non vanishing vector, and that for all $n \leq \bar{n}-1$, $b_n$ is vanishing; $\lambda_j$ are analytic functions of $\lambda$, that is
$$\lambda_j = \lambda_0 + \sum_{n \leq \bar{n}} c_n^{(j)} \lambda^n + O((\lambda)^{\bar{n}+1}) \tag{5.73}$$
and for any fixed $j$ the sequence $c_n^j$ is a bounded sequence. Inserting (5.73) in the Beta function, using analyticity and equating the coefficient of $\lambda^n$ with $n \leq \bar{n}-1$ we get
$$c_n^{(j-1)} = c_n^{(j)} + \sum_{k=j+1}^{0} d_{j,k}^n + O(\gamma^{\vartheta j}) \tag{5.74}$$
where the last sum represents the contribution of $D_{j,k}$, so that
$$|d_{j,k}^n| \leq \gamma^{-\vartheta(k-j)} D_n \sup_{2 \leq m \leq n-1} |\bar{c}_m^{(j)} - \bar{c}_m^{(k)}| \tag{5.75}$$
where we have used that $\vec{D}_{j,k}$ is at least quadratic in the running coupling constants, and $D_{\bar{n}}$ is a suitable constant (in $j$). Note that
$$|c_n^{(j-1)} - \bar{c}_n| \leq \bar{D}_{\bar{n}} \sum_{j'=-\infty}^{j} [(\sum_{k=j'+1}^{0} \gamma^{-\vartheta(k-j')} \sup_{2 \leq m \leq n-1} |c_m^{(j')} - c_m^{(k)}|) + \gamma^{\vartheta j'}] \tag{5.76}$$
The above inequality implies by induction that, for $n \leq \bar{n}-1$
$$\sup_{2 \leq m \leq n} |c_m^{(k)} - \vec{c}_m| \leq C^n \gamma^{\frac{\vartheta}{2} k} \tag{5.77}$$
for a suitable $C$; assume in fact that it is true for $k \geq j$ and from (5.74) we get
$$|c_n^{(j-1)} - \vec{c}_n| \leq C^{n-1} \bar{D}_{\bar{n}} \tag{5.78}$$
$$\sum_{j'=-\infty}^{j} [\sum_{k=j'+1}^{0} \gamma^{-\vartheta(k-j')}(\gamma^{\frac{\vartheta}{2}k} + \gamma^{\frac{\vartheta}{2}j'}) + \gamma^{\vartheta j'}] \leq K C^{n-1} \bar{D}_{\bar{n}} \gamma^{\frac{\vartheta}{2}(j-1)}$$
so that (5.77) holds for $j-1$ if $C \geq K \bar{D}_{\bar{n}}$. On the other hand (5.77) implies
$$|d_{j,k}^n| \leq \bar{C}^n \gamma^{\frac{\vartheta}{2}(j-1)} \tag{5.79}$$
Writing now the analogous of (5.74) for $n = \bar{n}$ we get
$$c_{\bar{n}}^{(j-1)} = c_{\bar{n}}^{(j)} + b_{j,\bar{n}} + d_{j,k}^{\bar{n}} \tag{5.80}$$
which can be rewritten as
$$c_{\bar{n}}^{(j-1)} = c_{\bar{n}}^{(j)} + b_{\bar{n}} + O(\gamma^{\frac{\vartheta}{2}j}) \tag{5.81}$$
so that $c_{\bar{n}}^{(j)}$ is necessarily a diverging as $j \to -\infty$, and this is a contradiction.

A similar argument can be repeated for $\beta_{j,\xi,L}$.

## 5.7 Non-perturbative Adler-Bardeen theorem

If we remove the infrared cut-off in the WI (5.15) we get, after some manipulations

$$-i p_\mu \langle j_{\mathbf{p}}^{5,\mu}; \psi_{\mathbf{k}} \bar\psi_{\mathbf{k-p}} \rangle_N = \gamma^5 [\langle \psi_{\mathbf{k-p}} \bar\psi_{\mathbf{k-p}} \rangle_N - \langle \psi_{\mathbf{k}} \bar\psi_{\mathbf{k}} \rangle_N] + \Delta_N^5(\mathbf{k}, \mathbf{k-p}) \quad (5.82)$$

where

$$\Delta_N^5(\mathbf{k}, \mathbf{k-p}) = \int d\mathbf{k}' C_{\mathbf{k}',\mathbf{p}}^{\mu,N} < \bar\psi_{\mathbf{k}'} \gamma^\mu \gamma^5 \psi_{\mathbf{k'-p}}; \psi_{\mathbf{k}} \bar\psi_{\mathbf{k-p}} >_{h,N} \quad (5.83)$$

with

$$C_{\mathbf{k},\mathbf{p}}^{\mu,N} = ([\chi_N(\mathbf{k})]^{-1} - 1)\mathbf{k}^\mu - ([\chi_N(\mathbf{k-p})]^{-1} - 1)(\mathbf{k}^\mu - \mathbf{p}^\mu) \quad (5.84)$$

By comparing (5.83) with (1.11), derived formally from the non regularized formal functional integral, we see that the difference is given by the term $\Delta_N^5$. Such term has been analyzed in §5.2 where we have proven that,

$$\lim_{N \to \infty} \Delta_N^0(\mathbf{k}, \mathbf{k-p}) = \frac{e}{4\pi} v(\mathbf{p}) \mathbf{p}_\mu \langle j_{\mathbf{p}}^{5,\mu}; \psi_{\mathbf{k}} \bar\psi_{\mathbf{k-p}} \rangle \quad (5.85)$$

and using that

$$ev(\mathbf{p}) \langle j_{\mathbf{p}}^{5,\mu}; \psi_{\mathbf{k}} \bar\psi_{\mathbf{k-p}} \rangle = \varepsilon_{\mu,\nu} \langle A_{\mathbf{p}}^\mu; \psi_{\mathbf{k}} \bar\psi_{\mathbf{k-p}} \rangle \quad (5.86)$$

we get

$$-i p_\mu \langle j_{\mathbf{p}}^{5,\mu}; \psi_{\mathbf{k}} \bar\psi_{\mathbf{k-p}} \rangle \quad (5.87)$$
$$= \gamma^5 [\langle \psi_{\mathbf{k-p}} \bar\psi_{\mathbf{k-p}} \rangle - \langle \psi_{\mathbf{k}} \bar\psi_{\mathbf{k}} \rangle] + \frac{e}{2\pi}(-i p_\mu) i \varepsilon^{\mu,\nu} < A_{\nu,\mathbf{p}}; \psi_{\mathbf{k}} \bar\psi_{\mathbf{k+p}} >$$

and we see that (5.87) and (1.11) differs for the last term in (5.87), which is called *axial anomaly*.

Note that the anomaly coefficient is linear in $e$, that is the anomaly is not renormalized by higher order corrections. This result is then a non-perturbative version of the Adler-Bardeen theorem in $d = 2$, see Refs.[32],[33], obtained by cancellation of Feynman graphs. We will see in the following chapter that this property requires in an essential way the fast decaying property of the boson propagator and it can be violated for non soft boson interactions.

## 5.8 Further remarks

The vanishing of beta function, at a purely perturbative level, was proved in Refs. [34],[35] using the properties of local invariance of the theory. The first non-perturbative proof of the vanishing of the Beta function was given in Ref. [36],[37], by a contradiction argument starting from certain properties of some exactly solvable models. More exactly one analyzes the Luttinger model by a multiscale analysis, and the only way for which the Schwinger functions can have the same properties of the ones found by the (operatorial) exact solution Ref. [18] is the validity of (5.1) (as we will see, the beta function of the Luttinger model or of infrared QED2 coincide up to irrelevant term).

Exact solutions are quite rare and essentially only at low dimension, so that one would like to achieve such a property remaining in a purely functional integral approach, avoiding methods valid only on $d = 2$. The conjecture that a non perturbative proof of vanishing of the beta function would follow from certain local symmetries was finally proved in [31] and it is the one presented here, with some simplifications with respect to the original one.

# Chapter 6

# Thirring and Gross-Neveu Models

## 6.1 The Thirring model

In the previous chapters we have considered a regularized version of QED2 and we have shown how to remove the ultraviolet and the fermionic momentum ultraviolet cut-off $N \to \infty$, uniformly in the fermionic mass. By a slight extension of the previous analysis, it would be not hard to show that also the bosonic ultraviolet cut-off could be removed, by performing the limit $\bar{K} \to \infty$.

A peculiarity of QED2 with photon mass is that it has not ultraviolet divergences in the perturbative expansion and, as a consequence, one can choose *finite* bare parameters. In this respect, QED2 is deeply different from QED4, in which ultraviolet divergences must be reabsorbed in bare parameters which are *diverging* when regularizations are removed. In this chapter we will consider QFT models which, despite still 2-dimensional, requires diverging bare parameters.

We consider the QFT with Generating Function

$$\mathcal{W}_{K,N}(J,\phi) = \tag{6.1}$$
$$\log \int P_Z(d\psi^{(\leq N)}) P(dA) e^{\int d\mathbf{x}[eZ_1\bar{\psi}_\mathbf{x}(A_{\mu,\mathbf{x}}\gamma_\mu)\psi_\mathbf{x} + J_{\mu,\mathbf{x}}A_{\mu,\mathbf{x}} + \phi_\mathbf{x}\bar{\psi}_\mathbf{x} + \bar{\phi}_\mathbf{x}\psi_\mathbf{x}]}$$

in which the fermionic propagator is given by

$$g^{(\leq N)}(\mathbf{x}-\mathbf{y}) = \frac{1}{Z}\int d\mathbf{p}\frac{-i\slashed{p} + Z_4 m}{\mathbf{p}^2 + Z_4^2 m^2}e^{-i\mathbf{p}(\mathbf{x}-\mathbf{y})}\chi_N(\mathbf{p}) \tag{6.2}$$

and the boson propagator is

$$v_K(\mathbf{x}-\mathbf{y}) = \int \frac{d\mathbf{p}}{(2\pi)^2}e^{-i\mathbf{p}(\mathbf{x}-\mathbf{y})}\chi_K(\mathbf{p}) \tag{6.3}$$

with $||v_K|| \leq C\gamma^{2K}$ and such that

$$|v_K(\mathbf{k})| \leq \frac{\gamma^{2K}}{1+(\gamma^K|\mathbf{x}|)^M} \qquad (6.4)$$

This model has a perturbative structure much more similar to $d=4$ gauge models, as it is renormalizable with divergence index is $2 - \frac{f}{2} - b$, if $b, f$ are the external bosonic and fermionic lines (to be compared with $4 - \frac{3}{2}f - b$ for QED$_4$); remember that with the choice (3.7) the theory is superrinormalizable and the index is $2 - n - f/2$, if $n$ is the perturbative order.

If the bosons are integrated out, a purely fermionic theory is obtained which can be considered a regularization of the *massive Thirring model*, describing the self interaction of a fermion fields via a current-current interaction

$$e^{\mathcal{W}_{K,N}(J,\phi)} = \int P_Z(d\psi^{(\leq N)}) \qquad (6.5)$$

$$e^{\frac{1}{2}\int d\mathbf{x}d\mathbf{y} v_K(\mathbf{x}-\mathbf{y})[Z_1 e\bar{\psi}_\mathbf{x}\gamma_\mu\psi_\mathbf{x}+J_{\mu,\mathbf{x}}][Z_1 e\bar{\psi}_\mathbf{y}\gamma_\mu\psi_\mathbf{y}+J_{\mu,\mathbf{y}}]+\int d\mathbf{x}[\phi_\mathbf{x}\bar{\psi}_\mathbf{x}+\bar{\phi}_\mathbf{x}\psi_\mathbf{x}]}$$

There are two ultraviolet cut-offs and different properties will be found, in the case (6.3), depending if the fermionic or the bosonic cut-off is removed first, that is in the limit $K \to \infty, N \to \infty$ or $N \to \infty, K \to \infty$. We will call $\lambda = -2e^2$.

## 6.2 Removing the fermionic ultraviolet cut-off before the bosonic one

The analysis of this case is essentially identical to the one discussed in detail for QED2, up to some trivial modifications (for details, Ref. [27],[38]).

**Theorem 6.1.** *If $e$ is small enough, and choosing*

$$Z = Z_1 = \gamma^{-\eta K}(1+O(e^2)), \quad Z_4 = \gamma^{\eta_1 K}(1+O(e^2)) \qquad (6.6)$$

*with $\eta = ae^4 + O(e^6)$, $\eta_1 = be^2 + O(e^4)$, $a, b > 0$ suitable constants, the 2-point Schwinger function verifies in the limit* $\lim_{K\to\infty} \lim_{N\to\infty}$

$$|\langle\bar{\psi}_\mathbf{x}\psi_\mathbf{y}\rangle| \leq \frac{C}{|\mathbf{x}-\mathbf{y}|^{1+\eta}} e^{-c\sqrt{\kappa\mu^{1+\eta_1'}}|\mathbf{x}-\mathbf{y}|} \qquad (6.7)$$

*with $\eta_1' = (1-\eta_1)^{-1} - 1$.*

*In the massless case $m = 0$*

$$\langle\bar{\psi}_\mathbf{x}\psi_\mathbf{y}\rangle = \frac{g(\mathbf{x}-\mathbf{y})}{|\mathbf{x}-\mathbf{y}|^\eta}(1+R(e)) \qquad (6.8)$$

with $|R(e)| \leq Ce^2$ Moreover the following axial Ward Identity holds, in the massless case and in the limit $K \to \infty, N \to \infty$

$$-i p^\mu \langle j^5_{\mu,\mathbf{p}} \psi_\mathbf{k} \bar\psi_{\mathbf{k}-\mathbf{p}} \rangle_{K,N} = \gamma^5 \langle \psi_{\mathbf{k}-\mathbf{p}} \bar\psi_{\mathbf{k}-\mathbf{p}} \rangle_{K,N} -$$
$$\gamma^5 \langle \psi_\mathbf{k} \bar\psi_\mathbf{k} \rangle_{K,N} + \frac{e}{2\pi} \varepsilon_{\mu,\nu} i p_\mu < A_{\nu,\mathbf{p}} \psi_\mathbf{k} \bar\psi_{\mathbf{k}-\mathbf{p}} >_{K,N} \quad (6.9)$$

with $j^5_\mu = Z \bar\psi \gamma_\mu \gamma_5 \psi$, $j^5 = Z_4 \bar\psi \gamma_5 \psi$.

Note that the bare wave function and mass must be chosen with a singular power law dependence from the ultraviolet cut-off $K$, in order to have non trivial Schwinger functions. According to power counting, there are *four* marginal or relevant monomials (namely $\bar\psi\psi, j^\mu j^\mu, A^\mu j^\mu, \bar\psi \partial \psi$), but all the ultraviolet divergences can be reabsorbed in *two* bare parameters $(Z, Z_4)$ only, as a consequence of the Ward Identities. In particular, the wave funcion renormalization $Z$ can be choosen as equal to the charge renormalization $Z_1$, as in perturbative QED4.

The proof of Theorem 6.1 is based an an analysis essentially identical to the one described in the previous chapters. The fields $\psi^{(N)}, \psi^{(N-1)}, .., \psi^{(K)}$ are integrated following the procedure described in chapt.3, obtaining (see Lemma 3.4), if $\lambda = -2e^2$

$$\frac{Z_K}{Z} = 1 + O(\lambda) \quad \frac{Z_K^{(1)}}{Z} = 1 + O(\lambda) \quad \frac{Z_K^{(4)}}{Z^{(4)}} = 1 + O(\lambda) \quad (6.10)$$

The integration of the fields $\psi^{(K-1)}, \psi^{(K-2)}, ..$ is done as described in chapt. 4, obtaining, from §4.5 for $h \leq K$

$$Z_h = \gamma^{-\eta h}(1 + O(\lambda)) \quad \frac{Z_h^{(2)}}{Z_h} = (1 + O(\lambda)) \quad Z_h^{(2)} = \gamma^{\eta_1 h}(1 + O(\lambda)) \quad (6.11)$$

Hence for having finite renormalizations at scales $O(1)$ (that is the "laboratory" scales) we must choose the bare renormalizations as in (6.6). Note that the second of (6.10) follows from the second of (5.1), while the fact that $\lambda_h = \lambda_K + O(\lambda_K^2)$ follows from the first of (5.1).

The short and large distance behaviour of the 2-point Schwinger function is anomalous and expressed by a non-universal critical index. Finally the chiral anomaly is linear in $e$, that is the anomaly is not renormalized by higher orders corrections.

The massless WI are given by, from chapt.5

$$\partial^\mu_\mathbf{z} \langle j^\mu_\mathbf{z}; \psi_\mathbf{x} \bar\psi_\mathbf{y} \rangle = a[\delta(\mathbf{z} - \mathbf{x}) - \delta(\mathbf{z} - \mathbf{y})] \langle \psi_\mathbf{x} \bar\psi_\mathbf{y} \rangle \quad (6.12)$$

$$\partial^\mu_\mathbf{z} \langle j^{5,\mu}_\mathbf{z}; \psi_\mathbf{x} \bar\psi_\mathbf{y} \rangle = \bar a[\delta(\mathbf{z} - \mathbf{x}) - \delta(\mathbf{z} - \mathbf{y})] \gamma^5 \langle \psi_\mathbf{x} \bar\psi_\mathbf{y} \rangle \quad (6.13)$$

with
$$a^{-1} = 1 - \frac{\lambda}{4\pi} \quad \bar{a}^{-1} = 1 + \frac{\lambda}{4\pi} \tag{6.14}$$

In the massless case we can write the the *Schwinger-Dyson* equation

$$<\psi_{\mathbf{k}}\bar{\psi}_{\mathbf{k}}>_{K,N} = \chi_N(\mathbf{k})\frac{-i\,\mathbf{k}}{\mathbf{k}^2}[Z^{-1} - e^2\int\frac{d\mathbf{p}}{(2\pi)^2}\widehat{v}_K(\mathbf{p})\gamma_\mu <j_{\mu,\mathbf{p}}\psi_{\mathbf{k}}\bar{\psi}_{\mathbf{k}-\mathbf{p}}>_{K,N}] \tag{6.15}$$

which can be combined with the WI and its analogous for the current obtaining

$$<\psi_{\mathbf{k}}\bar{\psi}_{\mathbf{k}}>_{K,N} = \mathcal{B}_{1,K,N}(\mathbf{k}) + \frac{\chi_N(\mathbf{k})}{Z}\frac{-i\,\mathbf{k}}{\mathbf{k}^2}$$
$$-e^2\frac{[\bar{a}-a]}{2}\chi_N(\mathbf{k})(\frac{-i\,\mathbf{k}}{\mathbf{k}^2})\int\frac{d\mathbf{p}}{(2\pi)^2}i\widehat{v}_K(\mathbf{p})\frac{\gamma_\mu\mathbf{p}_\mu}{\mathbf{p}^2}<\psi_{\mathbf{k}-\mathbf{p}}\bar{\psi}_{\mathbf{k}-\mathbf{p}}>_{K,N} \tag{6.16}$$

where $\mathcal{B}_{1,K,N}(\mathbf{k})$ is a term depending on the *integrated difference between the WI with or without cut-off*. It is an immediate consequence of the analysis in chapt. 5 that

$$\lim_{K\to\infty}\lim_{N\to\infty}\mathcal{B}_{1,K,N}(\mathbf{k}) = 0 \tag{6.17}$$

## 6.3 Removing the bosonic ultraviolet cut-off before the fermionic one

We consider now (6.2) with bosonic propagator (6.3) and we assume the bosonic cut-off is removed first; this is equivalent to consider (13.19) assuming that the removal of the fermionic cut-off is done starting from a local current-current or Thirring interaction; in the limit $K\to\infty$ the functional integral is given by

$$e^{\mathcal{W}^N(\phi,J)} = \tag{6.18}$$
$$\int P_Z(d\psi^{(\le N)})e^{-\int d\mathbf{x}\frac{\lambda}{4}Z_1^2(\bar{\psi}_{\mathbf{x}}\gamma_\mu\psi_{\mathbf{x}})(\bar{\psi}_{\mathbf{x}}\gamma_\mu\psi_{\mathbf{x}}) + \int d\mathbf{x}(\bar{\eta}_{\mathbf{x}}\psi_{\mathbf{x}} + \bar{\psi}_{\mathbf{x}}\eta_{\mathbf{x}})}$$

The analysis of such functional integral is essentially identical to the one presented in chapt. 5 (for more details, see Ref.[39]) and the results can be summarized in the following theorem.

**Theorem 6.2.** *If $\lambda$ is small enough (uniformly in m), by choosing the bare parameters as*

$$Z = Z_1 = \gamma^{-\eta N}, \quad Z_4 = \gamma^{\eta_1 N} \tag{6.19}$$

with $\eta, \eta_1$ analytic functions of $e$ and $\eta = a\lambda^2 + O(e^6)$, $\eta_1 = -b\lambda + O(e^4)$, $a, b > 0$ suitable constants, the limit the 2-point Schwinger function verifies in the limit $\lim_{N \to \infty} \lim_{K \to \infty}$

$$|\langle \bar\psi_\mathbf{x} \psi_\mathbf{y} \rangle| \leq \frac{C}{|\mathbf{x}-\mathbf{y}|^{1+\eta}} e^{-c\sqrt{\kappa \mu^{1+\eta'_1}}|\mathbf{x}-\mathbf{y}|} \qquad (6.20)$$

with $\eta'_1 = (1-\eta_1)^{-1} - 1$.

In the massless case $m = 0$

$$\langle \bar\psi_\mathbf{x} \psi_\mathbf{y} \rangle = \frac{g(\mathbf{x}-\mathbf{y})}{|\mathbf{x}-\mathbf{y}|^\eta}(1 + R(\lambda)) \qquad (6.21)$$

with $|R(\lambda)| \leq C|\lambda|$. Moreover the following axial Ward Identity holds, in the massless case and in the limit $N \to \infty, K \to \infty$ if $\lambda = -2e^2$

$$\mathbf{p}^\mu \langle j^5_{\mu,\mathbf{p}} \psi_\mathbf{k} \bar\psi_{\mathbf{k}-\mathbf{p}} \rangle_{K,N} = \gamma^5 \langle \psi_{\mathbf{k}-\mathbf{p}} \bar\psi_{\mathbf{k}-\mathbf{p}} \rangle_{K,N} - \qquad (6.22)$$

$$\gamma^5 \langle \psi_\mathbf{k} \bar\psi_\mathbf{k} \rangle_{K,N} + [\frac{e}{2\pi} + Ae^3 + O(e^4)]\varepsilon_{\mu,v} i p_\mu < A_{\nu,\mathbf{p}} \psi_\mathbf{k} \bar\psi_{\mathbf{k}-\mathbf{p}} >_{K,N}$$

with $A > 0$ and $j^5_\mu = Z_2 \bar\psi \gamma_\mu \gamma_5 \psi$, $j^5 = Z_4 \bar\psi \gamma_5$.

Note that the anomaly non-renormalization depends from the order in which the cut-offs are removed. The anomaly is not renormalized if the fermionic cut-off is removed first, while the anomaly have higher order corrections if the bosonic cut-off is removed first.

The above theorem is proved by a multiscale analisys identical to the one in chapt.4, with $N$ replacing $K$ and $\mathcal{R}\mathcal{V}^N = 0$. Hence the beta function for $\lambda_h$ coincides with $\beta_{\lambda,L,j}$ with $N$ replacing $K$ and up to $O(\gamma^{-\vartheta(N-j)})$ terms (for the short memory property) so that from Theorem 5.1

$$|\beta_{\lambda,L,j}(\lambda_j, \ldots, \lambda_j)| \leq C\lambda_j^2 \gamma^{-\vartheta(N-j)} \qquad (6.23)$$

from which

$$|\lambda_{-\infty} - \lambda_j| \leq c_1 |\lambda|^2 \gamma^{-\vartheta(N-j)} \qquad (6.24)$$

which means that for any fixed scale $j$ the effective coupling $\lambda_j$ has no flow (it is constant in $j$) when the ultraviolet cut-off is removed ($N \to \infty$). On the other hand $\lambda_j$ has a bounded but non trivial flow from the cut-off scale $N$ to $O(1)$ scales. Moreover

$$Z_j = c_h \gamma^{-\eta j} \qquad \mu_j = \widehat{c}_j \gamma^{-\eta_1 j} \qquad (6.25)$$

with

$$|c_j - 1| \leq C\lambda^2 \qquad |\widehat{c}_j - 1| \leq C\lambda^2 \ . \qquad (6.26)$$

When the ultraviolet cut-off is removed, the Schwinger functions at fixed coordinates depend only from $\lambda_{-\infty} \equiv \lambda_0$; this means that the physical observables depend only on the dressed parameters (which are fixed from the experiments).

Note that the Schwinger functions are the same, when expressed in terms of $\lambda_{-\infty}$, if the bosonic cut-off is removed before or after the fermionic one (but they are different when expressed in terms of the bare parameters). On the contrary, the bare parameters depend on the details of the regularization, in agreement with the QFT phylosophy in which the QFT are unobservable.

The WI can be eqivalently written in the following form

$$\partial_{\mathbf{z}}^{\mu} \langle j_{\mathbf{z}}^{\mu}; \psi_{\mathbf{x}} \bar{\psi}_{\mathbf{y}} \rangle = a[\delta(\mathbf{z}-\mathbf{x}) - \delta(\mathbf{z}-\mathbf{y})]\langle \psi_{\mathbf{x}} \bar{\psi}_{\mathbf{y}} \rangle \qquad (6.27)$$

$$\partial_{\mathbf{z}}^{\mu} \langle j_{\mathbf{z}}^{5,\mu}; \psi_{\mathbf{x}} \bar{\psi}_{\mathbf{y}} \rangle = \bar{a}[\delta(\mathbf{z}-\mathbf{x}) - \delta(\mathbf{z}-\mathbf{y})]\gamma^5 \langle \psi_{\mathbf{x}} \bar{\psi}_{\mathbf{y}} \rangle \qquad (6.28)$$

with

$$a^{-1} = 1 - \frac{\lambda}{4\pi} + c_+ \lambda^2 + R_1(\lambda) \quad \bar{a}^{-1} = 1 + \frac{\lambda}{4\pi} + c_+ \lambda^2 + R_2(\lambda) \qquad (6.29)$$

We discuss briefly why in this case the anomaly is renormalized by higher orders corrections, referring for more details to [39]. In the analogous of the analysis in §5.3 the $\mathcal{L}$ operation acts on the kernels $G_{2,1}^h$ as replacing them as with their values at zero momenta. We get then two running coupling constants $\nu_{\pm,j}$, one corresponding to $\bar{J}\psi_\omega^+ \psi_\omega^-$ and the other to $\bar{J}\psi_{-\omega}^+ \psi_{-\omega}^-$; we can write, for $h+1 \leq j \leq N-1$,

$$\nu_{\pm,j-1} = \nu_{\pm,j} + \beta_{\pm,\nu}^j(\lambda_j, \nu_j.., \lambda_N, \nu_N), \qquad (6.30)$$

with

$$\beta_{\pm,\nu}^j(\lambda_j, \nu_j.., \lambda_N, \nu_N) = \beta_{\pm,\nu}^j(\lambda_j, .., \lambda_N) + \sum_{j'=j}^{N} \nu_{\pm,j'} \beta_{\pm,\nu}^{j,j'}(\lambda_j, .., \lambda_N) \qquad (6.31)$$

and, given a positive $\vartheta < 1/4$,

$$|\beta_{\pm,\nu}^j(\lambda_j, .., \lambda_N)| \leq C\bar{\lambda}_j \gamma^{-2\vartheta(N-j)} \quad , |\beta_{\pm,\nu}^{j,j'}(\lambda_j, .., \lambda_N)| \leq C\bar{\lambda}_j^2 \gamma^{-2\vartheta|j-j'|} \qquad (6.32)$$

We fix $\nu_{\pm,N}$ so that

$$\nu_{\pm,N} = -\sum_{j=h+1}^{N} \beta_{\pm,\nu}^j(\lambda_j, \nu_j.., \lambda_N, \nu_N) \qquad (6.33)$$

By a fixed point argument, one can show that, if $\bar{\lambda}_h$ is small enough, it is possible to choose $\nu_{\pm,N}$ so that

$$|\nu_{\omega,j}| \leq c_0 \bar{\lambda}_h \gamma^{-\vartheta(N-j)} \qquad (6.34)$$

for any $h+1 \leq j \leq N$. Moreover, from an explicit computation of (6.30), we get $\nu_- = \frac{\lambda}{4\pi} + O(\lambda^2)$ and $\nu_+ = c_+\lambda^2 + O(\lambda^3)$ with $c_+ < 0$.

Finally by combining the Schwinger-Dyson equation with the WI and its

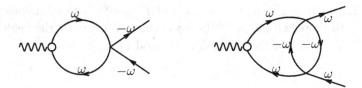

Fig. 6.1  Lowest order contributiond to $\nu_\pm$

analogous for the current at finite cut-off obtaining

$$<\psi_\mathbf{k}\bar\psi_\mathbf{k}>_{K,N} = \mathcal{B}_{2,K,N}(\mathbf{k}) + \frac{\chi_N(\mathbf{k})}{Z}\frac{-i\,\rlap{/}{\mathbf{k}}}{\mathbf{k}^2} \qquad (6.35)$$
$$-e^2\frac{[\bar a_2 - a_2]}{2}\chi_N(\mathbf{k})(\frac{-i\,\rlap{/}{\mathbf{k}}}{\mathbf{k}^2})\int\frac{d\mathbf{p}}{(2\pi)^2}i\hat v_K(\mathbf{p})\frac{\gamma_\mu \mathbf{p}_\mu}{\mathbf{p}^2}<\psi_{\mathbf{k}-\mathbf{p}}\bar\psi_{\mathbf{k}-\mathbf{p}}>_{K,N}$$

and $a_2^{-1} = 1 - \frac{e^2}{2\pi} - Ae^4 + O(e^6)$, $\bar a_2^{-1} = 1 + \frac{e^2}{2\pi} - Ae^4 + O(e^6)$; again $\mathcal{B}_{2,K,N}(\mathbf{k})$ is a term depending on the integrated difference between the WI with or without cut-off, but in this case it is not vanishing at all, but it holds

$$\mathcal{B}_{2,K,N}(\mathbf{k}) = \sigma\frac{1}{Z}\chi_N(\mathbf{k})(\frac{-i\,\rlap{/}{\mathbf{k}}}{\mathbf{k}^2}) + \rho <\psi_\mathbf{k}\bar\psi_\mathbf{k}> + H_{K,N}(\mathbf{k}) \qquad (6.36)$$

with $\rho = \bar A e^4 + O(e^6)$, with $\bar A > 0$, $\sigma = O(e^6)$ and $\lim_{N\to\infty}\lim_{K\to\infty} H_{K,N}(\mathbf{k}) = 0$.

The same conditions ensuring the validity of the anomaly non-renormalization ensures the validity of a closed equation for the 2-point Schwinger function, in the limit in which cut-offs are removed, which is equal to the one obtained combining the limiting Schwinger-Dyson equations with the WI. On the contrary, if the bosonic cut-off is removed first the 2-point function verifies a *different* closed equation.

## 6.4 The Gross-Neveu model

We consider a regularized version of the Gross-Neveu, describing fermions with a local, current-current interaction

$$e^{\mathcal{W}_{N,L}(J^A,\phi)} = \int P(d\psi^{\leq N}) \qquad (6.37)$$

$$e^{-\frac{\lambda_N}{4}\int d\mathbf{x}(\sum_{i=1}^n \bar\psi_{i,\mathbf{x}}\gamma^\mu\psi_{i\mathbf{x}})(\sum_{i=1}^n \bar\psi_{i,\mathbf{x}}\gamma^\mu\psi_{i,\mathbf{x}})+\sum_{i=1}^n\int d\mathbf{x}[\phi_{i,\mathbf{x}}\bar\psi_{i,\mathbf{x}}+\bar\phi_{i,\mathbf{x}}\psi_{i,\mathbf{x}}]}$$

where $i = 1,..,n$, $n \geq 2$ is a color index and $P(d\psi^{\leq N})$ is the fermionic integration with propagator $\delta_{i,j}g^{(\leq N)}(\mathbf{x}-\mathbf{y})$. One can integrate the fields $\psi^{(N)}, \psi^{(N-1)}, .., \psi^{(j)}$ as for the Thirring model and we arrive to

$$e^{\mathcal{W}(\phi,J)} = e^{-L^2 E_j}\int P_{\widetilde Z_j}(d\psi^{(\leq j)}) e^{-\mathcal{V}^{(j)}(\sqrt{Z_j}\psi^{(\leq j)})+\mathcal{B}^{(j)}(\sqrt{Z_j}\psi^{(\leq j)},\phi,J)}$$
(6.38)

We proceed in this way up to the scale $h^*$ such that $\gamma^{h^*} = m$.

The crucial difference with respect the $n = 1$ case is that the beta function for $\lambda_h$ is non vanishing; it is found

$$\lambda_{h-1} = \lambda_h - \frac{2}{\pi}(n-1)\lambda_h^2 + O(\lambda_h^3) \qquad (6.39)$$

$$\frac{m_{h-1}}{m_h} = 1 + b\lambda_h + O(\lambda_h^2) \qquad (6.40)$$

$$\frac{Z_{h-1}}{Z_h} = 1 + c\lambda_h^2 + O(\lambda_h^2) \qquad (6.41)$$

From the above equations we see that

$$0 < \lambda_h \leq \frac{\lambda_0}{1 + \frac{2}{\pi}(n-1)\lambda_0 h} \qquad (6.42)$$

It is then possible to choose

$$\lambda_N = O(\frac{\lambda_0}{1 + 2\pi^{-1}(n-1)\lambda_0 N}) \quad Z_N = 1 + O(\lambda) \quad m_N = O(m\lambda_N) \quad (6.43)$$

such that $\lambda_0 = \lambda$, $Z_0 = 1$, $m_0 = m$ and the expansion for Schwinger functions is convergent for $0 \leq \lambda \leq \varepsilon$.

Note that the effective interaction becomes weaker and weaker for large momentum scales; this phenomenon is called *asymptotic freedom*. This result, proved in Refs. [24], [26], has been the first example of non-perturbative construction of a renormalizanble QFT.

# Chapter 7

# Axioms Verification and Wilson Fermions

## 7.1 Osterwalder-Schrader axioms

We have seen several examples, in $d=2$, of functional integrals in which the renormalization can be performed non-perturbatively and the cut-offs can be removed. As we said in chapt 1, one has still to prove that such functions verify the Osterwalder-Schrader axioms, see Ref.[5], guaranting that a QFT can be reconstructed starting from such functions. It is time then to see precisely which are such axioms in $d=2$.

One introduces a set of test functions in the following way. For any $n \in \mathbb{N}$, setting $\mathbf{x} = (x^{(1)}, \ldots, x^{(n)})$, let $\mathcal{S}(\mathbb{R}^{2n})$ be the space of the complex functions on $\mathbb{R}^{2n}$, with labels, $\underline{\omega} = (\omega_1, \ldots, \omega_n)$, $\underline{\sigma} = (\sigma_1, \ldots, \sigma_n)$, s.t., for any integer $m$, and any $f_{n,\underline{\omega},\underline{\sigma}}(\mathbf{x}) \in \mathcal{S}(\mathbb{R}^{2n})$, the Schwartz norm

$$||f_{n,\underline{\omega},\underline{\sigma}}||_m = \max_{r: r_j \leq m} \sup_{x^{(j)} \in \mathbb{R}^4} \frac{|\partial_1^{r_1} \cdots \partial_n^{r_n} f_{n,\underline{\omega},\underline{\sigma}}(\mathbf{x})|}{1 + \sum_{i=1}^n |x^{(i)}|^{r_i}} \qquad (7.1)$$

is finite. Let $\mathcal{S}_{\neq}(\mathbb{R}^{2n})$ be the space of the functions in $\mathcal{S}(\mathbb{R}^{2n})$ which vanish, together with all their partial derivatives, if $x^{(i)} = x^{(j)}$ for some $1 \leq i < j \leq n$; and let $\mathcal{S}_{<}(\mathbb{R}^{2n})$ be the space of the functions in $\mathcal{S}_{\neq}(\mathbb{R}^{2n})$ which vanish, together with all their partial derivatives, if the ordering of the times $x_0^{(1)}, \ldots, x_0^{(n)}$ is different from $0 < x_0^{(1)} < x_0^{(2)} < \cdots < x_0^{(n)}$.

Let the "space translation", $\tau_y$, for $y = (0, y_1)$, be defined as

$$(\tau_y f)_{n,\underline{\omega},\underline{\sigma}}(\mathbf{x}) = f_{n,\underline{\omega},\underline{\sigma}}(\tau_y \mathbf{x}) \qquad (7.2)$$

with $\tau_y \mathbf{x} = (x^{(1)} + y, \ldots, x^{(n)} + y)$.

Let the "time reflection" be defined as

$$(\Theta f)_{n,\underline{\omega},\underline{\sigma}}(\mathbf{x}) = f^*_{n,\underline{\omega}^*,\underline{\sigma}^*}(\vartheta_0 \mathbf{x}), \qquad (7.3)$$

with $\vartheta_0 \underline{x} = (\vartheta_0 x^{(1)}, \ldots, \vartheta_0 x^{(n)})$, where $\vartheta_0(x_0, x_1) = (-x_0, x_1)$; $f^*(x^{(1)}, \ldots, x^{(n)})$ is the complex conjugate of $f(x^{(n)}, \ldots, x^{(1)})$; and the labels $\underline{\omega}^*$ and $\underline{\sigma}^*$ are defined respectively to be $\omega_n, \ldots, \omega_1$ and $-\sigma_n, \ldots, -\sigma_1$. The Osterwalder–Schrader (OS) axioms are the following:

E1.) $S^{(n)}_{\underline{\omega},\underline{\sigma}}(\underline{x})$ is a distribution on $\mathcal{S}_<(\mathbb{R}^{(2n)})$; indeed, for any integer $m$, there exist two constants $c_m, C_m > 0$ s.t.

$$||S^{(n)}_{\underline{\omega},\underline{\sigma}}||_m = \sup_{f \in \mathcal{S}_<(\mathbb{R}^{(2n)})} \frac{\left(S^{(n)}_{\underline{\omega},\underline{\sigma}}, f\right)}{||f||_m} \leq C_m (n!)^{c_m} . \qquad (7.4)$$

E2.) $S^{(n)}_{\underline{\omega},\underline{\sigma}}(\underline{x})$ is invariant under the Euclidean group of translation and rotation of all the coordinates.

E3.) $S^{(n)}_{\underline{\omega},\underline{\sigma}}(\underline{x})$ is antisymmetric under the exchange of the $x^{(i)}, \omega_i, \sigma_i$ respectively with $x^{(j)}, \omega_j, \sigma_j$, for any $1 \leq i < j \leq n$.

E4.) For any sequence of "time ordered" test functions, $\left\{ f_{n,\underline{\omega},\underline{\sigma}}(\underline{x}) \in \mathcal{S}_<(\mathbb{R}^{(2n)}) \right\}_{n \geq 0, \underline{\omega}, \underline{\sigma}}$, the Schwinger functions are "reflection inveriant":

$$S^{(n)}_{\underline{\omega},\underline{\sigma}}\left((\Theta f)_{n,\underline{\omega},\underline{\sigma}}\right) = S^{(n)}_{\underline{\omega},\underline{\sigma}}(f_{n,\underline{\omega},\underline{\sigma}}) \qquad (7.5)$$

and "reflection positive":

$$\sum_{m,\underline{\omega}',\underline{\sigma}'} \sum_{n,\underline{\omega},\underline{\sigma}} S^{(m+n)}_{\underline{\omega}',\underline{\sigma}';\underline{\omega},\underline{\sigma}}((\Theta f)_{m,\underline{\omega}',\underline{\sigma}'} \otimes f_{n,\underline{\omega},\underline{\sigma}}) \geq 0 . \qquad (7.6)$$

E5.) For any $f_{n,\underline{\omega},\underline{\sigma}} \in \mathcal{S}_<(\mathbb{R}^{(2n)})$ and $g_{m,\underline{\omega}',\underline{\sigma}'} \in \mathcal{S}_<(\mathbb{R}^{(2m)})$, it holds:

$$\lim_{|y| \to \infty} S^{(m+n)}_{\underline{\omega}',\underline{\sigma}';\underline{\omega},\underline{\sigma}}\left((\Theta g)_{m,\underline{\omega}',\underline{\sigma}'} \otimes (\tau_y f)_{n,\underline{\omega},\underline{\sigma}}\right) = 0 . \qquad (7.7)$$

The property E1) ensures that $S^{(n)}_{\underline{\omega},\underline{\sigma}}$ does not grow too quikly with $n$; property E2) ensures the proper law of transformations. E3) is the most interesting axiom: it states that the expectation function of a function of the fields multiplied by the same function after reflection om any hyperplane, abd complex conjugaton, must be positive. This property ensures, after continuation, the locality axiom and the positivity of the Hilbert space. Finally E4) ensures the antisymmetry and E6) states that the Schwinger functions asymptotically factorizes; such last property ensures the unicity of the vacuum in the Wigthman axioms.

The *Osterwalder-Schrader reconstruction theorem* is the following.

**Theorem 7.1.** *Any set of functions satisfying E1)-E5) determine a unique Quantum Field Theory in the sense of Wigthman, see Ref.[4].*

## 7.2 Lattice regularization and fermion doubling

For definiteness we discuss the verification of the axioms in the case of the Thirring model (6.18); similar results could be proved for QED2 or the Gross-Neveu model. The properties E2) and E3) are obvious by construction; the properties E1) and E5) can be proved by a bound on the generic Schwinger functions which follow in a strigthforward way from the bounds introduced previously; we refer to Ref.[39] for a detailed derivation. The most interesting property is E5), as reflection positivity is broken by the momentum cut-off introduced to regularize the theory, and it is quite difficult to verify directly that is restored when cut-off is removed.

Of course there are several possible ways for introducing the cut-offs. The one followed before, the momentum regularization, is the closest to the formal continuum limit (the regularized propagator is linear in $\mathbf{k}$) and this is convenient under many respects, for instance in the derivation of WI and closed equation for the two point Schwinger function. However such a choice has the big disavantage that the crucial property of *positive definiteness* is quite difficul to prove; such a property is however automatically fulfilled with a *lattice regularization*, see Ref.[40].

The simplest way for introducing a lattice regularization consists in replacing $k, k_0$ in the propagator with $a^{-1}\sin ka$ and $a^{-1}\sin k_0 a$, that is introducing a regularized fermionic propagator

$$g_\omega(\mathbf{k}) = \frac{1}{-ia^{-1}\sin k_0 a + a^{-1}\sin ka} \tag{7.8}$$

Such regularization suffers of the well known *fermion doubling* problem: the fermion propagator has *four* poles instead of a single one, that is $(0,0), (0, \pi/a), (\pi/a, 0), (\pi/a, \pi/a)$. In the continuum limit $a \to 0$ this means that there are four fermion state per field component and such extra unwanted fermions influence possibly the physical behaviour in a non trivial way.

Several solutions have been proposed; we will follow here the Wilson formulation of adding a term to the free action, called *Wilson term*, to cancel the unwanted poles. Then in the Wilson lattice regularization the fermionic integration is

$$P_{Z_a}(d\psi) = \exp[-\frac{Z_a}{L^2}\sum_{\omega,\omega'=\pm}\sum_{\mathbf{k}\in\mathcal{D}_{\bar{a}}}(\widehat{r}^{-1}(\mathbf{k}))_{\omega',\omega}\widehat{\psi}^+_{\mathbf{k},\omega'}\widehat{\psi}^-_{\mathbf{k},\omega}]\prod_{\mathbf{k}\in\mathcal{D}_a}\prod_{\omega=\pm}\frac{d\widehat{\psi}^+_{\mathbf{k},\omega}d\widehat{\psi}^-_{\mathbf{k},\omega}}{\mathcal{N}_a(\mathbf{k})} \tag{7.9}$$

where the covariance $\hat{r}_{\omega,\omega'}(k)$ is s.t.:

$$\hat{r}(\mathbf{k}) = \frac{1}{\mu_a^2(\mathbf{k}) - e_+(\mathbf{k})e_-(\mathbf{k})} \begin{pmatrix} e_+(\mathbf{k}) & -\mu_a - c_a(\mathbf{k}) \\ -\mu_a - c_a(\mathbf{k}) & e_-(\mathbf{k}) \end{pmatrix}_{\omega',\omega}$$

with $k_0 = (m_0 + 1/2)2\pi/L$, $k = (m + 1/2)2\pi/L$, $n, n_0 = -L/2\bar{a}, 1, \ldots, L/2\bar{a} - 1$ and

$$e_\omega(\mathbf{k}) = -i\frac{\sin(k_0 a)}{\sin\frac{\pi a}{L}} + \omega\frac{\sin(ka)}{\sin\frac{\pi a}{L}} \qquad c_a(\mathbf{k}) = \frac{1 - \cos(k_0 a)}{a} + \frac{1 - \cos(ka)}{a};$$
(7.10)

and $\bar{\mathcal{N}}_a$ is the normalization. The generating function, in the Thirring case, is given by

$$\int P_{Z_a}(d\psi) \exp\left\{ -\lambda_a Z_a^2 V(\psi) + (a^{-1}\nu_a) Z_a N(\psi) \right\}$$
(7.11)

$$\exp\left\{ Z_a^{(2)} \sum_\omega \int d\mathbf{x}\, J_{\mathbf{x},\omega} \psi_{\mathbf{x},\omega}^+ \psi_{\mathbf{x},\omega}^- + \sum_\omega \int d\mathbf{x}\, [\varphi_{\mathbf{x},\omega}^+ \psi_{\mathbf{x},\omega}^- + \psi_{\mathbf{x},\omega}^+ \varphi_{\mathbf{x},\omega}^-] \right\}$$

where $N(\psi) = \sum_{\omega=\pm} \int d\mathbf{x}\, \psi_{\mathbf{x},\omega}^+ \psi_{\mathbf{x},-\omega}^-$.

The above regularization has the crucial property of preserving Reflection Positivity; moreover, the fermion doubling problem is solved, as the term $(1-\cos(k_0 a))/a + (1-\cos(ka))/a$ has the effect that, in the massless case, only one pole is present. On the other hand, even if the bare mass $\mu_a$ is vanishing, both the oddness of the propagator and the *global* symmetry $\psi_{\mathbf{x},\omega}^\pm \to e^{\pm i\alpha_\omega} \psi_{\mathbf{x},\omega}^\pm$ are not true, and this leads to the generation through the interaction of a mass; to get an interacting massless theory, a mass counterterm $\nu_a$ has then to be introduced. We call $S^{N,L,h,a,(m;n)}$ the Schwinger function with the momentum regularization and $\bar{S}^{a,L(m;n)}$ the Schwinger functions with lattice regularization; then the following result hold.

**Theorem 7.2.** *Given $\lambda$ small enough and $\mu > 0$, there exist functions $\lambda_a(\lambda), \nu_a(\lambda), \mu_a \equiv \mu g_a(\lambda), Z_a \equiv Z_a(\lambda)$ verifying*

$$Z_a = a^{-\eta_z}(1 + O(\lambda^2)) \quad , \quad \mu_a = \mu a^{-\eta_\mu}(1 + O(\lambda)),$$
(7.12)

*with $\eta_z = a_z \lambda^2 + O(\lambda^4)$, $\eta_\mu = a_\mu \lambda + O(\lambda^2)$, $a_z, a_\mu > 0$ such that the limit*

$$\lim_{L, a^{-1} \to \infty} \bar{S}_{\underline{\sigma};\underline{\omega},\underline{\varepsilon}}^{a,(m;n)} \equiv \bar{S}_{\underline{\sigma};\underline{\omega},\underline{\varepsilon}}^{(m;n)},$$
(7.13)

*exist at non coinciding points.*

*Moreover, given $N > 0$, let $a_N = \pi(4\gamma^{N+1})^{-1}$; then*

$$\lim_{N \to \infty} [S_{\underline{\sigma};\underline{\omega},\underline{\varepsilon}}^{N,(m;n)}(\mathbf{z};\mathbf{x}) - \bar{S}_{\underline{\sigma};\underline{\omega},\underline{\varepsilon}}^{a_N,(m;n)}(\mathbf{z};\mathbf{x})] = 0.$$
(7.14)

The above result says that in the limit of removed cutoffs, the two *different regularizations* of the Thirring model gives the *same* Schwinger function, if the "bare" paramenters are suitably chosen; of course the bare parameters must be chosen different by $O(\lambda^2)$ and coincide at the infrared scale. This menas that both the regularizations produce Schwinger functions which are reflection positive, when the cut-offs are removed.

## 7.3 Integration of the doubled fermions

In order to prove Theorem 7.2 we have to adapt the previous renormalized expansion to the case of lattice fermions; we just sketch the main necessary modifications.

To begin with, we define $\bar{f}(\mathbf{k})$ so that

$$C_N^{-1}(\mathbf{k}) + \bar{f}(\mathbf{k}) = 1 \tag{7.15}$$

where $C_N^{-1}(\mathbf{k}) = \sum_{j=-\infty}^{N} f_j(\mathbf{k})$; since $C_N^{-1}(\mathbf{k}) = 0$ for $|\mathbf{k}| \geq \gamma^{N+1} = \pi/(4a)$, the support of the function $\bar{f}(\mathbf{k})$ is given by the set $\{\mathbf{k} : |\mathbf{k} - \pi/a| \leq 3\pi/4a\}$. Therefore, it is possible to decompose the propagator $\hat{r}_{\omega,\omega'}(\mathbf{k})$, as

$$\hat{r}_{\omega,\omega'}(\mathbf{k}) = \hat{r}_{\omega,\omega'}^{(\leq N)}(\mathbf{k}) + \hat{r}_{\omega,\omega'}^{(N+1)}(\mathbf{k}) \tag{7.16}$$

with $\hat{r}_{\omega,\omega'}^{(\leq N)}(\mathbf{k}) = C_N^{-1}(\mathbf{k})\hat{r}_{\omega,\omega'}(\mathbf{k})$ and $\hat{r}_{\omega,\omega'}^{(N+1)}(\mathbf{k}) = \bar{f}(\mathbf{k})\hat{r}_{\omega,\omega'}(\mathbf{k})$. Note that

$$|r_{\omega,\omega'}^{(N+1)}(\mathbf{x})| \leq C\gamma^N \frac{C_P}{1 + (\gamma^N|\mathbf{x} - \mathbf{y}|)^P} \tag{7.17}$$

since the function $\bar{f}(\mathbf{k})$ is a $C^\infty$ function, with a compact support of size $a^{-2}$, and $[1 - \cos(k_0 a) + 1 - \cos(ka)]/a \geq \tilde{C}a^{-1} = C\gamma^N$ on its support.

We can therefore write the Generating function as

$$\int P_{Z_{N+1},\hat{\mu}_{N+1},C_N}(d\psi^{(\leq N)}) P_{Z_{N+1},\hat{\mu}_{N+1},\bar{f}}(d\psi^{(N+1)}). \tag{7.18}$$

$$\exp\left\{-\lambda_{N+1} Z_{N+1}^2 V(\psi) + \nu_{N+1} Z_{N+1} N(\psi)\right\} \tag{7.19}$$

$$\cdot \exp\left\{Z_{N+1}^{(2)} \sum_\omega \int d\mathbf{x}\, J_{\mathbf{x},\omega}\psi^+_{\mathbf{x},\omega}\psi^-_{\mathbf{x},\omega} + \sum_\omega \int d\mathbf{x}\, [\varphi^+_{\mathbf{x},\omega}\psi^-_{\mathbf{x},\omega} + \psi^+_{\mathbf{x},\omega}\varphi^-_{\mathbf{x},\omega}]\right\} \tag{7.20}$$

where $Z_{N+1} = Z_a$, $Z_{N+1}^{(2)} = Z_a^{(2)}$, $\lambda_{N+1} = \lambda_a$, $\nu_{N+1} = 4\pi^{-1}\nu_a$, $\mu_{N+1}(\mathbf{k}) = \mu_a(\mathbf{k})$ and $\psi = \psi^{(\leq N)} + \psi^{(N+1)}$. and the integration of $P_{Z_{N+1},\hat{\mu}_{N+1},\bar{f}}(d\psi^{(N+1)})$ gives no troubles, thanks to the bound (7.17).

## 7.4 Lattice fermions

We have now to adapt the integration procedure in chapt. 4, in order to take into account of the presence of the lattice. The main difference is given by the presence in the effective potential of a term proportional to $N(\psi)$ and in the renormalized free measure of a non diagonal term proportional to $[1-\cos(k_0 a)+1-\cos(ka)]/a$. Hence, after the fields of scale $N+1, N, .., j+1$ are integrated, (7.19) can be written as

$$e^{-L^2 \bar{E}_j} \int P_{\widetilde{Z}_j, \widehat{\mu}_j, C_j}(d\psi^{(\leq j)}) e^{-\bar{\mathcal{V}}^{(j)}(\sqrt{Z_j}\psi^{(\leq j)})+\bar{\mathcal{B}}^{(j)}(\sqrt{Z_j}\psi^{(\leq j)}, \varphi, J)}, \quad (7.21)$$

where $\int P_{\widetilde{Z}_j, \widehat{\mu}_j, C_j}$ is the integration with propagator given by $\widetilde{Z}_j(\mathbf{k})^{-1}\widehat{r}_{\omega,\omega'}^{(\leq j)}(\mathbf{k})$, whose expression is obtained from that of $Z_a^{-1}\widehat{r}_{\omega,\omega'}^{(\leq N)}(\mathbf{k})$ by replacing $Z_a, \mu_a(\mathbf{k}), C_N(\mathbf{k})$ with $\widetilde{Z}_j(\mathbf{k}), \widehat{\mu}_j(\mathbf{k}), C_j(\mathbf{k})$; $\bar{\mathcal{V}}^{(j)}, \bar{\mathcal{B}}^{(j)}$ very similar to the analogous quantities in chapt.4, up to some obvious modifications that we now discuss.

The function $\widehat{\mu}_j(\mathbf{k})$, as we shall explain better below, is the sum of a term proportional to $[1-\cos(k_0 a)+1-\cos(ka)]/a$ and another one proportional to $\mu$, that we shall call $\widetilde{\mu}_j(\mathbf{k})$, as it will play the same role of the function with the same name in the continuum model. This allows us to define all localization $\mathcal{L}_j$ as the analogue of the ones in chap.4

$$\mathcal{L}_0 \widehat{W}_{2,\underline{\omega}}^{(h)}(\mathbf{k}) = \frac{1}{4} \sum_{\eta,\eta'=\pm 1} \widehat{W}_{2,\underline{\alpha},\omega}^{(h)}(\bar{\mathbf{k}}_{\eta\eta'})$$

$$\mathcal{L}_1 \widehat{W}_{2,\omega,\omega'}^{(h)}(\mathbf{k}) = \frac{1}{4} \sum_{\eta,\eta'=\pm 1} \widehat{W}_{2,\omega,\omega'}^{(h)}(\bar{\mathbf{k}}_{\eta\eta'}) \left[ \eta \frac{\sin k_0 a}{\sin \frac{\pi a}{L}} + \eta' \frac{\sin ka}{\sin \frac{\pi a}{L}} \right] \quad (7.22)$$

with $\mathbf{k}_{\eta,\eta'} = (\eta\pi/L, \eta'\pi/L)$. The above expressions of course reduces to the one in chapt.4 when the lattice step is sent to zero and the volume box is sent to infinity. However, since the operator $\mathcal{P}_0$ does not cancel the non diagonal part of the propagator, which is even in the momentum (while the diagonal one is odd), extra terms are produced by the action of the $\mathcal{L}$ operator with respect to the ones in chapt.4. It follows that

$$\mathcal{L}\widehat{W}_{2,\omega,\omega}^{(j)} = \mathcal{L}_1 \mathcal{P}_0 \widehat{W}_{2,\omega,\omega}^{(j)}, \quad \mathcal{L}\widehat{W}_{2,\omega,-\omega}^{(j)} = \mathcal{L}_0 \mathcal{P}_0 \widehat{W}_{2,\omega,-\omega}^{(j)} + \mathcal{L}_0 \mathcal{P}_1 \widehat{W}_{2,\omega,-\omega}^{(j)}, \quad (7.23)$$

and this implies that we can write

$$\mathcal{L}\mathcal{V}^{(j)}(\psi^{[h,j]}) = z_j F_\zeta^{[h,j]} + (s_j + \gamma^j n_j) F_\sigma^{[h,j]} + l_j F_\lambda^{[h,j]} \quad (7.24)$$

where $\gamma^j n_j = \mathcal{L}_0 \mathcal{P}_0 \widehat{W}_{2,\omega,-\omega}^{(j)}$, while $s_j = \mathcal{L}_0 \mathcal{P}_1 \widehat{W}_{2,\omega,-\omega}^{(j)}$. The proof of (7.23) follows by induction from the remark that $\widehat{W}_{2,\omega,\omega}^{(j)}$ is given by the sum of graphs with

1) either an even number of $\nu$ vertices, an even number of non diagonal propagators and an odd number of diagonal propagators;

2) or an odd number of $\nu$ vertices, an odd number of non diagonal propagators and an odd number of diagonal propagators. Moreover $\widehat{W}^{(j)}_{2,\omega,-\omega}$ is given by the sum of graphs with

3) either an even number of $\nu$ vertices, an odd number of non diagonal propagators and an even number of diagonal propagators;

4) or an odd number of $\nu$ vertices, an even number of non diagonal propagators and an even number of diagonal propagators.

The renormalization of the free measure is done exactly as in chapt.4, that is we do not put the term proportional to $n_j$ in the free measure, but we define a new running coupling constant $\nu_j = n_j(Z_j/Z_{j-1})$. It follows that the rescaled potential $\widehat{\mathcal{V}}^{(j)}(\psi^{(\leq j)})$ differs from that of chapt. 4 because its local part contains the term $\gamma^j \nu_j F_\sigma^{(\leq j)}$, that is it is equal to $\lambda_j F_\lambda^{[h,j]}$ | $\gamma^j \nu_j F_\sigma^{(\leq j)}$ One then performs the integration with respect to $\psi^{(j)}$, whose propagator is of the form

$$\frac{1}{\widetilde{Z}_j(\mathbf{k})}\widehat{r}^{(j)}_{\omega,\omega'}(\mathbf{k}) = \frac{1}{\widetilde{Z}_j(\mathbf{k})} \frac{\widetilde{f}_j(\mathbf{k})}{e_+(\mathbf{k})e_-(\mathbf{k}) - \mu_j^2(\mathbf{k})} \begin{pmatrix} e_-(\mathbf{k}) & -\widehat{\mu}_j(\mathbf{k}) \\ -\widehat{\mu}_j(\mathbf{k}) & e_+(\mathbf{k}) \end{pmatrix}_{\omega,\omega'} \quad (7.25)$$

with $\widehat{\mu}_j(\mathbf{k}) = \widetilde{\mu}_j(\mathbf{k}) + [Z_{N+1}/\widetilde{Z}_j(\mathbf{k})][1 - \cos(k_0 a) + 1 - \cos(ka)]/a$, $\widetilde{\mu}_j(\mathbf{k})$ being a function equal to $\mu$ for $j = N+1$.

In order to control the expansion we have to prove that $\bar{\lambda}_h \equiv \max_{h \leq j \leq N+1} |\lambda_j|$ and $\bar{\nu}_h \equiv \max_{h \leq j \leq N+1} |\nu_j|$ stay small, if $\lambda_a$ is small enough and $\nu_a$ is suitably chosen. This can be proved by noting that the propagator (7.25) can be written as

$$\widetilde{Z}_j(\mathbf{k})^{-1}\widehat{r}^{(j)}_{\omega,\omega'}(\mathbf{k}) = \widehat{g}^{(j)}_{\omega,\omega'}(\mathbf{k}) + \widehat{g}^{R,(j)}_{\omega,\omega'}(\mathbf{k}) , \quad (7.26)$$

where $\widehat{g}^{(j)}_{\omega,\omega'}(\mathbf{k})$ has *exactly* the same form as the single scale propagator appearing in the multiscale integration in chapt.4 and

$$|g^{R,(j)}_{\omega,\omega'}(\mathbf{x},\mathbf{y})| \leq \gamma^{-(N-j)}\gamma^j \frac{C_P}{1 + |\gamma^j|\mathbf{x}-\mathbf{y}||^P} . \quad (7.27)$$

By (7.26) (7.27) we see that the single scale propagators of the lattice model are equal to those of the continuum model, of course with different $\widetilde{Z}_j(\mathbf{k})$ and $\widetilde{\mu}_j(\mathbf{k})$ functions, up to a correction which is vanishing in the limit $N \to \infty$.

By using the above decomposition, the flow equation for $\lambda_j$ can be written, for $j \leq N+1$, as

$$\lambda_{j-1} = \lambda_j + \widetilde{\beta}_\lambda^j(\lambda_N, ..., \lambda_j) + r_\lambda^j(\lambda_a, \lambda_N, ..., \lambda_j) \quad (7.28)$$

$$+ \sum_{k \geq j} \nu_k \widetilde{\beta}_\lambda^{j,k}(\lambda_a, \nu_a, \lambda_N, \nu_N, ..., \lambda_j, \nu_j) \quad (7.29)$$

where we have included the sum over all trees with at least one $\nu$-endpoint in the last term in the r.h.s. of (7.29) and we have split the sum of all trees with no $\nu$-endpoint as $\widetilde{\beta}_\lambda^j + r_\lambda^j$, where $\widetilde{\beta}_\lambda^j$ contains the trees with propagator $g_{\omega,\omega'}^{(j)}$, $j \leq N$ (the decomposition (7.26) is used), while all other terms are included in $r_\lambda^j$. By the short memory property

$$|\widetilde{\beta}_\lambda^{j,k}| \leq C\bar{\lambda}_j \gamma^{-(k-j)/4} \quad , \quad |r_\lambda^j| \leq C\bar{\lambda}_j^2 \gamma^{-(N-j)/4} . \quad (7.30)$$

On the other hand the only difference between $\widetilde{\beta}_\lambda^j(\lambda_N, ..., \lambda_j)$ and the function $\beta_{j,\lambda,L}(\lambda_N, ..., \lambda_j)$ defined in (4.76) (with $N$ replacing $K$) comes from the fact that in the continuum model the delta function of conservation of momenta is $L^2 \delta_{k,0} \delta_{k_0,0}$, while in the lattice model is $L^2 \sum_{n,m \in Z^2} \delta_{k,2\pi n/a} \delta_{k_0,2\pi m/a}$. However, the difference between the two delta functions has no effect on the local part $\mathcal{L}\mathcal{V}^N$, because of the compact support of $\psi^{\leq N}$ and only slightly affects the non local terms. To see that, let us consider a particular tree $\tau$ and a vertex $v \in \tau$ of scale $h_v$ with $2n$ external fields of space momenta $\mathbf{k}_i$; the conservation of momentum implies that $\sum_i \varepsilon_i \mathbf{k}_i = \mathbf{m} \frac{2\pi}{a}$, with $m$ an arbitrary integer. On the other hand, $\mathbf{k}_i$ is of order $\gamma^{h_v}$ for any $i$, hence $m$ can be different from 0 only if $n$ is of order $\gamma^{N-h_v}$. Since the number of endpoints following a vertex with $2n$ external fields is greater or equal to $n-1$ and there is a small factor (of order $\bar{\lambda}_j$) associated with each endpoint, we get an improvement, in the bound of the terms with $|m| > 0$, with respect to the others, of a factor $\exp(-C\gamma^{N-h_v})$. Hence, by using the remark preceding (7.30), it is easy to show that

$$\widetilde{\beta}_{\lambda,j}(\lambda_N, ..., \lambda_j) = \beta_{j,\lambda,L}(\lambda_N, ..., \lambda_j) + \widehat{\beta}_{\lambda,j}(\lambda_N, ..., \lambda_j) . \quad (7.31)$$

where $\widehat{\beta}_{\lambda,j}(\lambda_N, ..., \lambda_j) \leq C\bar{\lambda}_j^2 \gamma^{-(N-j)/4}$ for a suitable constant $C$ and $\beta_{\lambda,L,j}(\lambda_N, ..., \lambda_j)$ is the beta function for the continuum model, verifying the crucial bound (5.1) with $N$ replacing $K$.

In the same way the flow equation for $\nu_j$

$$\nu_{j-1} = \gamma \nu_j + \beta_\nu^{(j)}(\lambda_a, \nu_a; \lambda_N, \nu_N; ...; \lambda_j, \nu_j) \quad (7.32)$$

can be written, for $j \leq N+1$, as

$$\nu_{j-1} = \gamma \nu_j + \beta_\nu^{(1,j)}(\lambda_a, \lambda_N, ..., \lambda_j) + \quad (7.33)$$

$$+ \sum_{k \geq j} \nu_k \widetilde{\beta}_\nu^{(j,k)} \left( \lambda_a, \nu_a, \lambda_N, \nu_N, ..., \lambda_j, \nu_j \right) \tag{7.34}$$

where $\beta_\nu^{(j,1)}$ is a sum over trees with no endpoints of type $\nu$. By using the decomposition (5.1), the parity properties of $g_{\omega,\omega'}^{(j)}(\mathbf{x}, \mathbf{y})$ we get the bounds

$$|\beta_\nu^{(1,j)}| \leq C |\lambda_a| \gamma^{-(N-j)/4} \quad , \quad |\widetilde{\beta}_\nu^{(j,k)}| \leq C |\lambda_a| \gamma^{-(k-j)/4} \tag{7.35}$$

From the above properties we can show that it is possible to choose $\nu_a = \nu_{N+1}$ so that $\nu_j = O(\lambda_a \gamma^{-(N-j)/8})$ and $\lambda_j$ stays close to $\lambda_a$.

**Lemma 7.1.** *For any given $\lambda_{N+1}$ small enough, it is always possible to fix $\nu_{N+1}$ so that, for any $j \leq N+1$,*

$$|\nu_j| \leq C |\lambda_a| \gamma^{-(N-j)/8} \quad , \quad |\lambda_j - \lambda_a| \leq C \lambda_a^2 . \tag{7.36}$$

*Proof.* We consider the Banach space $\mathcal{M}_\xi$ of sequences $\underline{\nu} = \{\nu_j\}_{j \leq N+1}$ such that

$$\|\underline{\nu}\|_\xi = \sup_{j \leq N+1} \gamma^{(N-j)/8} |\nu_j| \leq \xi |\lambda_a| , \tag{7.37}$$

with $\xi$ to be fixed later. From (7.29) and (5.1) it follows that there exists $\varepsilon_0$ such that, if both $|\lambda_a|$ and $\xi |\lambda_a|$ are smaller than $\varepsilon_0$, then, for any $\underline{\nu}$, $\underline{\nu}' \in \mathcal{M}_\xi$,

$$|\lambda_j(\underline{\nu}) - \lambda_a| \leq C \lambda_a^2 \quad , \quad |\lambda_j(\underline{\nu}) - \lambda_j(\underline{\nu}')| \leq C |\lambda_a| \|\underline{\nu} - \underline{\nu}'\|_\xi . \tag{7.38}$$

We want to show that it is possible to choose $\nu_{N+1}$ so that $\underline{\nu} \in \mathcal{M}_\xi$. Note that, if $\underline{\nu} \in \mathcal{M}_\xi$, $\lim_{j \to -\infty} \nu_j = 0$; by some simple algebra, this implies that

$$\nu_j = -\sum_{k \leq j} \gamma^{k-j-1} \beta_\nu^{(k)} \left( \lambda_a, \nu_a; \lambda_N, \nu_N; ...; \lambda_j, \nu_j \right) . \tag{7.39}$$

Hence, we look for a fixed point of the operator $\mathbf{T} : \mathcal{M}_\xi \to \mathcal{M}_\xi$ defined as

$$\mathbf{T}(\underline{\nu})_j = -\sum_{k \leq j} \gamma^{k-j-1} \beta_\nu^{(k)} \left( \lambda_a, \nu_a, \lambda_N(\underline{\nu}), \nu_N, .., \lambda_j(\underline{\nu}), \nu_j \right) . \tag{7.40}$$

By (7.4) we find

$$|\mathbf{T}(\underline{\nu})_j| \leq \sum_{k \leq j} C |\lambda_a| \gamma^{-(j-k)} \gamma^{-(N-k)/8} \leq c_0 |\lambda_a| \gamma^{-(N-j)/8} . \tag{7.41}$$

Hence the operator $\mathbf{T} : \mathcal{M}_\xi \to \mathcal{M}_\xi$ leaves $\mathcal{M}_\xi$ invariant, if $\xi \geq c_0$ and $\lambda_a$ is sufficiently small, and it is also a contraction since $|\mathbf{T}(\underline{\nu})_j - \mathbf{T}(\underline{\nu}')_j| \leq C |\lambda_a| \|\underline{\nu} - \underline{\nu}'\|_\xi$. It follows that there is a unique fixed point in $\mathcal{M}_\xi$, satisfying the flow equation. ∎

An important consequence of the above analysis is that we get, as $N \to \infty$, exactly the same expansion in terms of trees, containing only $\lambda$ endpoints with a fixed coupling constant $\widetilde{\lambda}_{-\infty}(\lambda_a) = \lim_{j \to -\infty} \lambda_j$; in fact, the trees containing at least one $\nu$ vertex vanish in this limit.

By a fixed point argument, one can show that we can fix $\lambda_a$ so that $\widetilde{\lambda}_{-\infty}(\lambda_a)$ has the same value as $\lambda_{-\infty}(\lambda)$ in the continuum model; this remark completes the proof of Theorem 7.1.

# Chapter 8

# Infraed QED4 with Large Photon Mass

## 8.1 Regularization

Up to now we has seen just $d = 1 + 1$ models. The problem with $QED4$ is that the effective coupling increases at large scales, so that the ultraviolet cut-off cannot be removed. It is generally believed that a non-perturbative construction of QED4, with no cut-offs, is possible only considering it as part of the Electroweak theory, which is asymptotically free; on the contrary in QED4 the presence of the Landau pole apparently forbids the removal of the ultraviolet cut-off. Nevertheless, one can consider QED4 with a fixed ultraviolet cut-off, also called *infrared QED4*. The presence of the ultraviolet cut-off violates Ward Identities but one can ask to what degree they are restored at momenta far from the ultraviolet cut-off; however in the bosonic approach, the WI are quite difficult to evaluate.

We consider a $QED_4$ with an ultraviolet cut-off and a large photon mass; the generating function is

$$e^{\bar{\mathcal{W}}_{N,L}(J^A,\phi)} = \int P(d\psi)P(dA)e^{\int d\mathbf{x}[e\bar{\psi}_\mathbf{x}(A_{\mu,\mathbf{x}}\gamma_\mu)\psi_\mathbf{x} + J^A_{\mu,\mathbf{x}}A_{\mu,\mathbf{x}} + \phi_\mathbf{x}\bar{\psi}_\mathbf{x} + \bar{\phi}_\mathbf{x}\psi_\mathbf{x}]}$$
(8.1)

where $\mathbf{x}_\mu = n_\mu a$, $\mu = 0,1,2,3$ with $L/a$ integer and $n_\mu = -L/2a, 1, \ldots, L/2a - 1$, the fermionic integration is

$$P(d\psi^{(\leq N)}) = \mathcal{N}^{-1}\mathcal{D}\psi \exp[-\frac{Z_N}{L^4}\sum_{\mathbf{k}\in\mathcal{D}}\bar{\psi}_\mathbf{k}\chi_N^{-1}(\mathbf{k})(i\,\slashed{k} + mI)\psi_\mathbf{k}] \quad (8.2)$$

where $\slashed{k} = \gamma_\mu k_\mu$ and $\chi_N(\mathbf{k})$ is a non-vanishing smooth cut-off function, never vanishing and selecting momenta $|\mathbf{k}| \leq \gamma^N$. Finally $A_{\mu,\mathbf{x}}$ is an euclidean boson field defined by the Gaussian measure $P(dA)$ with covariance

$v_{\mu,\nu}(\mathbf{x} - \mathbf{y}) = \delta_{\mu,\nu} v(\mathbf{x} - \mathbf{y})$. We will consider the case

$$v(\mathbf{x} - \mathbf{y}) = \frac{1}{L^4} \sum_{\mathbf{p} \in \mathcal{D}} e^{-i\mathbf{p}(\mathbf{x}-\mathbf{y})} \frac{\chi_N(\mathbf{p})}{\mathbf{p}^2 + M_N^2} \qquad (8.3)$$

with $M_N = \gamma^N M$.

The above model is a regularization of *infrared QED4*, with photon mass $M$ and in the Feynman gauge. We will keep the ultraviolet cut-off fixed (for instance $N = 0$) and we are interested in the continuum limit (which is trivial as $N$ is fixed) and in the infinite volume limit; such a limit is non trivial for the infrared singularities associated to the massless fermionic propagators.

By the same methods applied in $d = 1 + 1$ the following theorem follows (more details can be found Ref.[41]).

**Theorem 8.1.** *There exists $\varepsilon_0$, independent from $N, m$, such that for $|e| \leq \varepsilon_0$:*

*1) the limit*

$$\lim_{L, a^{-1} \to \infty} \langle \psi_{\mathbf{x}_1}; ...; \psi_{\mathbf{x}_n}; \bar{\psi}_{\mathbf{y}_1}; ...; \bar{\psi}_{\mathbf{y}_n}; A_{\mu_1, \mathbf{z}_1}; ...; A_{\mu_m, \mathbf{z}_m} \rangle_{N, L, a} \qquad (8.4)$$

*exists at non coinciding points.*

*2) The following identity is true, for some finite nonvanishing $\mathbf{p}, \mathbf{k}, \mathbf{k}-\mathbf{p}$,*

$$i p_\mu \langle A_{\mu, \mathbf{p}}; \psi_\mathbf{k} \bar{\psi}_{\mathbf{k}-\mathbf{p}} \rangle_N = e_0 \hat{v}(\mathbf{p}) [\langle \psi_{\mathbf{k}-\mathbf{p}} \bar{\psi}_{\mathbf{k}-\mathbf{p}} \rangle_N - \langle \psi_\mathbf{k} \bar{\psi}_\mathbf{k} \rangle_N](1 + H_N(\mathbf{k}, \mathbf{p})) \qquad (8.5)$$

*with*

$$e_0 = e(1 - c_+ e^2 + O(e^4)) \qquad (8.6)$$

*where $c_+$ is a constant and*

$$|H_N(\mathbf{k}, \mathbf{p})| \leq C|e|(\kappa \gamma^{-N})^{\frac{1}{8}} \qquad (8.7)$$

*for a suitable constant $C$ and $\kappa = \max(|\mathbf{k}|, |\mathbf{p}|, |\mathbf{k} - \mathbf{p}|)$.*

Note that $\varepsilon_0$ is *independent* from the fermionic mass; this makes the proof of (8.4) non trivial, as one has to solve the infrared problem associated to the singularity of the fermionic propagator at vanishing momenta in the massless case. The violation of the WI due to the cut-offs has been absorbed in a redefinition of the electric charge, up to corrections which are small at momentum scales far from the ultraviolet scale. If we fix $e_0$, the dressed charge, to its physical value, the *bare charge e* depends on the details of regularization.

## 8.2 Tree expansion

Also in $d=4$ we can integrate the photon fields as in chapter 3; we call $\lambda = -2e^2$. $\mathcal{W}_{N,L}$ is invariant under the Euclidean transformation

$$\psi'(\mathbf{x}') = S(\Lambda)\psi(\mathbf{x}) \qquad \bar\psi'(\mathbf{x}') = \bar\psi(\mathbf{x})S(\Lambda)^{-1} \qquad \mathbf{x}'_\nu = \Lambda_{\mu,\nu}\mathbf{x}_\nu \qquad (8.8)$$

with $S(\Lambda)\gamma^\nu S(\Lambda)^{-1} = (\Lambda^{-1})_{\mu,\nu}\gamma_\nu$. The local quadratic invariant terms are then only $\bar\psi_\mathbf{x}\psi_\mathbf{x}$ and $\bar\psi_\mathbf{x}\,\not{\!\partial}\psi_\mathbf{x}$.

The functional integral is analyzed through a multiscale integration procedure; the starting point is to write the fermionic propagator as

$$g(\mathbf{x}-\mathbf{y}) = \sum_{h=-\infty}^{N} g^{(h)}(\mathbf{x}-\mathbf{y}) \qquad (8.9)$$

with, by integrating by parts, for any positive integer $K$

$$|g^{(h)}(\mathbf{x}-\mathbf{y})| \le \gamma^{3h}\frac{C_K}{1+(\gamma^h|\mathbf{x}-\mathbf{y}|)^K} \qquad (8.10)$$

In a similar way the following bound for the boson propagator is obtained:

$$|v(\mathbf{x}-\mathbf{y})| \le \gamma^{2N}\frac{C_K}{1+(\gamma^N|\mathbf{x}-\mathbf{y}|)^K} \qquad (8.11)$$

We integrate the field $\psi^{(\le N)}$ and we split the effective potential $\mathcal{V}^{(N-1)}$ as $\mathcal{L}\mathcal{V}^{(N-1)} + \mathcal{R}\mathcal{V}^{(N-1)}$. The action of $\mathcal{L}$ on the kernels $W_{2l}^{(N-1)}$ of $\mathcal{V}^{(N-1)}$ is $\mathcal{L}W_{2l}^{(N-1)} = 0$ if $l > 1$, while if $l = 1$

$$\mathcal{L}\int d\mathbf{x}d\mathbf{y} W_2^{(N-1)}(\mathbf{x},\mathbf{y})\bar\psi_\mathbf{x}\psi_\mathbf{y} = \int d\mathbf{x}d\mathbf{y} W_2^{(N-1)}(\mathbf{x},\mathbf{y})[\bar\psi_\mathbf{x}\psi_\mathbf{x}+(\mathbf{x}_\mu-\mathbf{y}_\mu)\bar\psi_\mathbf{x}\partial_\mu\psi_\mathbf{x}] \qquad (8.12)$$

Note that by parity $\int d\mathbf{y} W_2^{(N)}(\mathbf{x},\mathbf{y}) = 0$ and by Euclidean invariance $\int d\mathbf{y}(\mathbf{x}_\mu - \mathbf{y}_\mu)W_2^{(N)}(\mathbf{x},\mathbf{y}) = \gamma_\mu S$ with $S$ a scalar so that

$$\mathcal{L}\mathcal{V}^{(N-1)} = z_{N-1}\int d\mathbf{x}\bar\psi_\mathbf{x}\,\not{\!\partial}\psi_\mathbf{x} \qquad (8.13)$$

At the end of the iterative procedure we obtain

$$\mathcal{W}_{N,L}(J,\phi) = -L^4 E_N + \sum_{m^\phi+n^J\ge 1} S_{2m^\phi,n^J}(\phi,J) \qquad (8.14)$$

where $E_N$ and $S_{2m^\phi,n^J}(\phi,J)$ can be written as sum of trees defined as in chapt 4.

By using that $|\int d\mathbf{r} v(\mathbf{r})| \le C\gamma^{-2N}$ and assuming $|z_k| \le C|\lambda|, |Z_k^{(1)}-1| \le C|\lambda|$ we find

$$\int d\mathbf{x}_{v_0}|W_{\tau,\mathbf{P},T}(\mathbf{x}_{v_0})| \le L^4 \gamma^{-h[-4+\frac{3|P_{v_0}|}{2}-2m_{4,v_0}+n^J_{v_0}]} \qquad (8.15)$$

$$\prod_{v \text{ not e.p.}} \left\{ \frac{1}{s_v!} C^{\sum_{i=1}^{s_v} |P_{v_i}| - |P_v|} \gamma^{-[-4 + \frac{3|P_v|}{2} - 2m_{4,v} + n_v^J + z_v]} \right\} [\prod_v \gamma^{-2N\bar{m}_{4,v}}]$$
(8.16)

Finally by using the identity

$$\gamma^{h2m_{4,v_0}} \prod_v \gamma^{(h_v - h_{v'})2m_{4,v}} = \prod_v \gamma^{h_v 2\bar{m}_{4,v}}$$
(8.17)

where $m_{4,v}$ are the $\lambda$-endpoints internal to the cluster $v$, and $\bar{m}_{4,v}$ are the $\lambda$-endpoints internal to the cluster $v$ and not to any smaller cluster, we obtain

$$\int d\mathbf{x}_{v_0} |W_{\tau, \mathbf{P}, T}(\mathbf{x}_{v_0})| \leq L^4 \gamma^{-h[-4 + \frac{3|P_{v_0}|}{2} + n_{v_0}^J]}$$

$$C^n |\lambda|^{\bar{n}} \prod_{v \text{ not e.p.}} \left\{ \frac{1}{s_v!} \gamma^{-[-4 + \frac{3|P_v|}{2} + n_v^J + z_v]} \right\} [\prod_v \gamma^{-2(N - h_v)\bar{m}_{4,v}}] \quad (8.18)$$

Proceeding as in chapt 4, we see that the sum over the scales can be done.

We have finally to show that $z_h$ is bounded for any $h$. By construction $z_h$ verifies an equation of the form

$$z_{h-1} = z_h + \beta_z^{(h)}(z_{h+1}, .., z_N, \lambda)$$
(8.19)

and $\beta_z^{(h)}(z_{h+1}, .., z_N, \lambda)$ is sum of trees with at least one end-point of type $\lambda$, as the trees with only $z$ end-points are vanishing at zero momenta, for the compact support properties of the propagators. Calling $\bar{h}$ the scale of a $\lambda$ end-point, the factor $\gamma^{-2(N-\bar{h})}$ in (8.18) and the fact that $4 - \frac{3|P_v|}{2} + z_v > 0$ implies that, for a suitable constant $\vartheta > 0$

$$|\beta_z^{(h)}(z_{h+1}, .., z_N, \lambda)| \leq C\lambda \gamma^{2\vartheta(h-N)}$$
(8.20)

and by iteration

$$|z_h| \leq \sum_{k=h}^{N} |\beta_z^{(h)}| \leq C|\lambda|$$
(8.21)

This implies that the kernels $W_{2l}^{(h)}$ of the effective potential verify the bound

$$\frac{1}{L^4} \int d\underline{\mathbf{x}} |W_{2l}^{(h)}(\mathbf{x})| \leq C^l \gamma^{-h(-4+3l)} |\lambda|^{l-1}$$
(8.22)

The corrections to the WI can be bounded in a similar way to chapt.6 and the bound (8.7) immediately follows.

# PART 3
# Lattice Statistical Mechanics

# Chapter 9

# Universality in Generalized Ising Models

## 9.1 The nearest neighbor Ising model

We start now the discussion of several statistical physics models; despite apparently totally unrelated with the QFT models discussed previously, they will reveal, after some manipulations, a mathematical structure remarkably similar, so that the renormalization methods developed before can be applied also to them and give detailed informations about their properties. We start recalling the 2d nerest neighbor Ising model, which, as discussed in Chapt 1, plays a central role in statistical physics.

If $\Lambda$ is a square subset of $\mathbb{Z}^2$ with side $L$ and periodic boundary conditions and $\mathbf{x} = (x_0, x)$, the hamiltonian of the nearest-neighbor Ising model is

$$H_I = J \sum_{\mathbf{x} \in \Lambda} [\sigma_{x,x_0}\sigma_{x,x_0+1} + \sigma_{x,x_0}\sigma_{x+1,x_0}] = \sum_{\mathbf{x} \in \Lambda} H_{\mathbf{x}} \qquad (9.1)$$

and periodic boundary conditions are imposed. The partition function at inverse temperature $\beta$ is given by

$$Z_I = \sum_{\{\sigma_{\mathbf{x}} = \pm\}} e^{-\beta H} \qquad (9.2)$$

and $f = \lim_{L \to \infty} \frac{1}{L^2} \log Z$ is the *free energy*. The specific heat, by (1.24), is proportional to $\sum_{\mathbf{x},\mathbf{y} \in \Lambda} <H_{I,\mathbf{x}}; H_{I,\mathbf{y}}>_T$ where

$$<O(\mathbf{x}); O(\mathbf{y})> = \frac{1}{Z} \sum_{\{\sigma_{\mathbf{x}}\}} O(\mathbf{x}) O(\mathbf{y}) e^{-\beta H} \qquad (9.3)$$

and $<O(\mathbf{x}); O(\mathbf{y})>_T = <O(\mathbf{x})O(\mathbf{y})> - <O(\mathbf{x})><O(\mathbf{y})>$.

One can rewrite the sum appearing at the exponent in (9.2) as $\sum_b \tilde{\sigma}_b$, where $\sum_b$ is the sum over the bonds linking nearest neighbor sites of $\Lambda$ and

$\tilde\sigma_b$ is the product of the spin variables over the two extremes of $b$. If we expand the exponential in power series we find:

$$Z = \sum_{\{\sigma_\mathbf{x}\}} \prod_b (\cosh\beta J + \tilde\sigma_b \sinh\beta J) = (\cosh\beta J)^B \sum_{\{\sigma_\mathbf{x}\}} \prod_b (1 + \tilde\sigma_b \tanh\beta J) \quad (9.4)$$

where $B$ is the number of bonds of $\Lambda$. Developing the product, we are led to a sum of terms proportional to $\tilde\sigma_{b_1}\cdots\tilde\sigma_{b_k}$. To every term $\tilde\sigma_{b_1}\cdots\tilde\sigma_{b_k}$ we can associate a distinct figure on the lattice, given by the geometric set of lines $b_1,\ldots,b_k$. If $\tilde\sigma_{b_1}\cdots\tilde\sigma_{b_k}$ contains $\sigma_\mathbf{x}^n$ for some $\mathbf{x}$ and $n$ an odd integer, such term gives a vanishing contribution to the sum in (9.4); hence the figures giving non vanishing contributions are just the closed multipolygons $\gamma$, that is polygons or union of polygons with a point (but not sides) in common. The partition function can be then rewritten as:

$$Z_I = (\cosh\beta J)^B 2^S \sum_\gamma (\tanh\beta J)^{|\gamma|} . \quad (9.5)$$

where $S$ is the total number of sites, the sum is over all the multipolygon $\gamma$ with length $|\gamma|$. If open boundary conditions are assumed, only multipolygons *not* winding up the lattice are allowed. In the case of periodic boundary conditions the representation is the same, but the polygons are allowed to wind up the lattice.

Starting from (9.5), $Z$ can be written as the sum of four Grassmann integrals with different boundary conditions, following the analysis of Refs. [42],[43],[12],[44] (see App.A.)

$$Z_I = -Z_{+,+} + Z_{+,-} + Z_{-,+} + Z_{-,-} \quad (9.6)$$

$$Z_{\varepsilon,\varepsilon'} = (\cosh\beta J)^B 2^S \frac{1}{2} \int \prod_{\mathbf{x}\in\Lambda} dH_\mathbf{x} d\bar H_\mathbf{x} dV_\mathbf{x} d\bar V_\mathbf{x} e^{S_{\varepsilon,\varepsilon'}} \quad (9.7)$$

where

$$\begin{aligned}S_{\varepsilon,\varepsilon'} = &\sum_{\mathbf{x}\in\Lambda} \tanh\beta J[\bar H_{x,x_0} H_{x+1,x_0} + \bar V_{x,x_0} V_{x,x_0+1}]\\ &+ \sum_{\mathbf{x}\in\Lambda} [\bar H_{x,x_0} H_{x,x_0} + \bar V_{x,x_0} V_{x,x_0} + \bar V_{x,x_0}\bar H_{x,x_0} + V_{x,x_0}\bar H_{x,x_0}\\ &+ H_{x,x_0}\bar V_{x,x_0} + V_{x,x_0} H_{x,x_0}]\end{aligned} \quad (9.8)$$

and $H_\mathbf{x},\bar H_\mathbf{x},V_\mathbf{x},\bar V_\mathbf{x}$ are *Grassmann variables* such that

$$\bar H_{x,x_0+L} = \varepsilon \bar H_{x,x_0} \qquad \bar H_{x+L,x_0} = \varepsilon' \bar H_{x,x_0}$$
$$H_{x,x_0+L} = \varepsilon H_{x,x_0} \qquad H_{x+L,x_0} = \varepsilon' H_{x,x_0} \quad (9.9)$$

and identical relations hold for the variables $V, \bar{V}$.

If $J$ is not constant but it depends on the bounds one obtains a similar formula in which $S_{\varepsilon,\varepsilon'}$ is given by

$$S_{\varepsilon,\varepsilon'} = \sum_{\mathbf{x}} \tanh \beta J_{1;x,x_0;x+1,x_0} \bar{H}_{x,x_0} H_{x+1,x_0} +$$

$$\tanh \beta J_{2;x,x_0;x,x_0+1} \bar{V}_{x,x_0} V_{x,x_0+1}] + \sum_{\mathbf{x}} [\bar{H}_{x,x_0} H_{x,x_0} + \bar{V}_{x,x_0} V_{x,x_0} +$$

$$\bar{V}_{x,x_0} \bar{H}_{x,x_0} + V_{x,x_0} \bar{H}_{x,x_0} + H_{x,x_0} \bar{V}_{x,x_0} + V_{x,x_0} H_{x,x_0}] \quad (9.10)$$

and the factor $(\cosh J)^B$ is replaced by $\prod_b \cosh J_b$, where the product is over all the possible nearest neighbor bounds. We will call $Z_I(J_{\mathbf{x},\mathbf{x}'})$ the Ising model partition function with non constant $J$.

Note that the exponent is quadratic in the fields, that is one has a Gaussian Grassman integral which can be exactly computed. Indeed $S_{\varepsilon,\varepsilon'}(t)$ can be also written as, if $t = \tanh \beta J$:

$$S_{\varepsilon,\varepsilon'}(t) = \sum_{\mathbf{k} \in \mathcal{D}_{\varepsilon,\varepsilon'}} \left[ t \widehat{\bar{H}}_{\mathbf{k}} \widehat{\bar{H}}_{-\mathbf{k}} e^{ik} + \right. \quad (9.11)$$

$$\left. t\widehat{\bar{V}}_{\mathbf{k}} \widehat{V}_{-\mathbf{k}} e^{ik_0} + \widehat{\bar{H}}_{\mathbf{k}} \widehat{H}_{-\mathbf{k}} + \widehat{\bar{V}}_{\mathbf{k}} \widehat{V}_{-\mathbf{k}} + \widehat{\bar{V}}_{\mathbf{k}} \widehat{H}_{-\mathbf{k}} + \widehat{V}_{\mathbf{k}} \widehat{\bar{H}}_{-\mathbf{k}} + \widehat{H}_{\mathbf{k}} \widehat{\bar{V}}_{-\mathbf{k}} + \widehat{V}_{\mathbf{k}} \widehat{\bar{H}}_{-\mathbf{k}} \right]$$

where $\mathbf{k} = (k, k_0)$ and $D_{\varepsilon,\varepsilon'}$ is the set of $\mathbf{k}$'s such that

$$k = \frac{2\pi n_1}{L} + \frac{(\varepsilon' - 1)\pi}{L} \quad k_0 = \frac{2\pi n_0}{L} + \frac{(\varepsilon - 1)\pi}{L} \quad (9.12)$$

with $-[L/2] \le n_0 \le [(L-1)/2]$, $-[L/2] \le n_1 \le [(L-1)/2]$, $n_0, n_1 \in \mathbb{Z}$.

Let us say that $\mathbf{k} > 0$ if its first component $k_0$ is $> 0$. Then we can rewrite (9.12) as, if $\widehat{H}_{\mathbf{k}}, \widehat{V}_{\mathbf{k}}$ are defined in (2.27):

$$\sum_{\mathbf{k}>0} \left[ t\widehat{\bar{H}}_{\mathbf{k}} \widehat{H}_{-\mathbf{k}} e^{ik} - t\widehat{H}_{\mathbf{k}} \widehat{\bar{H}}_{-\mathbf{k}} e^{-ik} + t\widehat{\bar{V}}_{\mathbf{k}} \widehat{V}_{-\mathbf{k}} e^{ik_0} - t\widehat{V}_{\mathbf{k}} \widehat{\bar{V}}_{-\mathbf{k}} e^{-ik_0} + \quad (9.13) \right.$$

$$\widehat{\bar{H}}_{\mathbf{k}} \widehat{H}_{-\mathbf{k}} - \widehat{H}_{\mathbf{k}} \widehat{\bar{H}}_{-\mathbf{k}} + \widehat{\bar{V}}_{\mathbf{k}} \widehat{V}_{-\mathbf{k}} - \widehat{V}_{\mathbf{k}} \widehat{\bar{V}}_{-\mathbf{k}} + \widehat{\bar{V}}_{\mathbf{k}} \widehat{H}_{-\mathbf{k}} - \widehat{\bar{H}}_{\mathbf{k}} \widehat{V}_{-\mathbf{k}} + \widehat{V}_{\mathbf{k}} \widehat{\bar{H}}_{-\mathbf{k}}$$

$$\left. -\widehat{\bar{H}}_{\mathbf{k}} \widehat{V}_{-\mathbf{k}} + \widehat{H}_{\mathbf{k}} \widehat{\bar{V}}_{-\mathbf{k}} - \widehat{V}_{\mathbf{k}} \widehat{H}_{-\mathbf{k}} + \widehat{V}_{\mathbf{k}} \widehat{H}_{-\mathbf{k}} - \widehat{H}_{\mathbf{k}} \widehat{V}_{-\mathbf{k}} \right] = \sum_{\mathbf{k}>0} \bar{\Psi}^T_{\mathbf{k}} M_{\mathbf{k}} \Psi_{-\mathbf{k}}$$

where $\bar{\Psi}^T_{\mathbf{k}} = (\widehat{\bar{H}}_{\mathbf{k}}, \widehat{H}_{\mathbf{k}}, \widehat{\bar{V}}_{\mathbf{k}}, \widehat{V}_{\mathbf{k}})$ and the matrix $M_{\mathbf{k}}$ is defined as:

$$M_{\mathbf{k}} = \begin{pmatrix} 0 & 1+te^{ik} & -1 & -1 \\ -(1+te^{-ik}) & 0 & 1 & -1 \\ 1 & -1 & 0 & 1+te^{ik_0} \\ 1 & 1 & -(1+te^{-ik_0}) & 0 \end{pmatrix}$$

Then, unless for a sign,

$$\int \prod_{x\in\Lambda} d\overline{H}_x dH_x d\overline{V}_x dV_x e^{S_{\varepsilon,\varepsilon'}(t)} = \tag{9.14}$$

$$\prod_{\mathbf{k}>0} \left[ \int d\widehat{\overline{H}}_\mathbf{k} d\widehat{\overline{H}}_{-\mathbf{k}} d\widehat{H}_\mathbf{k} d\widehat{H}_{-\mathbf{k}} d\widehat{\overline{V}}_\mathbf{k} d\widehat{\overline{V}}_{-\mathbf{k}} d\widehat{V}_\mathbf{k} d\widehat{V}_{-\mathbf{k}} \cdot e^{\Psi_\mathbf{k}^T M_\mathbf{k} \Psi_{-\mathbf{k}}} \right]$$

and we see that the r.h.s. of (9.15) is equal, by (2.12), to $\prod_{\mathbf{k}>0} \det M_\mathbf{k}$. Now, developing the determinant of $M_\mathbf{k}$ along its first row we find:

$$\det M_\mathbf{k} = -(1+te^{ik})\det\begin{pmatrix} -(1+te^{-ik}) & 1 & -1 \\ 1 & 0 & 1+te^{ik_0} \\ 1 & -(1+te^{-ik_0}) & 0 \end{pmatrix}$$

$$+\det\begin{pmatrix} -(1+te^{-ik}) & 0 & -1 \\ 1 & -1 & 1+te^{ik_0} \\ 1 & 1 & 0 \end{pmatrix} + \det\begin{pmatrix} -(1+te^{-ik}) & 0 & 1 \\ 1 & -1 & 0 \\ 1 & 1 & -(1+te^{-ik_0}) \end{pmatrix}$$

More explicitely the last expression is equal to

$$-(1+te^{ik})\Big[-(1+te^{-ik})|1+te^{ik_0}|^2 + (1+te^{ik_0}) + (1+te^{-ik_0})\Big] -$$

$$\Big[-(1+te^{-ik})(1+te^{ik_0}) - 2\Big] + \Big[-(1+te^{-ik})(1+te^{-ik_0}) + 2\Big] \tag{9.15}$$

so that

$$\det M_\mathbf{k} = \Big[1+t^2+2t\cos k\Big]\Big[1+t^2+2t\cos k_0\Big] - 4t(\cos k + \cos k_0)$$
$$-4t^2\cos k\cos k_0 = (1+t^2)^2 - 2t(1-t^2)(\cos k + \cos k_0) \tag{9.16}$$

In the limit $L \to \infty$, the four Grassmann integrals in (12.2) have the same limit so that the free energy $f$ is given by

$$-\beta f = \lim_{L\to\infty} \frac{1}{L^2}\log Z_I = \log(2\cosh^2\beta J) +$$

$$\frac{1}{2}\int_{-\pi}^{\pi}\frac{dk}{2\pi}\int_{-\pi}^{\pi}\frac{dk_0}{2\pi}\log\{(1+t^2)^2 - 2t(1-t^2)(\cos k + \cos k_0)\} \tag{9.17}$$

$$= \frac{1}{2}\int_{-\pi}^{\pi}\frac{dk}{2\pi}\int_{-\pi}^{\pi}\frac{dk_0}{2\pi}\log\Big\{4\big[\cosh^2 2\beta J - \sinh 2\beta J(\cos k + \cos k_0)\big]\Big\}$$

Note that the argument of the logarithm in the last expression is always $\geq 0$ and it vanishes if and only if $\beta = \beta_c$ with $\sinh 2\beta_c J = 1$, that is the equation for the critical temperature. In the following we shall also write this condition in the equivalent form

$$\tanh \beta_c J = \sqrt{2} - 1 \tag{9.18}$$

The specific heat has the form

$$C_v \simeq -C \log|\beta - \beta_c| + C_2 \tag{9.19}$$

that is it has a logarithmic singularity at $\beta_c$, which is a signal of the occurring of a phase transition at $\beta_c$. It is also possible to compute the correlations and one sees that they are exponentially vanishing for large distances except at $\beta = \beta_c$ when they have a weaker power law decay. For instance for large distances at $\beta \neq \beta_c$

$$|<\sigma_\mathbf{x}\sigma_{\mathbf{x}'};\sigma_\mathbf{y}\sigma_{\mathbf{y}'}>_T| \leq C \frac{e^{-\kappa m(\beta)|\mathbf{x}-\mathbf{y}|}}{|\mathbf{x}-\mathbf{y}|^2} \tag{9.20}$$

where $m(\beta) = O(|\beta - \beta_c|)$ is the correlation legnth, while if $\beta = \beta_c$ and $\mathbf{x}, \mathbf{x}'$ nearest neighbor

$$<\sigma_\mathbf{x}\sigma_{\mathbf{x}'};\sigma_\mathbf{y}\sigma_{\mathbf{y}'}>_T| \simeq \frac{1}{|\mathbf{x}-\mathbf{y}|^2} \tag{9.21}$$

## 9.2 Heavy and light Majorana fermions

It is convenient to manipulate further the above Grassman integral to write it in a way showing a remarkable analogy with the models of QFT in $2d$ we have seen before. We perform the change of variables

$$\overline{H}_\mathbf{x} + iH_\mathbf{x} = e^{i\frac{\pi}{4}}\psi_\mathbf{x} - e^{i\frac{\pi}{4}}\chi_\mathbf{x} \qquad \overline{H}_\mathbf{x} - iH_\mathbf{x} = e^{-i\frac{\pi}{4}}\overline{\psi}_\mathbf{x} - e^{-i\frac{\pi}{4}}\overline{\chi}_\mathbf{x}$$
$$\overline{V}_\mathbf{x} + iV_\mathbf{x} = \psi_\mathbf{x} + \chi_\mathbf{x} \qquad \overline{V}_\mathbf{x} - iV_\mathbf{x} = \overline{\psi}_\mathbf{x} + \overline{\chi}_\mathbf{x} \tag{9.22}$$

so that, if $S_{\varepsilon,\varepsilon'} = \sum_\mathbf{x} S_{\mathbf{x},\varepsilon,\varepsilon'}$

$$S_{\mathbf{x},\varepsilon,\varepsilon'} = S^{(\psi)}_{\mathbf{x},\varepsilon,\varepsilon'} + S^{(\chi)}_{\mathbf{x},\varepsilon,\varepsilon'} + Q_{\mathbf{x},\varepsilon,\varepsilon'} \tag{9.23}$$

where

$$S^{(\psi)}_{\mathbf{x},\varepsilon,\varepsilon'} = \frac{t}{4}[\psi_\mathbf{x}(\partial_1 - i\partial_0)\psi_\mathbf{x} + \overline{\psi}_\mathbf{x}(\partial_1 + i\partial_0)\overline{\psi}_\mathbf{x}] + \tag{9.24}$$
$$+ \frac{t}{4}[-i\overline{\psi}_\mathbf{x}(\partial_1\psi_x + \partial_0\psi_\mathbf{x}) + i\psi_\mathbf{x}(\partial_1\overline{\psi}_\mathbf{x} + \partial_0\overline{\psi}_\mathbf{x})] + i(\sqrt{2} - 1 - t)\overline{\psi}_\mathbf{x}\psi_\mathbf{x}$$

with the definitions

$$\partial_1\psi_\mathbf{x} = \psi_{x+1,x_0} - \psi_\mathbf{x} \qquad \partial_0\psi_x = \psi_{x,x_0+1} - \psi_\mathbf{x} \tag{9.25}$$

Moreover

$$S^{(\chi)}_{\mathbf{x},\varepsilon,\varepsilon'} = \frac{t}{4}[\chi_\mathbf{x}(\partial_1 - i\partial_0)\chi_\mathbf{x} + \overline{\chi}_\mathbf{x}(\partial_1 + i\partial_0)\overline{\chi}_\mathbf{x}] + \tag{9.26}$$
$$\frac{t}{4}[-i\overline{\chi}_\mathbf{x}(\partial_1\chi_\mathbf{x} + \partial_0\chi_\mathbf{x}) + i\chi_\mathbf{x}(\partial_1\overline{\chi}_\mathbf{x} + \partial_0\overline{\chi}_\mathbf{x})] - i(\sqrt{2} + 1 + t)\overline{\chi}_\mathbf{x}\chi_\mathbf{x}$$

and

$$Q_{\mathbf{x},\varepsilon,\varepsilon'} = \frac{t}{4}\{-\psi_\mathbf{x}(\partial_1\chi_\mathbf{x} + i\partial_0\chi_\mathbf{x}) - \overline{\psi}_\mathbf{x}(\partial_1\overline{\chi}_\mathbf{x} - i\partial_0\overline{\chi}_\mathbf{x}) -$$
$$-\chi_\mathbf{x}(\partial_1\psi_\mathbf{x} + i\partial_0\psi_\mathbf{x}) - \overline{\chi}_\mathbf{x}(\partial_1\overline{\psi}_\mathbf{x} - i\partial_0\overline{\psi}_\mathbf{x}) + i\overline{\psi}_\mathbf{x}(\partial_1\chi_\mathbf{x} - \partial_0\chi_\mathbf{x}) + \quad (9.27)$$
$$+i\psi_\mathbf{x}(-\partial_1\overline{\chi}_\mathbf{x} + \partial_0\overline{\chi}_\mathbf{x}) + i\overline{\chi}_\mathbf{x}(\partial_1\psi_\mathbf{x} - \partial_0\psi_\mathbf{x}) + i\chi_\mathbf{x}(-\partial_1\overline{\psi}_\mathbf{x} + \partial_0\overline{\psi}_\mathbf{x})\}$$

If we define, if $\phi$ denotes either $\psi$ or $\chi$

$$P(d\phi) = \mathcal{N}_\phi^{-1} \prod_{\mathbf{k}\in D_{\varepsilon,\varepsilon'}} d\phi_\mathbf{k} d\bar{\phi}_\mathbf{k} \exp[\frac{t}{2L^2} \sum_{\mathbf{k}\in D_{\varepsilon,\varepsilon'}} \xi_\mathbf{k}^T A_\phi(\mathbf{k})\xi_{-\mathbf{k}}] \quad (9.28)$$

where

$$A_\phi(\mathbf{k}) = \begin{pmatrix} i\sin k + \sin k_0 & -im_\phi(\mathbf{k}) \\ im_\phi(\mathbf{k}) & i\sin k - \sin k_0 \end{pmatrix} \quad \xi^\mathbf{T}_\mathbf{k} = (\phi_\mathbf{k}, \bar{\phi}_\mathbf{k}) \quad \xi^\mathbf{T}_{-\mathbf{k}} = (\phi_{-\mathbf{k}}, \bar{\phi}_{-\mathbf{k}})$$

with $m_\phi$ defined, *differently* for $\phi = \psi$ (choose $-t$) and for $\phi = \chi$ (choose $+t$), by

$$\frac{t}{2}m_\phi(\mathbf{k}) = (\sqrt{2}-1\mp t) + \frac{t}{2}(\cos k_0 + \cos k - 2). \quad (9.29)$$

then we can write

$$Z^{\varepsilon,\varepsilon'} = (\cosh\beta J)^B 2^S \frac{1}{2} \int P_{\varepsilon,\varepsilon'}(d\psi) P_{\varepsilon,\varepsilon'}(d\chi) e^{Q(\chi,\psi)}. \quad (9.30)$$

Fermionic variables with a free measure of the form $P(d\phi)$ are usually called *Majorana fermions*, to distinguish them from the ones with a measure like (4.9) which are called *Dirac fermions*. Note that $m^\chi(0) \equiv m^\chi$ and $m^\psi(0) \equiv m^\psi$ can be seen as the mass of the $\chi$ or $\psi$ fermion; while $m_\chi$ is bounded away from zero for any $\beta$, $m_\psi(0)$ vanishes at $\beta = \beta_c$. In this sense, one can think to $\chi$ and $\psi$ as heavy and light Majorana fermions.

Defining

$$\langle\phi_\mathbf{x}\phi_\mathbf{y}\rangle = g^\phi_{+,+}(\mathbf{x},\mathbf{y}) \quad \langle\bar{\phi}_\mathbf{x}\bar{\phi}_\mathbf{y}\rangle = g^\phi_{-,-}(\mathbf{x},\mathbf{y}) \quad \langle\bar{\phi}_\mathbf{x}\phi_\mathbf{y}\rangle = g^\phi_{+,-}(\mathbf{x},\mathbf{y}) \quad (9.31)$$

we see that, for any integer $P$

$$|g^\chi_{\omega,\omega'}(\mathbf{x},\mathbf{y})| \leq \frac{C_P}{1+(m_\chi|\mathbf{x}-\mathbf{y}|)^P} \quad (9.32)$$

A similar bound holds for $g^\psi_{\omega,\omega'}$.

We can integrate out the $\chi$-fields so obtaining

$$\int P(d\psi) \int P(d\chi) e^{Q(\chi,\psi)} = \int \overline{P}(d\psi) e^{L^2\mathcal{N}} \quad (9.33)$$

proceeding in the following way. If $\xi_k^T = = (\psi_k, \overline{\psi}_k, \chi_k, \overline{\chi}_k)$, we can write

$$\int P(d\psi) \int P(d\chi) e^{Q(\chi,\psi)} = \frac{1}{\mathcal{N}} \exp\{-\frac{t}{4L^2} \sum_k \xi_k^T C_k \xi_{-k}\} \qquad (9.34)$$

with $C_k$ given by

$$C_k = \begin{pmatrix} -i\sin k - \sin k_0 & -im_{\psi,k}^{(j)} & i\sin k - \sin k_0 & i(\cos k - \cos k_0) \\ im_{\psi,k}^{(j)} & -i\sin k + \sin k_0 & -i(\cos k - \cos k_0) & i\sin k + \sin k_0 \\ i\sin k - \sin k_0 & i(\cos k - \cos k_0) & -i\sin k - \sin k_0 & -im_{\chi,k}^{(j)} \\ -i(\cos k - \cos k_0) & i\sin k + \sin k_0 & im_{\chi,k}^{(j)} & -i\sin k + \sin k_0 \end{pmatrix}$$

The determinant $B^{(j)}(\mathbf{k}) = \det C_k^{(j)}$ is equal to, calling $t_\psi = \sqrt{2} - 1$ and $t_\chi = -\sqrt{2} - 1$

$$B(\mathbf{k}) = \frac{16}{t^4}[2t[1-t^2](2-\cos k - \cos k_0) + (t-t_\psi)^2(t-t_\chi)^2] \qquad (9.35)$$

Using, for $l, m = 1, \ldots, 4$, the algebraic identity

$$\frac{1}{\mathcal{N}} \int [\prod_{\mathbf{k},i}(d\xi_k^{(j)})_i](\xi_{-\mathbf{k}'}^{(j)})_l(\xi_{\mathbf{k}'})_m \exp\{-\frac{t}{4L^2}\sum_k \xi_k^T C_k \xi_{-k}\} = \frac{4L^2}{t}(C_{\mathbf{k}'})_{lm}^{-1}$$

$$(9.36)$$

we find:

$$<\psi_{-\mathbf{k}}^{(j)}\psi_{\mathbf{k}}^{(j)}>_1 = \frac{4L^2}{t}\frac{c_{1,1}(\mathbf{k})}{B(\mathbf{k})} \quad , \quad <\overline{\psi}_{-\mathbf{k}}\psi_{\mathbf{k}}>_1 = \frac{4L^2}{t_\lambda}\frac{c_{-1,1}(\mathbf{k})}{B(\mathbf{k})} \quad ,$$

$$<\overline{\psi}_{-\mathbf{k}}^{(j)}\overline{\psi}_{\mathbf{k}}^{(j)}>_1 = \frac{4L^2}{t}\frac{c_{-1,-1}(\mathbf{k})}{B(\mathbf{k})} \qquad (9.37)$$

where, if $\varepsilon = \pm 1$

$$c_{\varepsilon,\varepsilon}(\mathbf{k}) = \frac{4}{t^2}\{2tt_\chi(-i\sin k \cos k_0 + \varepsilon \sin k_0 \cos k)$$
$$+[t^2 + t_\chi^2](i\sin k - \varepsilon \sin k_0)\}$$

$$c_{\varepsilon,-\varepsilon}(\mathbf{k}) = -i\varepsilon\frac{4}{t^2}\{-t(3t_\chi + t_\psi)\cos k \cos k_0$$
$$+[t^2 + 2t_\chi t_\psi + t_\chi^2](\cos k + \cos k_0) - (t(t_\psi + t_\chi) + 2\frac{t_\psi t_\chi^2}{t})\}$$

This implies

$$\bar{P}(d\psi) = \mathcal{N}^{-1}\prod_{\mathbf{k}\in\mathcal{D}} d\psi_\mathbf{k} d\overline{\psi}_\mathbf{k} \exp[-\frac{t_\psi}{L^2}\sum_{\mathbf{k}\in\mathcal{D}} \psi^T_\mathbf{k} T^{(1)}(\mathbf{k})\psi_{-\mathbf{k}}] \qquad (9.38)$$

where $T^{(1)}(\mathbf{k}) =$

$$\begin{pmatrix} \widetilde{Z}_1(i\sin k + \sin k_0) + \mu_{1,1}(\mathbf{k}) & -im_1 - i\mu_{1,2}(\mathbf{k}) \\ im_1 + i\mu_{1,2}(\mathbf{k}) & \widetilde{Z}_1(i\sin k - \sin k_0) + \mu_{2,2}(\mathbf{k}) \end{pmatrix}$$

with $\tilde{Z}_1 = 1 + O(m_\psi)$, $\psi^T_{\bf k} = (\psi_{\bf k}, \bar\psi_{\bf k})$, $\mu_{i,i}({\bf k}) = O({\bf k}^2)$ and non vanishing at ${\bf k} = \pi$, $\bar m_1 = \frac{m_\psi}{1-m_\psi/2}$.

Of course if we integrate over the $\psi$ variables we get again the explicit expression of the partition function of the Ising model (9.18). However (9.33) is quite interesting by itself; it says that the Ising model partition function is equivalent of a model of free Majorana fermion on a lattice, with mass $O(|\beta-\beta_c|)$; the critical temperature corresponds to the massless limit.

Note that $\mu_{i,i}({\bf k})$ have essentially the same form of the Wilson terms, which were added essentially by hand in chapt.7 in order to introduce a lattice regularization of $d = 2$ QFT. The exact solvability of the Ising model appears equivalent to the fact that the fermions are *non interacting*, that is the action is quadratic in the fermionic variables.

## 9.3  Generalized Ising models

From a physical point of view, there is no reason for which only nearest-neighbor spins should interact; much more reasonable is to assume that the interaction is short ranged, in the sense that it becomes weaker and weaker as more distant spins are considered. In the same way, it is also not very natural to exclude interactions involving four or a greater number of spins. Such considerations suggest to consider a more general Ising model with hamiltonian

$$H = H_I + \bar\lambda V \tag{9.39}$$

where $\bar\lambda$ is the coupling and $V$ have the form

$$\sum_{m=2}^{\infty} \int d{\bf x}_1 d{\bf x}_m v_m({\bf x}_1,..,{\bf x}_m) \prod_{i=1}^{m} \sigma_{{\bf x}_i}\sigma_{{\bf x}'_i} \tag{9.40}$$

where ${\bf x}, {\bf x}'$ are nearest neighbor and $v({\bf x}_1, ..., {\bf x}_m)$ short ranged. We will consider for definiteness an interaction of the form

$$V = \sum_{\bf x}[\sigma_{x+1,x_0}\sigma_{x,x_0+1} + \sigma_{x+1,x_0-1}\sigma_{x,x_0}] \tag{9.41}$$

which can be rewritten as

$$V = \sum_{\bf x}[\sigma_{x,x_0}\sigma_{x+1,x_0}\sigma_{x,x_0}\sigma_{x,x_0+1} + \sigma_{x+1,x_0-1}\sigma_{x+1,x_0}\sigma_{x,x_0}\sigma_{x+1,x_0}] \tag{9.42}$$

The rest of this chapter is devoted to the proof of the following theorem, proved in [45],[46] to which we refer for more details.

**Theorem 9.1.** *For $\bar{\lambda}$ small enough there exists a bounded function $\nu(\bar\lambda)$ such that, if*

$$\tanh \beta_c J = \sqrt{2} - 1 + \nu(\bar\lambda) \qquad (9.43)$$

*then for suitable constants $C_1, C_2$*

$$-C_1 \log|\beta - \beta_c| \leq |C_v| \leq -C_2 \log|\beta - \beta_c| \qquad (9.44)$$

*and, if $\mathbf{x}, \mathbf{x}'$ are nearest neighbor*

$$|<\sigma_{\mathbf{x}}\sigma_{\mathbf{x}'};\sigma_{\mathbf{y}}\sigma_{\mathbf{y}'}>_T| \leq C \frac{e^{-\kappa m(\beta)|\mathbf{x}-\mathbf{y}|}}{|\mathbf{x}-\mathbf{y}|^2} \qquad (9.45)$$

*where $m(\beta) = O(|\beta - \beta_c|)$ is the correlation length.*

The above result establishes a form of universality for the Ising model; the critical indices for the specific heat and the correlation length are insensitive to the perturbation. On the contrary the value of the critical temperature is not universal but it depends from the detail of the perturbation.

## 9.4 Fermionic representation of the generalized Ising model

Also the partition function of the model (9.39),(9.40) can be written as a Grassman integral; we show this for definiteness only in the case (9.42) but the same analysis can be repeated for the general case (9.41).

Noting that

$$\sigma_{x+1,x_0}\sigma_{x,x_0+1} = \sigma_{x,x_0}\sigma_{x+1,x_0}\sigma_{x,x_0}\sigma_{x,x_0+1} \qquad (9.46)$$

the partition function of (9.39) can be written as

$$Z = \sum_{\substack{\sigma=\pm 1 \\ \mathbf{x} \in \Lambda}} e^{-H_I(\sigma)} \prod_{\mathbf{x}} \{[1 + \tanh \bar\lambda \sigma_{x,x_0}\sigma_{x+1,x_0}\sigma_{x,x_0}\sigma_{x,x_0+1}]$$

$$[1 + \tanh \bar\lambda \sigma_{x,x_0}\sigma_{x+1,x_0}\sigma_{x+1,x_0-1}\sigma_{x+1,x_0})]\} \qquad (9.47)$$

If $Z_I$ is the partition function of the Ising model, we note that

$$\sigma_{x,x_0}\sigma_{x+1,x_0}e^{-H_I(\sigma)} = \frac{\partial}{\partial \beta J_{1;x,x_0;x+1,x_0}} Z_I(J_{\mathbf{x},\mathbf{x}'})|_{\{J_{\mathbf{x},\mathbf{x}'}\}=\{J\}} \qquad (9.48)$$

and from (9.10) this derivative gives an extra factor $\tanh \beta J + \operatorname{sech}^2 \beta J \bar H_{x,x_0} H_{x+1,x_0}$ in (9.7). In the same way

$$\sigma_{x,x_0}\sigma_{x,x_0+1}e^{-H_I(\sigma)} = \frac{\partial}{\partial \beta J_{2;x,x_0;x,x_0+1}} Z_I(J_{\mathbf{x},\mathbf{x}'})|_{\{J_{\mathbf{x},\mathbf{x}'}\}=\{J\}} \qquad (9.49)$$

and this derivative gives a factor $\tanh\beta J + \mathrm{sech}^2\beta J \bar{V}_{x,x_0} V_{x,x_0+1}$. We can write then, if $\delta_{+,+} = 1$ and $\delta_{+,-} = \delta_{-,+} = \delta_{-,-} = 2$

$$Z = \sum_{\varepsilon,\varepsilon'} (-1)^{\delta_{\varepsilon,\varepsilon'}} Z^{\varepsilon,\varepsilon'} \qquad (9.50)$$

where

$$Z^{\varepsilon,\varepsilon'} = (\cosh\beta J)^B 2^S \frac{1}{2} \int [\prod_{\mathbf{x}} dH_{\mathbf{x}} d\bar{H}_{\mathbf{x}} dV_{\mathbf{x}} d\bar{V}_{\mathbf{x}}] e^{S_{\varepsilon,\varepsilon'}}$$

$$\prod_{\mathbf{x}} \{[1 + \tanh\bar{\lambda}(t + \mathrm{sech}^2\beta J \bar{H}_{x,x_0} H_{x+1,x_0})(t + \mathrm{sech}^2\beta J \bar{V}_{x,x_0} V_{x,x_0+1})]$$

$$[1 + \tanh\bar{\lambda}(t + \mathrm{sech}^2\beta J \bar{H}_{x,x_0} H_{x+1,x_0})(t + \mathrm{sech}^2 J \bar{V}_{x+1,x_0-1} V_{x+1,x_0})]\}$$

The above expression can be rewritten as

$$\widehat{Z}^{\varepsilon,\varepsilon}_{2I} = (\cosh J)^B 2^S \frac{1}{2} \int [\prod_{\mathbf{x}} dH_{\mathbf{x}} d\bar{H}_{\mathbf{x}} dV_{\mathbf{x}} d\bar{V}_{\mathbf{x}} e^{S_{\varepsilon,\varepsilon}}] e^{\mathcal{V}} \qquad (9.51)$$

and, if $f = log(1 + \lambda\tanh^2\beta J)$

$$\mathcal{V} = \sum_{\mathbf{x}} [f + \widetilde{\lambda}[\bar{H}_{x,x_0} H_{x+1,x_0} + \bar{V}_{x,x_0} V_{x,x_0+1}] + \lambda\bar{H}_{x,x_0} H_{x+1,x_0} \bar{V}_{\mathbf{x}} V_{x,x_0+1}]$$

$$+ \sum_{\mathbf{x}} [f + \widetilde{\lambda}[\bar{V}_{x+1,x_0-1} V_{x+1,x_0} + \bar{H}_{x,x_0} H_{x+1,x_0}]$$

$$+ \lambda\bar{V}_{x+1,x_0-1} V_{x+1,x_0} \bar{H}_{\mathbf{x}} H_{x+1,x_0}] \qquad (9.52)$$

It is easy in fact to verify that

$$e^{f + \widetilde{\lambda}[\bar{H}_{\mathbf{x}} H_{x+1,x_0} + \bar{V}_{\mathbf{x}} V_{x,x_0+1}] + \lambda\bar{H}_{\mathbf{x}} H_{x+1,x_0} \bar{V}_{\mathbf{x}} V_{x,x_0+1}}$$

$$= (1 + \lambda\tanh^2\beta J)[1 + \widetilde{\lambda}[\bar{H}_{\mathbf{x}} H_{x+1,x_0} + \bar{V}_{\mathbf{x}} V_{x,x_0+1}]$$

$$+ (\lambda + (\widetilde{\lambda})^2)\bar{H}_{\mathbf{x}} H_{x+1,x_0} \bar{V}_{\mathbf{x}} V_{x,x_0+1}] \qquad (9.53)$$

hence the equality between (9.52) and (9.51) holds with the identification

$$\widetilde{\lambda}(1 + \tanh\bar{\lambda}\tanh^2\beta J) = \tanh\bar{\lambda}\mathrm{sech}^2\beta J \tanh\beta J$$

$$(1 + \tanh\bar{\lambda}\tanh^2\beta J)(\lambda + (\widetilde{\lambda})^2) = \tanh\bar{\lambda}\mathrm{sech}^4\beta J \qquad (9.54)$$

The conclusion of the above computations is that even the "physically harmless" inclusion of next to nearest interactions has the effect that the Grassmann integral is not Gaussian, so that exact solvability is lost. The exact solvability is then a rather delicate property, related to a certain simplifying assmption in the model, which is immediately lost when they are removed.

We shall consider for simplicity the partition function $Z^{-,-}$, i.e. the partition function in which the Grassmannian variables verify antiperiodic boundary conditions. The other partition functions in (9.50) admit similar expressions. Furthermore it will appear that the logarithm of $Z^{\varepsilon,\varepsilon'}$ divided by its expression for $\lambda = 0$ is insensitive to boundary conditions up to corrections which are exponentially small in the size $L$ of the system in the thermodynamic limit in which $L \to \infty$ (and if the limit $\beta \to \beta_c$ is performed after the thermodinamic limit).

## 9.5 Integration of the $\chi$-variables

We integrate the heavy $\chi$ fields

$$\int \bar{P}(d\psi)e^{L^2\mathcal{N}+\mathcal{V}^{(1)}(\psi)} = \int P(d\psi) \int P(d\chi)e^{Q(\chi,\psi)+\mathcal{V}(\psi,\chi)} \qquad (9.55)$$

where $\mathcal{N}$ is a constant, $\bar{P}(d\psi)$ is given by (9.38) and

$$\mathcal{V}^{(1)} = \sum_{n\geq 1}\sum_{\underline{\alpha},\underline{\omega},\underline{\varepsilon}}\sum_{\mathbf{x}_1,\dots,\mathbf{x}_{2n}} W_{\underline{\alpha},\underline{\varepsilon},2n}(\mathbf{x}_1,..,\mathbf{x}_{2n})\partial^{\alpha_1}\psi^{\varepsilon_1}_{\mathbf{x}_1,\omega_1}\dots\partial^{\alpha_{2n}}\psi^{\varepsilon_{2n}}_{\mathbf{x}_{2n},\omega_{2n}}$$

$$|\widehat{W}_{\underline{\alpha},\underline{\varepsilon},n}(\mathbf{k}_1,\dots\mathbf{k}_{n-1})| \leq L^2 C^n |\lambda|^{n/2}, \qquad n \geq 2 \qquad (9.56)$$

The term with $n = 1$ can be written as

$$\sum_\omega \sum_\mathbf{x} [i\omega\bar{\nu}_1\psi_\mathbf{x}\bar{\psi}_\mathbf{x} + \psi_\mathbf{x}(a_1\partial_0 + a_2\partial_1)\psi_\mathbf{x} + \bar{\psi}_\mathbf{x}(\bar{a}_1\partial_0 + \bar{a}_2\partial_1)\bar{\psi}_\mathbf{x}$$

$$+ \sum_{\mathbf{x}_1,\mathbf{x}_2}\sum_{\{\omega\}}\sum_{\alpha_1+\alpha_2\geq 2,\varepsilon_1,\varepsilon_2} W_{\underline{\alpha},\underline{\varepsilon}}(\mathbf{x}_1,\mathbf{x}_2)\partial^{\alpha_1}\psi^{\varepsilon_1}_{\mathbf{x}_1}\partial^{\alpha_2}\psi^{\varepsilon_2}_{\mathbf{x}_2} \qquad (9.57)$$

with $\bar{\nu}_1, a_i, \bar{a}_i = O(\lambda)$ and $|\widehat{W}_{\underline{\alpha},\underline{\varepsilon}}(\mathbf{k})| \leq L^2 C|\lambda|$

Calling

$$-\bar{\mathcal{V}}(\psi,\chi) = Q(\psi,\chi) + \tilde{\lambda}V(\psi,\chi) , \qquad (9.58)$$

the above result can be obtained from

$$\int P(d\chi)e^{-\bar{\mathcal{V}}(\psi,\chi)} = \sum_{n=0}^{\infty}\frac{(-1)^{n+1}}{n!}\mathcal{E}^T_\chi(\bar{\mathcal{V}};n) . \qquad (9.59)$$

where $\mathcal{E}^T_\chi$ is truncated expectation with respect to $P(d\chi)$. Also for $\mathcal{E}^T_\chi$ holds a formula similar to (2.118), namely

$$\mathcal{E}^T_\chi(\tilde{\chi}(P_{v_1}),\dots,\tilde{\chi}(P_{v_n})) = \sum_T \prod_{\ell\in T} g^\chi(\mathbf{x}(f^1_\ell) - \mathbf{x}(f^2_\ell))\int dP_T(\mathbf{t})\text{Pf}G^T(\mathbf{t})$$

$$(9.60)$$

where $T$ and $dP_T(\mathbf{t})$, $\ell \in T$ and $f_\ell^1, f_\ell^2$ the field labels associated to the points connected by $\ell$; finally if $2n = \sum_{i=1}^s |P_{v_i}|$, then $G^T(\mathbf{t})$ is a $(2n - 2s + 2) \times (2n - 2s + 2)$ antisymmetrix matrix, whose elements are given by $G^T_{f,f'} = t_{i(f),i(f')} g_\chi(f,f')$, where: $f, f' \notin F_T$ and $F_T = \cup_{\ell \in T}\{f_\ell^1, f_\ell^2\}$.

Formula (9.60) can be obtained exactly as in the proof of Lemma 3.1, starting from the formula (replacing (2.122))

$$\mathcal{E}_\chi\Big(\prod_{j=1}^s \widetilde{\chi}(P_j)\Big) = PfG = (-1)^n \int \mathcal{D}\chi \, \exp\Big[\frac{1}{2}(\chi, G\chi)\Big], \qquad (9.61)$$

where: the expectation $\mathcal{E}_\chi$ is w.r.t. $P(d\chi)$; if $2m = \sum_{j=1}^s |P_j|$, $G$ is the $2m \times 2m$ antisymmetric matrix with entries $G_{f,f'} = g^\chi(\mathbf{x}(f) - \mathbf{x}(f'))$; and

$$\mathcal{D}\chi = \prod_{j=1}^n \prod_{f \in P_j} d\chi_{\mathbf{x}(f)}^{\alpha(f)} \qquad (\chi, G\chi) = \sum_{f,f' \in \cup_i P_i} \chi_{\mathbf{x}(f)}^{\alpha(f)} G_{f,f'} \chi_{\mathbf{x}(f')}^{\alpha(f')}. \qquad (9.62)$$

By (9.32), (9.62) and the fact that $m_\chi = O(1)$, the bound (9.57) can be obtained by proceeding as in the proof of Lemma 3.1, using that $|PfG^T| = \sqrt{|\det G^T|}$ and bounding $\det G^T$ by Gram inequality.

As in the case of Wilson fermions discussed in the previous chapter, in which the interaction modifies the mass, we expect in this case that the interaction changes the value of the critical temperature; in order to take into account this fact it is convenient to write

$$\bar{P}(d\psi) = \widetilde{P}(d\psi) e^{-i\nu F_\nu(\psi)}, \qquad (9.63)$$

where $\widetilde{P}(d\psi)$ is equal to $\bar{P}(d\psi)$ with $m_1$ replaced by $\sigma = \frac{m_\psi + \bar{\nu}}{1 - m_\psi/2}$, $\nu = \frac{\bar{\nu}}{1 - m_\psi/2}$ and

$$F_\nu(\psi) = \sum_{\mathbf{x}} \bar{\psi}_{\mathbf{x}} \psi_{\mathbf{x}} \qquad (9.64)$$

## 9.6 Integration of the light fermions

We have now to analyze (9.55) which has indeed a form very similar to the functional integrals for QFT analyzed in the previous part. Despite their similarity, there are important differences.

First of all, in this case the lattice has a physical meaning, being related to the cristalline structure of solids, and it has not to be removed; in QFT the lattice is instead a mathematical artifact introduced to regularize the theory and the continuum limit must be taken. The main problem here is to take the thermodynamic limit $L \to \infty$, which is the analogue of removing the infrared cut-off in QFT; on the other hand the lattice provides a natural ultraviolet cut-off.

Another difference is that for each point of the lattice **x** only two Grassmann variables are associated, $\psi_\mathbf{x}, \bar\psi_\mathbf{x}$, while in QFT the spinorial nature of the fields has the effect that four indipendent variables are associated to any **x**, namely $\psi^\pm_{\pm,\mathbf{x}}$. This produces the crucial difference that in (9.55) local monomial quartic in the Grasmann fields are vanishing

$$\psi_\mathbf{x}\bar\psi_\mathbf{x}\psi_\mathbf{x}\bar\psi_\mathbf{x} = 0 \qquad (9.65)$$

for the anticommutativity of Grassman fields.

We perform then a multiscale integration of (9.55) very similar to the one described in chapt 4. Let us discuss first the integration of the partition function. Assume that we have integrated the scale $0, -1, .., h+1$ and we arrive at at

$$\int P(d\psi^{\le h})e^{\mathcal{V}^{(h)}} \qquad (9.66)$$

with

$$P(d\psi^{\le h}) = \mathcal{N}^{-1}\prod_\mathbf{k} d\psi_\mathbf{k} d\bar\psi_\mathbf{k} \exp[-t_\psi \frac{C_h(\mathbf{k})}{L^2}\sum_\mathbf{k} \psi^\mathbf{T}_\mathbf{k} T^{(h)}(\mathbf{k})\psi_{-\mathbf{k}}] \qquad (9.67)$$

and $T^{(h)}(\mathbf{k}) =$

$$\begin{pmatrix} i(\tilde Z_1 + a_h)\sin k + (\tilde Z_1 + b_h)\sin k_0) + \mu_{1,1} & -i\sigma - i\mu_{1,2} \\ i\sigma + i\mu_{1,2} & i(\tilde Z_1 + a_h)\sin k - (\tilde Z_1 + b_h)\sin k_0) + \mu_{2,2} \end{pmatrix}$$

with $\psi^\mathbf{T}_\mathbf{k} = (\psi_\mathbf{k}, \bar\psi_\mathbf{k})$ and $\mathcal{V}^h$ is a sum of monomials in the $\psi$. We define the localization as (as in chapt.7, with $L = \infty$ for definiteness)

$$\mathcal{L}W_4^h(\mathbf{k}_1, \mathbf{k}_2, \mathbf{k}_3) = W_4^h(0, 0, 0) \qquad (9.68)$$

$$\mathcal{L}W_2^h(\mathbf{k}) = W_2^h(0) + \sin k_0 \partial_0 W_2^h(0) + \sin k \partial_1 W_2^h(0) \qquad (9.69)$$

so that

$$\mathcal{L}\mathcal{V}^h = \gamma^h \nu_h \frac{1}{L^2}\sum_\mathbf{k} \bar\psi_\mathbf{k}\psi_{-\mathbf{k}} + \frac{1}{L^2}\sum_\mathbf{k}(\alpha_h \sin k_0 + \beta_h \sin k)\psi_\mathbf{k}\psi_{-\mathbf{k}}$$

$$\frac{1}{L^2}\sum_\mathbf{k}(\bar\alpha_h \sin k_0 + \bar\beta_h \sin k)\bar\psi_\mathbf{k}\bar\psi_{-\mathbf{k}} \qquad (9.70)$$

where we have used that the kernel of $\psi\bar\psi$ is even in **k**, and that there are no quartic terms as $\psi_\mathbf{x}\psi_\mathbf{x}\bar\psi_\mathbf{x}\bar\psi_\mathbf{x} = 0$. We write (9.66) as

$$\int P(d\psi^{\le h})e^{\mathcal{L}\mathcal{V}^{(h)} + \mathcal{R}\mathcal{V}^{(h)}} = \int \widehat P(d\psi^{\le h})e^{\sum_\mathbf{x} \gamma^h \nu_h \bar\psi_\mathbf{x}\psi_\mathbf{x} + \mathcal{R}\mathcal{V}^{(h)}} \qquad (9.71)$$

where $\hat{P}(d\psi^{\leq h})$ is identical to (9.67) with $a_{h-1}, b_{h-1}$ replacing $a_h, b_h$, with

$$a_{h-1} = a_h + C_h^{-1}\alpha_h \qquad b_{h-1} = b_h + C_h^{-1}\beta_h \qquad (9.72)$$

We can then write (9.71) as

$$\int P(d\psi^{\leq h-1}) \int P(d\psi^{(h)}) e^{\sum_x \gamma^h \nu_h \bar\psi_x \psi_x + \mathcal{R}\mathcal{V}^{(h)}} = \int P(d\psi^{(\leq h-1)}) e^{\mathcal{V}^{(h-1)}} \qquad (9.73)$$

and the procedure can be iterated.

The kernels can be written as sum of trees, with the following difference with respect to the ones in chapt 4:

1) The highest scale is $h = 2$
2) the end-points with scale $h < 1$ are associated $\gamma^h \nu_h$; to the end-points with scale 1 is associated $\mathcal{V}^{(1)}$.

In order to bound the kernels of $W^h$ we note that, assuming that $|a_h|, |b_h| \leq C|\lambda|$, it holds for any $h \leq 0$

$$|g^{(h)}_{\omega,\omega}(\mathbf{x} - \mathbf{y})| \leq \gamma^h \frac{C_N}{1 + (\gamma^h |\mathbf{x} - \mathbf{y}|)^N} \qquad (9.74)$$

so that the same bounds as in chapt.4 holds.

In order to control the flow of the running coupling constants, we start from the flow of $\nu_h$. Its beta function is given by

$$\nu_{k-1} = \gamma \nu_k + \beta_\nu^k(\nu_k; \ldots; \nu_1; \lambda) \qquad (9.75)$$

Since we want to fix $\nu_h$ in such a way that $\nu_{-\infty} = 0$, we must have:

$$\nu_1 = -\sum_{k=-\infty}^{1} \gamma^{k-2} \beta_\nu^k(\nu_k; \ldots; \nu_1; \lambda) \ . \qquad (9.76)$$

If we manage to fix $\nu_1$ we also get:

$$\nu_h = -\sum_{k \leq h} \gamma^{k-h-1} \beta_\nu^k(\nu_k; \ldots; \nu_1; \lambda) \ . \qquad (9.77)$$

We consider the Banach space $\mathcal{M}_\vartheta$ of sequences $\underline{\nu} = \{\nu_j\}_{j \leq N+1}$ such that, if $\vartheta$ is a constant

$$\|\underline{\nu}\|_\xi = \sup_{j \leq 0} \gamma^{-j\vartheta} |\nu_j| \leq \xi |\lambda| \ , \qquad (9.78)$$

We look for a fixed point of the operator $\mathbf{T} : \mathcal{M}_\vartheta \to \mathcal{M}_\vartheta$ defined as:

$$(\mathbf{T}\underline{\nu})_h = -\sum_{k \leq h} \gamma^{k-h-1} \beta_\nu^k(\nu_k; \ldots; \nu_1; \lambda) \ . \qquad (9.79)$$

Note that, if $|\lambda|$ is sufficiently small, then **T** leaves $\mathcal{M}_\vartheta$ invariant: in fact we can write

$$\beta_\nu^k(\nu_k;\ldots;\nu_1;\lambda) = \beta_{\nu,a}^k(\nu_k;\ldots;\nu_1) + \beta_{\nu,b}^k(\nu_k;\ldots;\nu_1;\lambda) \qquad (9.80)$$

where the first addend is the sum of trees with no $\lambda$-endpoint; of course $\beta_{\nu,a}^k(\nu_k;\ldots;\nu_1) = 0$ as it is given by chain graphs whose local part is vanishing as $g^{(k)}(0) = 0$. On the other hand

$$|\beta_{\nu,b}^k(\nu_k;\ldots;\nu_1;\lambda)| \leq c_1|\lambda|\gamma^{\vartheta h} \qquad (9.81)$$

by the short memory property, as they have necessarily and end-point at scale 0. Hence

$$|(\mathbf{T}\nu)_h| \leq \sum_{k \leq h} 2c_1|\lambda|\gamma^{(\vartheta/2)k}\gamma^{k-h} \leq c|\lambda|\gamma^{(\vartheta/2)h}, \qquad (9.82)$$

Furthermore, by using again the short memory property, we find that **T** is a contraction on $\mathcal{M}_\vartheta$:

$$|(\mathbf{T}\nu)_h - (\mathbf{T}\underline{\nu}')_h| \leq c \sum_{k \leq h} \gamma^{k-h-1} \sum_{k'=k}^{1} \gamma^{\vartheta(k-k')}|\lambda||\nu_{k'} - \nu'_{k'}|$$

$$\leq c''|\lambda|\gamma^{(\vartheta/2)h}||\underline{\nu} - \underline{\nu}'||_\vartheta \qquad (9.83)$$

hence $||(\mathbf{T}\nu) - (\mathbf{T}\underline{\nu}')||_\vartheta \leq c''|\lambda|||\underline{\nu} - \underline{\nu}'||_\vartheta$. Then, a unique fixed point $\underline{\nu}^*$ for **T** exists on $\mathcal{M}_\vartheta$.

We have finally to discuss the flow of $\vec{v}_h = (a_h, b_h)$; we can write

$$\vec{v}_{h-1} = \vec{v}_h + \beta_{\vec{v}}^h(\nu_h, .., \nu_1; \lambda) \qquad (9.84)$$

where in $\beta^h$ there is at least an endpint at scale 1 or a $|\nu_k| \leq C|\lambda|\gamma^{\vartheta k}$ (the contribution of chain graphs is vanishing) so that by the short memory property

$$|\beta_{\vec{v}}^h(\nu_h, .., \nu_1; \lambda)| \leq c_1|\lambda|\gamma^{\vartheta h} \qquad (9.85)$$

so that

$$|\vec{v}_h| \leq \sum_k c_1|\lambda|\gamma^{\vartheta h} \leq C|\lambda| \qquad (9.86)$$

The above iterative procedure can be iterated up to a scale $h^*$ defined as the minimal scale such that $\gamma^{h^*} \leq |\sigma|$, so that $\gamma^{h^*} = O(|t - \sqrt{2} + 1 - C_0^{-1}\nu|)$

## 9.7 Correlation functions and the specific heat

In the preceding sections we have found a convergent expansion for the free energy; in order to prove Theorem 9.1 we have do the same for the energy-energy correlation function (14.6) and the specific heat. The specific heat is essentially given by

$$\frac{1}{L^2} \sum_{\mathbf{x},\mathbf{y}\in\Lambda} <H_{I,\mathbf{x}}; H_{I,\mathbf{y}}>_T \tag{9.87}$$

where $H_I = \sum_{\mathbf{x}} H_{I,\mathbf{x}}$, and $H_I$ is the Ising model hamiltonian. We can write

$$<H_{I,\mathbf{x}}; H_{I,\mathbf{y}}>_{\Lambda,T} = \sum_{\varepsilon,\varepsilon'} (-1)^{\delta_{\varepsilon,\varepsilon'}} Z^{\varepsilon,\varepsilon'} \frac{\Omega_{\varepsilon,\varepsilon',\Lambda}(\mathbf{x}-\mathbf{y})}{\sum_{\varepsilon,\varepsilon'}(-1)^{\delta_{\varepsilon,\varepsilon'}} Z^{\varepsilon,\varepsilon'}} \tag{9.88}$$

where $\delta_{+,+} = -$ and $-1$ otherwise and

$$\Omega_{\varepsilon,\varepsilon',\Lambda}(\mathbf{x}-\mathbf{y}) = \frac{\int P_{\varepsilon,\varepsilon'}(dH,dV) e^{-\mathcal{V}} \partial_t S_{\mathbf{x},\varepsilon,\varepsilon'} \partial_t S_{\mathbf{y},\varepsilon,\varepsilon'}}{\int P_{\varepsilon,\varepsilon'}(dH,dV)] e^{-\mathcal{V}}} \tag{9.89}$$

Again the r.h.s. of (9.89) can be written as $\Omega_{\underline{\varepsilon},\Lambda}(\mathbf{x}-\mathbf{y}) = \frac{\partial}{\partial\phi(\mathbf{x})}\frac{\partial}{\partial\phi(\mathbf{y})}\mathcal{S}_{\underline{\varepsilon}}(\phi)|_{\phi=0}$ where, if $S_{\mathbf{x},\varepsilon,\varepsilon'}$ is the summand in (9.8)

$$e^{\mathcal{S}_{\underline{\varepsilon}}(\phi)} = \int P(d\psi) \int P(d\chi) e^{Q(\chi,\psi)-\mathcal{V}(\psi,\chi)} e^{\sum_{\mathbf{x}} \phi(\mathbf{x})\partial_t S_{\mathbf{x},\varepsilon,\varepsilon'}}. \tag{9.90}$$

We have then to slightly adapt the previous analysis of for the integration of $\mathcal{S}_{\underline{\varepsilon}}(\phi)$. One can proceed as before in order to integrate the massive $\chi$ fields and we iteratively integrate the $\psi$ fields. The action of $\mathcal{L}$ produces extra terms of the form $Z_h^{(1)} \int d\mathbf{k}d\mathbf{p} \phi_{\mathbf{p}} \psi_{\mathbf{k}}^{(\leq h)} \psi_{\mathbf{k}-\mathbf{p}}^{(\leq h)}$ and $Z_h^{(1)}$ have a bounded flow, for the same reasons discussed in the previous section. We obtain

$$\Omega(\mathbf{x},\mathbf{y}) = \sum_{h=h^*}^{0} \Omega^{(h)}(\mathbf{x},\mathbf{y}) \tag{9.91}$$

with

$$|\bar{\Omega}^{(h)}(\mathbf{x},\mathbf{y})| \leq \gamma^{2h} \frac{C_N}{1+(\gamma^h|\mathbf{x}-\mathbf{y}|)^N} \tag{9.92}$$

from which Theorem 9.1 immediately follows.

## Chapter 10

# Nonuniversality in Vertex or Isotropic Ashkin-Teller Models

## 10.1 Ashkin-Teller or Vertex models

We have seen in the previous chapter that any perturbation of the nearest neighbor Ising model, obtained adding small short range perturbations in the spins, does not change the critical properties of the specific heat or the asymptotic behavior of energy correlations. The reason why universality (at least for such quantities) holds appears, in this approach, quite subtle; when mapped in a fermionic system, all the local monomials in the fermions with degree higher than two are irrelevant. The only possible non irrelevant term is $\psi_\mathbf{x}\psi_\mathbf{x}\bar\psi_\mathbf{x}\bar\psi_\mathbf{x}$ which is indeed vanishing by the anticommutativity properties of Grasssman variables.

We consider in this and in the following chapter other spin lattice models, which can be still considered as perturbations of the Ising model (more exactly, they can be mapped in two copies of the Ising model), in which however universality can be violated.

The first model we consider is the *Ashkin-Teller* model, see Ref.[14], introduced as a generalization of the Ising model to a four component system. The assumption that the spins have only two values is physically unrealistic, as a magnetic dipole can have a continuum of pointing directions. A step in this direction is provided by the Ashkin-Teller model, in which each site of a bidimensional lattice is occupied by one of four kinds of atoms: $A, B, C, D$. Two neighbouring atoms interact with an energy: $\varepsilon_0$ for $AA, BB, CC, DD$; $\varepsilon_1$ for $AB, CD$; $\varepsilon_2$ for $AC, BD$; and $\varepsilon_3$ for $AD, BC$.

This Ashkin-Teller model can be expressed in terms of Ising spins; one associates to each site of the square lattice two spins variables, $\sigma_\mathbf{x}^{(1)}$ and $\sigma_\mathbf{x}^{(2)}$. If $(\sigma_\mathbf{x}^{(1)}, \sigma_\mathbf{x}^{(2)}) = (+,+)$ there is an atom $A$ associated to $\mathbf{x}$, if $(\sigma_\mathbf{x}^{(1)}, \sigma_\mathbf{x}^{(2)}) = (+,-)$ there is an atom $B$, if $(\sigma_\mathbf{x}^{(1)}, \sigma_\mathbf{x}^{(2)}) = (-,+)$ there is an atom $C$ and if $(\sigma_{\mathbf{x}}^{(1)}, \sigma_{\mathbf{x}}^{(2)}) = (-,-)$ there is an atom $D$.

The partition function is then given by $Z_\Lambda = \sum_{\sigma^{(1)},\sigma^{(2)}} e^{-H_\Lambda}$, where

$$H_\Lambda(\sigma^{(1)}, \sigma^{(2)}) = J^{(1)} H_I(\sigma^{(1)}) + J^{(2)} H_I(\sigma^{(2)}) - J^{(3)} V(\sigma^{(1)}, \sigma^{(2)}) + J^{(4)}$$

$$H_I(\sigma^{(j)}) = -\sum_{\mathbf{x} \in \Lambda}[y_1 \sigma_{\mathbf{x}}^{(j)} \sigma_{\mathbf{x}+\hat{e}_1}^{(j)} + y_2 \sigma_{\mathbf{x}}^{(j)} \sigma_{\mathbf{x}+\hat{e}_0}^{(j)}] \tag{10.1}$$

$$V_{AT}(\sigma^{(1)}, \sigma^{(2)}) = \sum_{\mathbf{x} \in \Lambda} [\sigma_{x,x_0}^{(1)} \sigma_{x+1,x_0}^{(1)} \sigma_{x,x_0}^{(2)} \sigma_{x+1,x_0}^{(2)} + \sigma_{x,x_0}^{(1)} \sigma_{x,x_0+1}^{(1)} \sigma_{x,x_0}^{(2)} \sigma_{x,x_0+1}^{(2)}]$$

where $H_I$ is the Ising model hamiltonian, $y_1 = y_2 = 1$, $\Lambda$ is a square subset of $\mathbb{Z}^2$ of side $L$ and

$$-J^{(1)} = (\varepsilon_0 + \varepsilon_1 - \varepsilon_2 - \varepsilon_3)/4 \quad -J^{(2)} = (\varepsilon_0 + \varepsilon_2 - \varepsilon_3 - \varepsilon_1)/4$$
$$-J^{(3)} = (\varepsilon_0 + \varepsilon_3 - \varepsilon_1 - \varepsilon_2)/4 \quad -J^{(4)} = (\varepsilon_0 + \varepsilon_1 + \varepsilon_2 + \varepsilon_3)/4 \tag{10.2}$$

The Ashkin-Teller model is not exactly solvable, except for some special choice of the parameters corrosponding to $J^{(3)} = 0$ in which reduces to two independent Ising models; in particular the specific heat has a log-singularity in corrispondence of the critical temperatures located at $\tanh J^{(1)} \beta_c = \sqrt{2} - 1$ and $\tanh J^{(2)} \beta_c = \sqrt{2} - 1$; in the case $J^{(1)} = J^{(2)}$ the Ashkin-Teller model is called isotropic and the two critical temperatures coincides. In this chapter we will discuss the isotropic model, and the anisotropic case will be discussed in the following chapter.

Another important lattice statistical mechanics model is the *8-Vertex model*, see Ref.[14], in which to each site of a bidimensional lattice one associates one among 8 possible vertices composed by four arrows pointing in or out the center. To each vertex is associated an energy, and $\varepsilon_A$ is the common energy of the first and second vertex, $\varepsilon_B$ the common energy of the third and fourth vertex and so on. Such a model has been introduced as a generalization of the "ice-type" models, describing crystals with hydrogen bounds which can be conveniently described by arrows placed on the bounds.

Also the 8V model can be exactly mapped, see Ref.[14], in two Ising models coupled by a four spin interaction bilinear in the energy densities of the two sublattices, with the following Hamiltonian

$$H_\Lambda(\sigma^{(1)}, \sigma^{(2)}) = J H_I(\sigma^{(1)}) + J H_I(\sigma^{(2)}) + \lambda V_{8V}(\sigma^{(1)}, \sigma^{(2)}) \tag{10.3}$$

where

$$V_{8V}(\sigma^{(1)}, \sigma^{(2)}) = -\lambda \sum_{\mathbf{x}} \Big\{ [\sigma_{x,x_0}^{(1)} \sigma_{x+1,x_0}^{(1)} \sigma_{x,x_0}^{(2)} \sigma_{x,x_0+1}^{(2)} +$$
$$\sigma_{x,x_0}^{(1)} \sigma_{x,x_0+1}^{(1)} \sigma_{x-1,x_0+1}^{(2)} \sigma_{x,x_0+1}^{(2)}] \Big\} \tag{10.4}$$

and with the identifications, if $Jy^{(1)} = J^{(1)}, Jy^{(2)} = J^{(2)}, a = e^{-\beta \varepsilon_A}, b = e^{-\beta \varepsilon_B}, c = e^{-\beta \varepsilon_C}, d = e^{-\beta \varepsilon_D}$

$$a = e^{\beta(J^{(1)}+J^{(2)}+\lambda)} \quad b = e^{\beta(-J^{(1)}-J^{(2)}+\lambda)} \tag{10.5}$$

$$c = e^{\beta(-J^{(1)}+J^{(2)}-\lambda)} \quad b = e^{\beta(J^{(1)}-J^{(2)}-\lambda)} \tag{10.6}$$

Contrary to the Ashkin-Teller model, the 8V model can be exactly solved and some critical exponents can be computed.

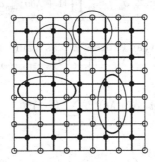

Fig. 10.1 The spins involved in the interaction of the models in (10.7). The heavy dots and lines or the light dots and lines mark the Ising lattices and the nearest neighbors Ising couplings. The ellipses symbolize the Ashkin–Teller four spins interactions ($\lambda a$–couplings) and the circles the Baxter four spins interactions ($\lambda b$) couplings.

The methods introduced in the previous chapters allow a detailed analyis of such model, at least in certain region of the parameters; remarkably, such analysis donot require any property of exact solvability.

For fixing the ideas we consider in this chapter a model with hamiltonian

$$H = JH_I(\sigma^{(1)}) + JH_I(\sigma^{(2)}) - \lambda a V_{AT} - \lambda b V_{8V} \tag{10.7}$$

reducing to the 8V or AT model for $b = 0$ or $a = 0$; from the analysis it will appear clear that similar results holds in a rather general class of models.

## 10.2 Fermionic representation

The partition function of the model (10.7) is

$$Z_{2I} = \sum_{\substack{\sigma_x^{(1)}=\pm 1 \\ x \in \Lambda}} \sum_{\substack{\sigma_x^{(2)}=\pm 1 \\ x=\Lambda}} e^{-H_I(\sigma^{(1)})} e^{-H_I(\sigma^{(2)})} e^{-V(\sigma^{(1)},\sigma^{(2)})} \tag{10.8}$$

Setting $\widehat{\lambda a} = \tanh(\lambda a), \widehat{\lambda b} = \tanh(\lambda b)$ we see that $Z_{2I}$ becomes $(\cosh \lambda a \cosh \lambda b)^{2S}$ times $\widehat{Z}_{2I}$ with

$$\widehat{Z}_{2I} = \sum_{\substack{\sigma^{(1)}=\pm 1 \\ x \in \Lambda}} \sum_{\substack{\sigma^{(2)}=\pm 1 \\ x \in \Lambda}} e^{-H_I(\sigma^{(1)})} e^{-H_I(\sigma^{(2)})}$$

$$\cdot \prod_{x \in \Lambda} [1 + \widehat{\lambda a} \sigma^{(1)}_{x,x_0} \sigma^{(1)}_{x+1,x_0} \sigma^{(2)}_{x,x_0} \sigma^{(2)}_{x+1,x_0}] \prod_{x \in \Lambda} [1 + \widehat{\lambda a} \sigma^{(1)}_{x,x_0} \sigma^{(1)}_{x,x_0+1} \sigma^{(2)}_{x,x_0} \sigma^{(2)}_{x,x_0+1}]$$

$$\cdot \prod_{x \in \Lambda} [1 + \widehat{\lambda b} \sigma^{(1)}_{x,x_0} \sigma^{(1)}_{x+1,x_0} \sigma^{(2)}_{x,x_0} \sigma^{(2)}_{x,x_0+1}] \prod_{x \in \Lambda} [1 + \widehat{\lambda b} \sigma^{(1)}_{x,x_0} \sigma^{(1)}_{x,x_0+1} \sigma^{(2)}_{x-1,x_0+1} \sigma^{(2)}_{x,x_0+1}]$$
(10.9)

Proceeding as in §9.4 we can express $\widehat{Z}_{2I}$ as a sum of sixteen partition functions labeled by $\gamma_1, \gamma_2 = (\varepsilon^{(1)}, \varepsilon'^{(1)}), (\varepsilon^{(2)}, \varepsilon'^{(2)})$ (corresponding to choosing each $\varepsilon$ and $\varepsilon'$ as $\pm$)

$$\widehat{Z}_{2I} = (\cosh \lambda a \cosh \lambda b)^{2S} \sum_{\gamma_1, \gamma_2} (-1)^{\delta_{\gamma_1} + \delta_{\gamma_2}} \widehat{Z}^{\gamma_1, \gamma_2}_{2I} \qquad (10.10)$$

with $\widehat{Z}_{2I}$ can be written as

$$\widehat{Z}^{\gamma_1,\gamma_2}_{2I} = \frac{(\cosh \beta J)^{2B} 2^{2S}}{4} \int \prod_{j=1}^{2} \Big( \big[ \prod_{x \in \Lambda} dH^{(\alpha_j)}_{\mathbf{x}} d\overline{H}^{(\alpha_j)}_{\mathbf{x}} dV^{(\alpha_j)}_{\mathbf{x}} d\overline{V}^{(\alpha_j)}_{\mathbf{x}} \big] e^{S^{(\alpha_j)}_{J,\gamma_j}} \Big) e^{-\mathcal{V}}$$
(10.11)

with

$$\mathcal{V} = \mathcal{V}_a + \mathcal{V}_b \qquad (10.12)$$

and, if $f_i = \log(1 + \widehat{\lambda[i]} \tanh^2 \beta J)$ and $[i] = a, b$

$$-\mathcal{V}_a = \sum_{x \in \Lambda} [2f_a + \widetilde{\lambda}_a [\overline{H}^{(1)}_{x,x_0} H^{(1)}_{x+1,x_0} + \overline{H}^{(2)}_{x,x_0} H^{(2)}_{x+1,x_0}] + \lambda_a \overline{H}^{(1)}_{x,x_0} H^{(1)}_{x+1,x_0} \overline{H}^{(2)}_{\mathbf{x}} H^{(2)}_{x+1,x_0}$$

$$+ \widetilde{\lambda}_a [\overline{V}^{(1)}_{x,x_0} V^{(1)}_{x,x_0+1} + \overline{V}^{(2)}_{x,x_0} V^{(2)}_{x,x_0+1}] + \lambda_a \overline{V}^{(1)}_{\mathbf{x}} V^{(1)}_{x,x_0+1} \overline{V}^{(2)}_{x,x_0} V^{(2)}_{x,x_0+1}]$$

$$-\mathcal{V}_b = \sum_{x \in \Lambda} [2f_b + \widetilde{\lambda}_b [\overline{H}^{(1)}_{x,x_0} H^{(1)}_{x+1,x_0} + \overline{V}^{(2)}_{x,x_0} V^{(2)}_{x,x_0+1}] + \lambda_b \overline{H}^{(1)}_{x,x_0} H^{(1)}_{x+1,x_0} \overline{V}^{(2)}_{\mathbf{x}} V^{(2)}_{x,x_0+1}]$$

$$+ \widetilde{\lambda}_b [\overline{V}^{(1)}_{x,x_0} V^{(1)}_{x,x_0+1} + \overline{H}^{(2)}_{x-1,x_0+1} H^{(2)}_{x,x_0+1}] + \lambda_b \overline{V}^{(1)}_{x,x_0} V^{(1)}_{x,x_0+1} \overline{H}^{(2)}_{x-1,x_0+1} H^{(2)}_{x,x_0+1}]$$
(10.13)

where

$$\widetilde{\lambda}_i (1 + \widehat{\lambda[i]} \tanh^2 \beta J) = \widehat{\lambda[i]} \operatorname{sech}^2 \beta J \tanh \beta J$$

$$(1 + \widehat{\lambda[i]} \tanh^2 \beta J)(\lambda_i + (\widetilde{\lambda}_i)^2) = \widehat{\lambda[i]} \operatorname{sech}^4 \beta J \qquad (10.14)$$

For small $\lambda$ it is $\widetilde{\lambda}_i = \lambda[i](\tanh J \operatorname{sech}^2 \beta J + O(\lambda))$, $\lambda_i = \lambda[i](\operatorname{sech}^4 \beta J + O(\lambda))$.

We shall consider for simplicity the partition function $\widehat{Z}_{2I}^{-,-,-,-} = i\widehat{Z}_{2I}^{-}$, i.e. the partition function in which all Grassmannian variables verify antiperiodic boundary conditions. The other fifteen partition functions in (10.10) admit similar expressions. The logarithm of $Z_{2I}^{\gamma_1,\gamma_2}$ divided by its expression for $\lambda = 0$ is insensitive to boundary conditions up to corrections which are exponentially small in the size $L$ of the system in the thermodynamic limit in which $L \to \infty$ so that it will turn out that it is sufficient to study just one of the sixteen partition functions and $\widehat{Z}_{2I}^{-,-,-,-}$ is chosen here (arbitrarily).

As in chap.9 we perform the following change of variables, $\alpha = 1, 2$

$$\overline{H}_{\mathbf{x}}^{(\alpha)} + iH_{\mathbf{x}}^{(\alpha)} = e^{i\frac{\pi}{4}}\psi_{\mathbf{x}}^{(\alpha)} - e^{i\frac{\pi}{4}}\chi_{\mathbf{x}}^{(\alpha)} \quad \overline{H}_{\mathbf{x}}^{(\alpha)} - iH_{\mathbf{x}}^{(\alpha)} = e^{-i\frac{\pi}{4}}\overline{\psi}_{\mathbf{x}}^{(\alpha)} - e^{-i\frac{\pi}{4}}\overline{\chi}_{\mathbf{x}}^{(\alpha)}$$

$$\overline{V}_{\mathbf{x}}^{(\alpha)} + iV_{\mathbf{x}}^{(\alpha)} = \psi_{\mathbf{x}}^{(\alpha)} + \chi_{\mathbf{x}}^{(\alpha)} \quad \overline{V}_{\mathbf{x}}^{(\alpha)} - iV_{\mathbf{x}}^{(\alpha)} = \overline{\psi}_{\mathbf{x}}^{(\alpha)} + \overline{\chi}_{\mathbf{x}}^{(\alpha)} \qquad (10.15)$$

which replaces the $H, V, \overline{H}, \overline{V}$ variables with "Majorana variables" $\psi^{(\alpha)}, \chi^{(\alpha)}$.

Subsequently we replace the Majorana variables with Dirac variables by setting

$$\psi_{1,\mathbf{x}}^{\mp} = \frac{1}{\sqrt{2}}(\psi_{\mathbf{x}}^{(1)} \pm i\psi_{\mathbf{x}}^{(2)}), \qquad \psi_{-1,\mathbf{x}}^{\mp} = \frac{1}{\sqrt{2}}(\overline{\psi}_{\mathbf{x}}^{(1)} \pm i\overline{\psi}_{\mathbf{x}}^{(2)}) \qquad (10.16)$$

$$\chi_{1,\mathbf{x}}^{\mp} = \frac{1}{\sqrt{2}}(\chi_{\mathbf{x}}^{(1)} \pm i\chi_{\mathbf{x}}^{(2)}), \qquad \chi_{-1,\mathbf{x}}^{\mp} = \frac{1}{\sqrt{2}}(\overline{\chi}_{\mathbf{x}}^{(1)} \pm i\overline{\chi}_{\mathbf{x}}^{(2)}).$$

The final expression is

$$\widehat{Z}_{2I}^{-} = \mathcal{N} \int P(d\psi) P(d\chi) e^{Q(\chi,\psi) - \mathcal{V}(\chi,\psi)}. \qquad (10.17)$$

where if $\phi$ denotes either $\psi$ or $\chi$

$$P(d\phi) = \mathcal{N}_\phi^{-1} \prod_{\mathbf{k}} \prod_{\omega=\pm 1} d\phi_{\mathbf{k},\omega}^+ d\phi_{\mathbf{k},\omega}^- \qquad (10.18)$$

$$\exp[\frac{t}{2L^2}\sum_{\mathbf{k}} \xi_{\mathbf{k}}^{(-),\mathbf{T}} A_\phi(\mathbf{k}) \xi_{\mathbf{k}}^{(+)}] \qquad (10.19)$$

where

$$A_\phi(\mathbf{k}) = \begin{pmatrix} i\sin k + \sin k_0 & -im_\phi(\mathbf{k}) \\ im_\phi(\mathbf{k}) & i\sin k - \sin k_0 \end{pmatrix} \quad \xi_{\mathbf{k}}^{\mathbf{T}} = (\phi_{\mathbf{k},1}^-, \phi_{\mathbf{k},-1}^-)$$

with $m_\phi$ defined, *differently* for $\phi = \psi$ (choose $-t$) and for $\phi = \chi$ (choose $+t$), by

$$\frac{t}{2}m_\phi(\mathbf{k}) = (\sqrt{2} - 1 \mp t) + \frac{t}{2}(\cos k_0 + \cos k - 2). \qquad (10.20)$$

Note the remarkable similarity with $P(\psi)$ with the integartion of lattice relatistic fermions in $d = 2$ discussed in chapt. 9.

Finally $Q(\chi, \psi)$ and $\mathcal{V}(\chi, \psi)$ are obtained respectively from (9.28) through the change of variables (10.15) and (10.17).

## 10.3 Anomalous behaviour

The main difference with respect to the case treated in the previous chapter is that the system can be mapped in terms of Dirac instead of Majorana fermions; as a consequence there local quartic monomials in the fields that with non negative dimension, namely $\psi^+_{+,\mathbf{x}}\psi^-_{+,\mathbf{x}}\psi^+_{-,\mathbf{x}}\psi^-_{-,\mathbf{x}}$. It holds the following result (originally proved in Ref.[47; 48], where more details can be found), if $H_\mathbf{x}(\sigma^{(1)},\sigma^{(2)})$ is the summand in (10.7)

**Theorem 10.1.** *If $a = 0$ or $b = 0$, for $\lambda$ small enough one can uniquely define $\nu(\lambda)$, bounded function in $\lambda$, so that the model is critical at $t = t_c = \sqrt{2}-1+\nu(\lambda)$. This means that, for $|t-t_c|$ strictly positive and small enough*

$$\lim_{|\Lambda|\to\infty} \langle H_\mathbf{x}(\sigma^{(1)},\sigma^{(2)}) H_\mathbf{y}(\sigma^{(1)},\sigma^{(2)}) \rangle_T = \Omega^a(\mathbf{x},\mathbf{y}) + \Omega^b(\mathbf{x},\mathbf{y}) \qquad (10.21)$$

*and the bounds, for any integer $N$*

$$|\Omega^a(\mathbf{x},\mathbf{y})| \leq \frac{1}{|\mathbf{x}-\mathbf{y}|^{2+2\eta_1}} \frac{C_N}{1+(\Delta|\mathbf{x}-\mathbf{y}|)^N} \qquad (10.22)$$

$$|\Omega^b(\mathbf{x},\mathbf{y})| \leq \frac{1}{|\mathbf{x}-\mathbf{y}|^{2+\vartheta}} \frac{C_N}{1+(\Delta|\mathbf{x}-\mathbf{y}|)^N} \qquad (10.23)$$

*hold, with $\vartheta > 0$ a constant and correlation length $\Delta^{-1}$ and critical indices $\eta_1, \eta_2$ given by*

$$\Delta = |t-t_c|^{1+\eta_2}, \quad \eta_1(\lambda) = -a_1(a+b)\lambda + O(\lambda^2) \quad \eta_2(\lambda) = -a_2(a+b)\lambda + O(\lambda^2) \qquad (10.24)$$

*with $a_1 > 0, a_2 > 0$ constants. Furthermore if $1 \leq |\mathbf{x}| \leq \Delta^{-1}$ the correlation is asymptotic to $\Omega^a$ in the sense that $\Omega^b$ is negligeble because*

$$\Omega^a(\mathbf{x},\mathbf{y}) = \frac{1+A(\mathbf{x}-\mathbf{y})}{(\mathbf{x}-\mathbf{y})^{2+2\eta_1}}, \qquad |A(\mathbf{x})| \leq C\left[|\lambda| + (\Delta|\mathbf{x}|)^{\frac{1}{2}}\right] \qquad (10.25)$$

*Finally the specific heat $C_v$ verifies*

$$C_1 \frac{1}{2\eta_1}[1-|\Delta|^{2\eta_1}] \leq C_v^\lambda \leq C_2 \frac{1}{2\eta_1}[1-|\Delta|^{2\eta_1}] \qquad (10.26)$$

*where $C_1, C_2$ are positive constants.*

Note that the logarithmic singularity of the specific heat of the Ising model is removed or changed in a power law (with a non universal critical index) depending on the sign of the interaction. Moreover also the critical index of the correlation length is changed; universality is violated in such models and the critical behaviour is expressed in term of non-universal critical indices.

## 10.4 Simmetry properties

We can integrate the massive fermions $\chi$ as in the previous chapter obtaining

$$\widehat{Z}_{2I}^{-} = \int P(d\psi) \int P(d\chi)e^{Q(\chi,\psi)}e^{-\mathcal{V}(\psi,\chi)} = \int \overline{P}(d\psi)e^{L^2\mathcal{N}^{(1)}-\mathcal{V}^{(1)}(\psi)} \tag{10.27}$$

where $\mathcal{N}^{(1)}$ is a constant such that the *effective potential* $\mathcal{V}^{(1)}(\psi)$ vanishes at $\psi = 0$ and $\overline{P}$ is suitably defined.

$$\mathcal{V}^{(1)} = \sum_{n \geq 1} \sum_{\underline{\alpha},\underline{\omega},\underline{\varepsilon}} \sum_{\mathbf{x}_1,\ldots,\mathbf{x}_{2n}} W_{\underline{\alpha},\underline{\omega},\underline{\varepsilon},2n}(\mathbf{x}_1,..,\mathbf{x}_{2n})\partial^{\alpha_1}\psi^{\varepsilon_1}_{\mathbf{x}_1,\omega_1}\ldots\partial^{\alpha_{2n}}\psi^{\varepsilon_{2n}}_{\mathbf{x}_{2n},\omega_{2n}}$$

$$|\widehat{W}_{\underline{\alpha},\underline{\omega},\underline{\varepsilon},n}(\mathbf{k}_1,\ldots\mathbf{k}_{n-1})| \leq L^2 C^n |\lambda|^n, \qquad n \geq 2 \tag{10.28}$$

The terms in (10.28) with $n=2$ can be written, for $l_1 = (\widehat{\lambda a} + \widehat{\lambda b})\text{sech}^4\beta J + O(\lambda^2)$ real, as

$$l_1 \sum_{\mathbf{x}} \psi^+_{1,\mathbf{x}}\psi^+_{-1,\mathbf{x}}\psi^-_{-1,\mathbf{x}}\psi^-_{1,\mathbf{x}} + \tag{10.29}$$

$$\sum_{\mathbf{x}_1,\ldots,\mathbf{x}_4} \sum_{\alpha_1+..\alpha_4 \geq 1,\underline{\varepsilon}} W_{\underline{\alpha},\underline{\omega},\underline{\varepsilon},2}(\mathbf{x}_1,..,\mathbf{x}_4)\partial^{\alpha_1}\psi^{\varepsilon_1}_{\mathbf{x}_1,\omega_1}\partial^{\alpha_2}\psi^{\varepsilon_2}_{\mathbf{x}_2,\omega_2}\partial^{\alpha_3}\psi^{\varepsilon_3}_{\mathbf{x}_3,\omega_3}\partial^{\alpha_4}\psi^{\varepsilon_4}_{\mathbf{x}_4,\omega_4}$$

The term with $n=1$ can be written as

$$\sum_{\omega}\sum_{\mathbf{x}}[i\omega\bar{\nu}_1\psi^+_{\mathbf{x},\omega}\psi^-_{\mathbf{x},-\omega} + \psi^+_{\mathbf{x},\omega}(i\omega a_1 \partial_0 + a_2\partial_1)\psi^-_{\mathbf{x},\omega}] +$$

$$\sum_{\mathbf{x}_1,\mathbf{x}_2}\sum_{\{\omega\}}\sum_{\alpha_1+\alpha_2 \geq 2, \varepsilon_1, \varepsilon_2} W_{\underline{\alpha},\underline{\omega},1,a}(\mathbf{x}_1,\mathbf{x}_2)\partial^{\alpha_1}\psi^{\varepsilon_1}_{\mathbf{x}_1,\omega_1}\partial^{\alpha_2}\psi^{\varepsilon_2}_{\mathbf{x}_2,\omega_2} \tag{10.30}$$

with $\bar{\nu}_1, a_1 a_2$ real and $|\widehat{W}_{\underline{\alpha},\underline{\omega},1,a}(\mathbf{k}_1)| \leq L^2 C|\lambda|$. Finally as in chap.9

$$\bar{P}(d\psi) = \tag{10.31}$$

$$\mathcal{N}^{-1}\prod_{\mathbf{k}\in\mathcal{D}}\prod_{\omega=\pm 1} d\psi^+_{\mathbf{k},\omega}d\psi^-_{\mathbf{k},\omega}\exp[-\frac{t_\psi Z_1}{L^2}\sum_{\mathbf{k}\in\mathcal{D}}\psi^+_{\mathbf{k},\omega}T^{(1)}_{\omega,\omega'}(\mathbf{k})\psi^-_{\mathbf{k},\omega'}]$$

where $T^{(1)}(\mathbf{k}) =$

$$\begin{pmatrix} \widetilde{Z}_1(i\sin k + \sin k_0) + \mu_{1,1}(\mathbf{k})Z_1^{-1} & -i\bar{m}_1 - i\mu_{1,2}(\mathbf{k})Z_1^{-1} \\ i\bar{m}_1 + i\mu_{1,2}(\mathbf{k})Z_1^{-1} & \widetilde{Z}_1(i\sin k - \sin k_0) + \mu_{2,2}(\mathbf{k})Z_1^{-1} \end{pmatrix}$$

with $\widetilde{Z}_1 = 1 + O(m_\psi)$, $Z_1 = 1$, $\bar{m}_1 = \frac{m_\psi}{1-m_\psi/2}$, $\mu_{i,j}(\mathbf{k})$ analytic functions in $\mathbf{k}$ of size $O(\mathbf{k}^2)$ with $\mu_{i,i}(\mathbf{k})$, $i=1,2$, odd and $\mu_{1,2}(\mathbf{k})$ even and real.

The proof of the above proposition is a repetition of the analysis in the previous chapter. The only difficulty and novelty is that a detailed analysis of the bilinear and quartic terms in $\mathcal{V}^{(1)}$ is necessary. In fact we have to show that the quadratic part can be written as in (10.30), saying that *there are no terms of the form* $\psi^\varepsilon_{\mathbf{x},\omega}\psi^{-\varepsilon}_{\mathbf{x},\omega}$, *or* $\psi^\varepsilon_{\mathbf{x},\omega}\psi^\varepsilon_{\mathbf{x},-\omega}$ *or* $\psi^\varepsilon_{\mathbf{x},\omega}\partial\psi^\varepsilon_{\mathbf{x},-\omega}$; despite the fact that such terms are absent in $\mathcal{V}$, they *could* be generated by the integration of the $\chi$ variables. This is not the case, as a consequence of symmetry properties verified by the model. Such symmetries are quite evident in tthe original spin representaion but quite involved in the fermionic language.

We start with noting that the formal action appearing in (10.13) is invariant under the following transformations.

1) *Parity*:
$$H^{(j)}_{\mathbf{x}} \to \overline{H}^{(j)}_{-\mathbf{x}}, \quad \overline{H}^{(j)}_{\mathbf{x}} \to -H^{(j)}_{-\mathbf{x}}, \quad V^{(j)}_{\mathbf{x}} \to \overline{V}^{(j)}_{-\mathbf{x}}, \quad \overline{V}^{(j)}_{\mathbf{x}} \to -V^{(j)}_{-\mathbf{x}}. \tag{10.32}$$
In terms of the variables $\widehat{\psi}^\alpha_{\omega,\mathbf{k}}$, this transformation is equivalent to $\widehat{\psi}^\alpha_{\omega,\mathbf{k}} \to i\omega\widehat{\psi}^\alpha_{\omega,-\mathbf{k}}$ (the same for $\chi$) and we shall call it *parity*.

2) *Complex conjugation*:
$$\psi^{(j)}_{\mathbf{x}} \to \overline{\psi}^{(j)}_{\mathbf{x}}, \quad \overline{\psi}^{(j)}_{\mathbf{x}} \to \psi^{(j)}_{\mathbf{x}}, \quad \chi^{(j)}_{\mathbf{x}} \to \overline{\chi}^{(j)}_{\mathbf{x}}, \quad \overline{\chi}^{(j)}_{\mathbf{x}} \to \chi^{(j)}_{\mathbf{x}}, \quad c \to c^*, \tag{10.33}$$
where $c$ is a generic constant appearing in the formal action and $c^*$ is its complex conjugate. In terms of the variables $\widehat{\psi}^\alpha_{\omega,\mathbf{k}}$, this transformation is equivalent to $\widehat{\psi}^\alpha_{\omega,\mathbf{k}} \to \widehat{\psi}^{-\alpha}_{-\omega,\mathbf{k}}$ (the same for $\chi$), $c \to c^*$ and we shall call it *complex conjugation*.

3) *Hole-particle*:
$$H^{(j)}_{\mathbf{x}} \to (-1)^{j+1} H^{(j)}_{\mathbf{x}}, \quad \overline{H}^{(j)}_{\mathbf{x}} \to (-1)^{j+1} \overline{H}^{(j)}_{\mathbf{x}},$$
$$V^{(j)}_{\mathbf{x}} \to (-1)^{j+1} V^{(j)}_{\mathbf{x}}, \quad \overline{V}^{(j)}_{\mathbf{x}} \to (-1)^{j+1} \overline{V}^{(j)}_{\mathbf{x}} \tag{10.34}$$
This transformation is equivalent to $\widehat{\psi}^\alpha_{\omega,\mathbf{k}} \to \widehat{\psi}^{-\alpha}_{\omega,-\mathbf{k}}$ (the same for $\chi$) and we shall call it *hole-particle*.

4) *Rotation*:
$$H^{(j)}_{x,x_0} \to i\overline{V}^{(j)}_{-x_0,-x}, \quad \overline{H}^{(j)}_{x,x_0} \to iV^{(j)}_{-x_0,-x}$$
$$V^{(j)}_{x,x_0} \to i\overline{H}^{(j)}_{-x_0,-x}, \quad \overline{V}^{(j)}_{x,x_0} \to iH^{(j)}_{-x_0,-x} \tag{10.35}$$
This transformation is equivalent to
$$\widehat{\psi}^\alpha_{\omega,(k,k_0)} \to -\omega e^{-i\omega\pi/4}\widehat{\psi}^\alpha_{-\omega,(-k_0,-k)}, \quad \widehat{\chi}^\alpha_{\omega,(k,k_0)} \to \omega e^{-i\omega\pi/4}\widehat{\chi}^\alpha_{-\omega,(-k_0,-k)} \tag{10.36}$$

and we shall call it *rotation*.

5) *Reflection*:

$$H^{(j)}_{x,x_0} \to i\overline{H}^{(j)}_{-x,x_0}, \quad \overline{H}^{(j)}_{x,x_0} \to iH^{(j)}_{-x,x_0}$$
$$V^{(j)}_{x,x_0} \to -i\overline{V}^{(j)}_{-x,x_0}, \quad \overline{V}^{(j)}_{x,x_0} \to iV^{(j)}_{-x,x_0} \tag{10.37}$$

This transformation is equivalent to $\widehat{\psi}^\alpha_{\omega,(k,k_0)} \to i\widehat{\psi}^\alpha_{-\omega,(-k,k_0)}$ (the same for $\chi$) and we shall call it *reflection*.

6) *The (1)$\longleftrightarrow$(2) symmetry*. In the $8V$ model we have the symmetry

$$H^{(1)}_{x,x_0} \to H^{(2)}_{x,x_0}, \overline{H}^{(1)}_{x,x_0} \to \overline{H}^{(2)}_{x,x_0}, V^{(1)}_{x,x_0} \to V^{(2)}_{x,x_0}, \overline{V}^{(1)}_{x,x_0} \to \overline{V}^{(2)}_{x,x_0}$$
$$H^{(2)}_{x,x_0} \to H^{(1)}_{x+1,x_0-1} \overline{H}^{(2)}_{x,x_0} \to \overline{H}^{(2)}_{x+1,x_0-1} \tag{10.38}$$
$$V^{(2)}_{x,x_0} \to V^{(1)}_{x+1,x_0-1}, \overline{V}^{(2)}_{x,x_0} \to \overline{V}^{(1)}_{x+1,x_0-1}$$

Let us check explicitly such symmetry on the quartic terms (on the quadratic one it is obvious). The quartic terms in $\mathcal{V}$ are

$$\sum_x [\bar{H}^{(1)}_{x,x_0} H^{(1)}_{x+1,x_0} \bar{V}^{(2)}_x V^{(2)}_{x,x_0+1} + \bar{V}^{(1)}_{x,x_0} V^{(1)}_{x,x_0+1} \bar{H}^{(2)}_{x-1,x_0+1} H^{(2)}_{x,x_0+1}] \tag{10.39}$$

and under the above transformation

$$\sum_x [\bar{H}^{(2)}_{x,x_0} H^{(2)}_{x+1,x_0} \bar{V}^{(1)}_{x+1,x_0-1} V^{(1)}_{x+1,x_0} + \bar{V}^{(2)}_{x,x_0} V^{(2)}_{x,x_0+1} \bar{H}^{(1)}_{x,x_0} H^{(1)}_{x+1,x_0}] \tag{10.40}$$

and making the shift $x, x_0 \to x-1, x_0+1$ we find

$$\sum_x [\bar{H}^{(2)}_{x-1,x_0+1} H^{(2)}_{x,x_0+1} \bar{V}^{(1)}_{x,x_0} V^{(1)}_{x,x_0+1} + \bar{V}^{(2)}_{x,x_0} V^{(2)}_{x,x_0+1} \bar{H}^{(1)}_{x-1,x_0-1} H^{(1)}_{x,x_0+1}] \tag{10.41}$$

from which invariance follows. This symmetry is equivalent to

$$\widehat{\psi}^{(1)}_\mathbf{k} \to \widehat{\psi}^{(2)}_\mathbf{k} \quad \widehat{\psi}^{(2)}_\mathbf{k} \to e^{i(k_0-k)}\widehat{\psi}^{(1)}_\mathbf{k} \tag{10.42}$$

(the same for $\chi$) and we shall call it (1)$\longleftrightarrow$(2) *symmetry*.

In the Ashkin-Teller model the symmetry is

$$H^{(1)}_{x,x_0} \to H^{(2)}_{x,x_0}, \quad \overline{H}^{(1)}_{x,x_0} \to \overline{H}^{(2)}_{x,x_0}, \quad V^{(1)}_{x,x_0} \to V^{(2)}_{x,x_0}, \quad \overline{V}^{(1)}_{x,x_0} \to \overline{V}^{(2)}_{x,x_0}$$
$$H^{(2)}_{x,x_0} \to H^{(1)}_{x,x_0}, \quad \overline{H}^{(2)}_{x,x_0} \to \overline{H}^{(2)}_{x,x_0}, \quad V^{(2)}_{x,x_0} \to V^{(1)}_{x,x_0}, \quad \overline{V}^{(2)}_{x,x_0} \to \overline{V}^{(1)}_{x,x_0}$$

which is equivalent to $\psi^\alpha_{\omega,\mathbf{k}} \to -i\alpha\psi^{-\alpha}_{\omega,\mathbf{k}}$.

It is easy to verify that the quadratic forms $P(d\chi)$, $P(d\psi)$ and $\overline{P}(d\psi)$ are separately invariant under the symmetries above. Then the effective action $\mathcal{V}^{(1)}(\psi)$ is still invariant under the same symmetries. Using the invariance of $\mathcal{V}^{(1)}$ under transformations (1)–(6), we now study in detail the structure of its quadratic and quartic terms.

*Quartic term.* Let us consider in (10.28) the term with $2n = 4$, $\alpha_1 = \alpha_2 = -\alpha_3 = -\alpha_4 = +$, $\omega_1 = -\omega_2 = \omega_3 = -\omega_4 = 1$; for simplicity of notation, let us denote it with

$$\sum_{\mathbf{k}_i} W(\mathbf{k}_1, \mathbf{k}_2, \mathbf{k}_3, \mathbf{k}_4) \widehat{\psi}^+_{1,\mathbf{k}_1} \widehat{\psi}^+_{-1,\mathbf{k}_2} \widehat{\psi}^-_{-1,\mathbf{k}_3} \widehat{\psi}^-_{1,\mathbf{k}_4} \delta(\mathbf{k}_1 + \mathbf{k}_2 - \mathbf{k}_3 - \mathbf{k}_4) \quad (10.43)$$

Under complex conjugation it becomes equal to

$$\sum_{\mathbf{k}_i} W^*(\mathbf{k}_1, \mathbf{k}_2, \mathbf{k}_3, \mathbf{k}_4) \widehat{\psi}^-_{-1,\mathbf{k}_1} \widehat{\psi}^-_{1,\mathbf{k}_2} \widehat{\psi}^+_{1,\mathbf{k}_3} \widehat{\psi}^+_{-1,\mathbf{k}_4} \delta(\mathbf{k}_3 + \mathbf{k}_4 - \mathbf{k}_1 - \mathbf{k}_2) \quad (10.44)$$

so that $W(\mathbf{k}_1, \mathbf{k}_2, \mathbf{k}_3, \mathbf{k}_4) = W^*(\mathbf{k}_3, \mathbf{k}_4, \mathbf{k}_1, \mathbf{k}_2)$.

*Quadratic terms.* We distinguish 4 cases (items (a)–(d) below).
a) Let us consider in (10.28) the term with $2n = 2$, $\alpha_1 = -\alpha_2 = +$ and $\omega_1 = -\omega_2 = \omega$; let us denote it with $\sum_{\omega,\mathbf{k}} W_\omega(\mathbf{k}) \widehat{\psi}^+_{\omega,\mathbf{k}} \widehat{\psi}^-_{-\omega,\mathbf{k}}$. Under parity it becomes

$$\sum_{\omega,\mathbf{k}} W_\omega(\mathbf{k})(i\omega)\widehat{\psi}^+_{\omega,-\mathbf{k}}(-i\omega)\widehat{\psi}^-_{-\omega,-\mathbf{k}} = \sum_{\omega,\mathbf{k}} W_\omega(-\mathbf{k}) \widehat{\psi}^+_{\omega,\mathbf{k}} \widehat{\psi}^-_{-\omega,\mathbf{k}}, \quad (10.45)$$

so that $W_\omega(\mathbf{k})$ is even in $\mathbf{k}$.
Under complex conjugation it becomes

$$\sum_{\omega,\mathbf{k}} W_\omega(\mathbf{k})^* \widehat{\psi}^-_{-\omega,\mathbf{k}} \widehat{\psi}^+_{\omega,\mathbf{k}} = -\sum_{\omega,\mathbf{k}} W_\omega(\mathbf{k})^* \widehat{\psi}^+_{\omega,\mathbf{k}} \widehat{\psi}^-_{-\omega,\mathbf{k}}, \quad (10.46)$$

so that $W_\omega(\mathbf{k})$ is purely imaginary.
Under hole-particle it becomes

$$\sum_{\omega,\mathbf{k}} W_\omega(\mathbf{k}) \widehat{\psi}^-_{\omega,-\mathbf{k}} \widehat{\psi}^+_{-\omega,-\mathbf{k}} = -\sum_{\omega,\mathbf{k}} W_{-\omega}(\mathbf{k}) \widehat{\psi}^+_{\omega,\mathbf{k}} \widehat{\psi}^-_{-\omega,\mathbf{k}}, \quad (10.47)$$

so that $W_\omega(\mathbf{k})$ is odd in $\omega$.

b) The term $\sum_{\omega,\alpha,\mathbf{k}} W^\alpha_\omega(\mathbf{k}) \widehat{\psi}^\alpha_{\omega,\mathbf{k}} \widehat{\psi}^\alpha_{-\omega,-\mathbf{k}}$ is such that $W^\alpha_\omega(\mathbf{k})$ is even in $\alpha$ and $\mathbf{k}$. Using hole-particle becomes

$$\sum_{\omega,\alpha,\mathbf{k}} W^\alpha_\omega(\mathbf{k}) \widehat{\psi}^\alpha_{\omega,\mathbf{k}} \widehat{\psi}^\alpha_{-\omega,-\mathbf{k}} = \sum_{\omega,\alpha,\mathbf{k}} W^\alpha_\omega(\mathbf{k}) \widehat{\psi}^{-\alpha}_{\omega,-\mathbf{k}} \widehat{\psi}^{-\alpha}_{-\omega,\mathbf{k}} \quad (10.48)$$

and using $(1)\longleftrightarrow(2)$ symmetry becomes in the AT case

$$\sum_{\omega,\alpha,\mathbf{k}} W_\omega^\alpha(\mathbf{k})\widehat{\psi}^{-\alpha}_{\omega,\mathbf{k}}\widehat{\psi}^{-\alpha}_{-\omega,-\mathbf{k}} = -\sum_{\omega,\alpha,\mathbf{k}} W_\omega^\alpha(\mathbf{k})\widehat{\psi}^{-\alpha}_{\omega,-\mathbf{k}}\widehat{\psi}^{-\alpha}_{-\omega,\mathbf{k}} \qquad (10.49)$$

so that $W_\omega^\alpha(\mathbf{k}) = 0$. In the 8V case, we note that $\widehat{\psi}^-_{+,\mathbf{k}}\widehat{\psi}^-_{-,-\mathbf{k}} = \frac{1}{2}[\widehat{\psi}^{(1)}_\mathbf{k}\bar{\psi}^{(1)}_{-\mathbf{k}} - \widehat{\psi}^{(2)}_\mathbf{k}\bar{\psi}^{(2)}_{-\mathbf{k}}] + [\widehat{\psi}^{(1)}_\mathbf{k}\bar{\psi}^{(2)}_{-\mathbf{k}} + \widehat{\psi}^{(2)}_\mathbf{k}\bar{\psi}^{(1)}_{-\mathbf{k}}]$, and the second term violates the hole-particle symmetry, while the first violates the $(1)\longleftrightarrow(2)$ symmetry.

c) Let us consider in (10.28) the term with $2n = 2$, $\alpha_1 = -\alpha_2 = +$, $\omega_1 = \omega_2 = \omega$ and let us denote it with $\sum_{\omega,\mathbf{k}} W_\omega(\mathbf{k})\widehat{\psi}^+_{\omega,\mathbf{k}}\widehat{\psi}^-_{\omega,\mathbf{k}}$. By using parity it becomes

$$\sum_{\omega,\mathbf{k}} W_\omega(\mathbf{k})\widehat{\psi}^+_{\omega,\mathbf{k}}\widehat{\psi}^-_{\omega,\mathbf{k}} = -\sum_{\omega,\mathbf{k}} W_\omega(\mathbf{k})\widehat{\psi}^+_{\omega,-\mathbf{k}}\widehat{\psi}^-_{\omega,-\mathbf{k}} , \qquad (10.50)$$

so that $W_\omega(\mathbf{k})$ is odd in $\mathbf{k}$.

d) The term $\sum_{\omega,\alpha,\mathbf{k}} W_\omega^\alpha(\mathbf{k})\widehat{\psi}^\alpha_{\omega,\mathbf{k}}\widehat{\psi}^\alpha_{\omega,-\mathbf{k}}$ is forbidden for the same reasons as in b).

## 10.5 Integration of the light fermions

The integration of the light fermions is essentially identical to the integration of the Wilson fermions in chapt. 7; indeed the analysis in the previous section ensures that the only terms with non negative dimension are

$$\psi^+_{\mathbf{x},\omega}\psi^-_{\mathbf{x},-\omega}; \qquad \psi^+_{\mathbf{x},\omega}\partial\psi^-_{\mathbf{x},\omega}; \qquad \psi^+_{\mathbf{x},\omega}\psi^-_{\mathbf{x},\omega}\psi^+_{\mathbf{x},-\omega}\psi^-_{\mathbf{x},-\omega} \qquad (10.51)$$

As in chapt.9, we write $\bar{P}(d\psi)$ as $P(d\psi^{(\le 0)})e^{-\nu F_\sigma(\psi^{\le 0})}$ with $F_\sigma = \frac{1}{L^2}\sum_{\mathbf{k}\in\mathcal{D}}\sum_{\omega=\pm 1} i\omega\widehat{\psi}^+_{\mathbf{k},\omega}\widehat{\psi}^-_{\mathbf{k},-\omega}$, and after the integration of $\psi^{(0)},...,\psi^{(h+1)}$ one arrives to an expression of the form

$$\int P_{Z_h,m_h}(d\psi^{(\le h)}) e^{-\mathcal{V}^{(h)}(\sqrt{Z_h}\psi^{(\le h)}) - L^2 E_h} , \qquad \mathcal{V}^{(h)}(0) = 0 , \qquad (10.52)$$

and

$$P_{Z_h,m_h}(d\psi^{(\le h)}) = \mathcal{N}_h^{-1} \prod_{\mathbf{k}} \prod_{\omega=\pm 1} \qquad (10.53)$$

$$d\psi^{(\le h)+}_{\mathbf{k},\omega} d\psi^{(\le h)-}_{\mathbf{k},\omega} \exp[-\frac{t_\psi Z_h}{L^2} \sum_{\mathbf{k}\in D_{-,-}} C_h(\mathbf{k})^{-1}\psi^{(\le h)+}_{\mathbf{k},\omega} T^{(h)}_{\omega,\omega'}(\mathbf{k})\psi^{(\le h)-}_{\mathbf{k},\omega'}]$$

where $T^{(h)}(\mathbf{k}) =$

$$\begin{pmatrix} t_\psi \widetilde{Z}_1(i\sin k + \sin k_0) + \mu_{1,1}(\mathbf{k})Z_h^{-1} & -im_h - i\mu_{1,2}(\mathbf{k})Z_h^{-1} \\ im_h + i\mu_{1,2}(\mathbf{k})Z_h^{-1} & \widetilde{Z}_1(i\sin k - \sin k_0) + \mu_{2,2}(\mathbf{k})Z_h^{-1} \end{pmatrix}$$

and $m_1 = \frac{m_\psi - \bar{\nu}}{1 - m_\psi/2}$. The localization operator is defined as in chapt. 7 and we get

$$\mathcal{L}\mathcal{V}^{(h)}(\psi^{(\leq h)}) = (s_h + \gamma^h n_h)F_\sigma^{(\leq h)} + l_h F_\lambda^{(\leq h)} + z_h F_\zeta^{(\leq h)}, \qquad (10.54)$$

where $s_1, z_1, a_1 = O(\lambda)$, $l_1 = (\widehat{\lambda a} + \widehat{\lambda b})\operatorname{sech}^4 \beta J + O(\lambda^2)$, $\nu_1 = \nu + O(\lambda)$ and

$$F_\sigma^{(\leq h)} = \frac{1}{L^2} \sum_{\mathbf{k} \in \mathcal{D}} \sum_{\omega = \pm 1} i\omega \widehat{\psi}_{\mathbf{k},\omega}^{(\leq h)+} \widehat{\psi}_{\mathbf{k},-\omega}^{(\leq h)-}$$

$$F_\lambda^{(\leq h)} = \frac{1}{L^8} \sum_{\mathbf{k}_1,\ldots,\mathbf{k}_4 \in \mathcal{D}} \widehat{\psi}_{\mathbf{k}_1,+1}^{(\leq h)+} \widehat{\psi}_{\mathbf{k}_2,-1}^{(\leq h)+} \widehat{\psi}_{\mathbf{k}_3,-1}^{(\leq h)-} \widehat{\psi}_{\mathbf{k}_4,+1}^{(\leq h)-} \delta(\mathbf{k}_1 - \mathbf{k}_2 + \mathbf{k}_3 - \mathbf{k}_4)$$

$$F_\zeta^{(\leq h)} = \frac{1}{L^2} \sum_{\mathbf{k} \in \mathcal{D}} \sum_{\omega = \pm 1} (i\sin k + \omega \sin k_0) \widehat{\psi}_{\mathbf{k},\omega}^{(\leq h)+} \widehat{\psi}_{\mathbf{k},\omega}^{(\leq h)-} \qquad (10.55)$$

where $\delta(\mathbf{k}) = 0$ if $\mathbf{k} \neq \mathbf{0}$ and $\delta(\mathbf{0}) = 1$. By putting $s_h F_\sigma^{(\leq h)} + z_h F_\zeta^{(\leq h)}$ in the free measure and rescaling the fields, as in chapt 4, we can rewrite (10.54) as

$$\int P_{Z_{h-1},m_{h-1}}(d\psi^{(\leq h)}) e^{-\widehat{\mathcal{V}}^{(h)}(\sqrt{Z_{h-1}}\psi^{(\leq h)}) - L^2 E_h}, \quad \mathcal{V}^{(h)}(0) = 0, \qquad (10.56)$$

where

$$\mathcal{L}\widehat{\mathcal{V}}^{(h)}(\psi^{(\leq h)}) = \gamma^h \nu_h F_\sigma^{(\leq h)} + \lambda_h F_\lambda^{(\leq h)}, \qquad (10.57)$$

After integrating the field $\psi^{(h)}$ we get an expression of the form (11.31) and the procedure can be iterated.

By repeating a fixed point argument similar to the one in the previous chapter we get that it is possible to choose $\nu$ so that $|\nu_h| \leq B|\lambda|\gamma^{\vartheta h}$.

We now consider the equation

$$\lambda_{h-1} = \lambda_h + \beta_\lambda^h(\lambda_h, \nu_h; \ldots; \lambda_1, \nu_1) \qquad (10.58)$$

Note that the analogue of (7.26) holds, namely

$$g_{\omega,\omega}^{(h)}(\mathbf{x} - \mathbf{y}) = g_{L,\omega}^{(h)}(\mathbf{x} - \mathbf{y}) + r_\omega^{(h)}(\mathbf{x} - \mathbf{y}) \qquad (10.59)$$

where

$$g_{L,\omega}^{(h)}(\mathbf{x} - \mathbf{y}) = \frac{1}{L^2} \sum_{\mathbf{k}} e^{-i\mathbf{k}(\mathbf{x} - \mathbf{y})} \widetilde{f}_h(\mathbf{k}) \frac{1}{ik + \omega k_0} \qquad (10.60)$$

and $r_\omega^{(h)}$ is the rest, satisfying the same bound as $g_{\omega,\omega}^{(h)}$, times a factor $\gamma^h$. This means that the propagator can be written to the same propagator appearing in QED2 plus a small correction; the above decomposition induces the following decomposition for $\beta_\lambda^h$:

$$\beta_\lambda^h(\lambda_h,\nu_h;\ldots;\lambda_1,\nu_1) = \beta_{\lambda,L,h}(\lambda_h,\ldots,\lambda_h) + $$
$$\sum_{k=h+1}^{1} D_\lambda^{h,k} + r_\lambda^h(\lambda_h,\ldots,\lambda_1) + \sum_{k \geq h} \nu_k \widetilde{\beta}_\lambda^{h,k}(\lambda_k,\nu_k;\ldots;\lambda_1,\nu_1) \quad (10.61)$$

where $\beta_{\lambda,L,h}$ collect the contributions obtained by trees with end-points associated to $\mathcal{LV}^h$, posing $r_\omega^{(k)} = 0$ and substituting the discrete $\delta$ function with $L^2 \delta_{\mathbf{k},0}$; hence $\beta_{\lambda,L,h}(\lambda_h,\ldots,\lambda_h)$ identical to (4.76), with $K=0$, so that

$$|\beta_{\lambda,L,h}| \leq c|\lambda|^2 \gamma^{\vartheta h}, \quad |D_\lambda^{h,k}| \leq c|\lambda|\gamma^{\vartheta(h-k)}|\lambda_k - \lambda_h|,$$
$$|r_\lambda^h| \leq c|\lambda|^2 \gamma^{\vartheta h}, \quad |\widetilde{\beta}_\lambda^{h,k}| \leq c|\lambda|\gamma^{\vartheta(h-k)} \quad (10.62)$$

Note that the first of (10.62) follows from Theorem 6.1 with $K=0$; this means that the flow of the running coupling constants can be controlled in such systems by the gauge symmetries hidden in such models.

An immediate consequence of (10.62) is that, by using (4.84) and by induction,

$$|\lambda_{h-1} - \lambda_h| \leq C\lambda^2 \gamma^{\vartheta h} + C\lambda^2 \sum_{k=h+1}^{1} \gamma^{-2\vartheta(k-h)} \gamma^{\vartheta h'} \leq C|\lambda|\gamma^{\vartheta(h-1)} \quad (10.63)$$

implying

$$\lambda_h = \lambda + O(\lambda^2) \quad Z_h = \gamma^{\eta h}(1+O(\lambda)) \quad \mu_h = \gamma^{\eta_\mu h}(1+O(\lambda)) \quad (10.64)$$

with $\eta = O(\lambda^2), \eta_\mu = O(\lambda)$, and the analyisis is essentially identical to the one in chapt.4.

## 10.6 The specific heat

As in chapt. 9 the specific heat can be written as

$$C_v = \frac{1}{\Lambda} \sum_{\mathbf{x},\mathbf{y}} \Omega_{\underline{\varepsilon},\Lambda}(\mathbf{x}-\mathbf{y}) \quad (10.65)$$

with

$$\Omega_{\underline{\varepsilon},\Lambda}(\mathbf{x}-\mathbf{y}) = \frac{\partial}{\partial \phi(\mathbf{x})} \frac{\partial}{\partial \phi(\mathbf{y})} \mathcal{S}_{\underline{\varepsilon}}(\phi)|_{\phi=0} \quad (10.66)$$

where, with the notation of (11.8)

$$e^{S_{\varepsilon}(\phi)} = \int P(d\psi) \int P(d\chi) e^{Q(\chi,\psi) - \mathcal{V}(\psi,\chi)} e^{\sum_{\mathbf{x}} \phi(\mathbf{x})[\partial_t S^1_{\mathbf{x},\varepsilon 1,\varepsilon' 1} + \partial_t S^2_{\mathbf{x},\varepsilon 2,\varepsilon' 2}]}.$$
(10.67)

Let us consider $S_{-,-,-,-}(\phi)$. One can integrate the massive $\chi$ fields and one finds, for $|\lambda| \leq \varepsilon$

$$e^{S(\phi)} = e^{L^2 \mathcal{N}} \int \overline{P}(d\psi) e^{-\mathcal{V}^{(1)}(\psi) + \mathcal{B}(\phi,\psi)}$$
(10.68)

where $\mathcal{N}$ is a normalization constant and

$$\mathcal{B}(\psi,\phi) = \sum_{m=1}^{\infty} \sum_{n=1}^{\infty} \sum_{\underline{\varepsilon},\underline{\alpha},\underline{\omega}} \sum_{\mathbf{x}_1} \cdots \sum_{\mathbf{x}_m} \sum_{\mathbf{y}_1} \cdots \sum_{\mathbf{y}_{2n}}$$
(10.69)

$$B_{m,2n,\underline{\varepsilon},\underline{\alpha},\underline{\omega}}(\mathbf{x}_1,\ldots,\mathbf{x}_m;\mathbf{y}_1,\ldots,\mathbf{y}_{2n}) \left[ \prod_{i=1}^{m} \phi(\mathbf{x}_i) \right] \left[ \prod_{i=1}^{2n} \partial^{\alpha_i} \psi^{\varepsilon_i}_{\mathbf{y}_i,\omega_i} \right]$$

where for $n \geq 2$

$$\sum_{\mathbf{y}_1,\ldots,\mathbf{y}_{2n}} |B_{m,2n,\underline{\alpha},\underline{\omega}}(\mathbf{x}_1,\ldots,\mathbf{x}_m;\mathbf{y}_1,\ldots,\mathbf{y}_{2n})| \leq C^n \varepsilon^{\frac{n}{2}}$$
(10.70)

and for $n = 1$

$$\sum_{\mathbf{x}} i\omega \phi(\mathbf{x}) \psi^+_{\mathbf{x},\omega} \psi^-_{\mathbf{x},-\omega} +$$
(10.71)

$$\sum_{\mathbf{y}_1,\mathbf{y}_2} \sum_{\mathbf{x}} \sum_{\{\varepsilon,\omega\}} \sum_{\alpha_1+\alpha_2 \geq 1} B_{1,\underline{\alpha},\underline{\omega}}(\mathbf{x};\mathbf{y}_1,\mathbf{y}_2) \phi(\mathbf{x}) \partial^{\alpha_1} \psi^{\varepsilon_1}_{\mathbf{y}_1,\omega_1} \partial^{\alpha_2} \psi^{\varepsilon_2}_{\mathbf{y}_2,\omega_2} + \widetilde{B}(\phi,\psi)$$

where $\sum_{\mathbf{y}_1,\mathbf{y}_2} |B_{1,2,\underline{\varepsilon},\underline{\alpha},\underline{\omega}}(\mathbf{x};\mathbf{y}_1,\mathbf{y}_2)| \leq C$ and $\widetilde{B}(\phi,\psi)$ contains the terms with $m \geq 2$. The symmetry considerations imply that the only possible local terms with $n = m = 1$ are of the form $\phi(\mathbf{x}) \psi^+_{\mathbf{x},1} \psi^-_{\mathbf{x},-1}$.

We write $\Omega_\Lambda(\mathbf{x},\mathbf{y}) = \Omega^a_\Lambda(\mathbf{x},\mathbf{y}) + \Omega^b_\Lambda(\mathbf{x},\mathbf{y})$, where the $\Omega^a_\Lambda(\mathbf{x},\mathbf{y})$ is given by the sum over trees belonging to $\mathcal{T}^2_{h,n}$ with endpoints $v$ to which are associated one of the terms in $\mathcal{LV}^{(h_v-1)}$ or $\mathcal{LB}^{(h_v-1)}$), and $\Omega^b_\Lambda(\mathbf{x},\mathbf{y})$ is the sum over the remaining trees. We can single out from $\Omega^a_\Lambda(\mathbf{x},\mathbf{y})$ the contribution from the trees with $n = 0$ so that

$$\Omega^a_\Lambda(\mathbf{x},\mathbf{y}) = \sum_{h,h'=h^*}^{1} \sum_{\omega=\pm 1} \left\{ \frac{(Z^{(1)}_{h \vee h'})^2}{Z_{h-1} Z_{h'-1}} [g^{(h)}_{\omega,\omega}(\mathbf{x}-\mathbf{y}) g^{(h')}_{-\omega,-\omega}(\mathbf{y}-\mathbf{x}) - \right.$$

$$\left. g^{(h)}_{+1,-1}(\mathbf{x}-\mathbf{y}) g^{(h')}_{-1,+1}(\mathbf{y}-\mathbf{x})] \right\} + \sum_{h=h^*}^{1} \left( \frac{Z^{(1)}_h}{Z_h} \right)^2 G^{(h),a}_\Lambda(\mathbf{x},\mathbf{y})$$
(10.72)

where $h \vee h' = \max\{h, h'\}$ and $g_{\omega_1,\omega_2}^{(h^*)}(\mathbf{x})$ has to be understood as $g_{\omega_1,\omega_2}^{(\leq h^*)}(\mathbf{x})$; moreover $(\frac{Z_h^{(1)}}{Z_h})^2 G_\Lambda^{(h)}(\mathbf{x})$ is given by the sum of trees with $n \geq 1$.

It holds that for $\lambda$ small enough, for any $N$ there exist a constant $N$ such that

$$|\partial_x^{m_1} \partial_{x_0}^{m_0} G_\Lambda^{(h),a}(\mathbf{x},\mathbf{y})| \leq \gamma^{(2+m_0+m_1)h}|\lambda_1|\frac{C_N}{1+(\gamma^h|\mathbf{x}-\mathbf{y}|)^N} \,. \qquad (10.73)$$

Moreover

$$Z_h^{(1)} = \gamma^{\eta_1 h}(1+O(\lambda)) \qquad (10.74)$$

with $\eta = a_1\lambda + O(\lambda^2)$. For $\Omega_\Lambda^b(\mathbf{x},\mathbf{y})$ the following bound holds

$$|\partial_x^{m_1} \partial_{x_0}^{m_0} \Omega_\Lambda^b(\mathbf{x},\mathbf{y})| \leq \sum_{h=h^*}^{1} \gamma^{(2+m_0+m_1+\tau)h}\frac{C_N}{1+(\gamma^h|\mathbf{x}-\mathbf{y}|)^N} \,, \qquad (10.75)$$

where $0 < \vartheta < 1$ is a constant; the extra factor $\gamma^{\vartheta h}$ in (10.75) (with respect to (10.73)) is due to the fact that the bound over all the trees which have at least one endpoint $v$ of fixed scale $h_v = 2$ can be improved by a factor $\gamma^{\vartheta h}$, by the short memory property. Proceeding as in §4.7, Theorem 10.1 follows.

# Chapter 11

# Universality-Nonuniversality Crossover in the Ashkin-Teller Model

## 11.1 The anisotropic AT model

We have considered in the previous chapter the isotropic Ashkin-Teller model and we have shown that its critical properties are different with respect to the ones of the Ising model; for instance the singularity of the specific heat is power-like instead as logarithmic. In this chapter we consider the *Anisotropic Ashkin Teller model*, whose hamiltonian can be written as

$$H_\Lambda(\sigma^{(1)}, \sigma^{(2)}) = J^{(1)} H_I(\sigma^{(1)}) + J^{(2)} H_I(\sigma^{(2)}) + V(\sigma^{(1)}, \sigma^{(2)}) \quad (11.1)$$

with $J^{(1)} \neq J^{(2)}$. It is convenient to introduce the variables $t^{(j)} = \tanh \beta J^{(j)}$, $j = 1, 2$ and

$$t = \frac{t^{(1)} + t^{(2)}}{2} \quad , \quad u = \frac{t^{(1)} - t^{(2)}}{2}. \quad (11.2)$$

The parameter $u$ measures the *anisotropy* of the system. We consider then the free energy or the specific heat as function of $t, u, J^{(3)}$.

In the $\lambda \equiv J^{(3)} = 0$ case, the model is exactly solvable as its hamiltonian is the sum of two indipendent Ising models hamiltonians. The model has *two* critical temperatures

$$t = t_c^\pm = \sqrt{2} - 1 \pm |u|. \quad (11.3)$$

and for $t$ close to $t_c^\pm$ the specific heat $C_v$ has a logarithmic divergence, namely if $0 < |t - t_c^\pm| \leq |u|/4$:

$$C_v = -C \log |t - t_c^\pm| (1 + f_0^\pm(t, u)) \quad (11.4)$$

where $C > 0$ is an $O(1)$ constant and $f_0^\pm(t, u)$ is a bounded function of $t, u$, vanishing for $t = t_c^\pm$. On the other hand when $u = 0$ the model has a single critical temperature and it reduces to the case treated in the previous chapter.

We consider now the case in which $\lambda$ is small with respect to $J^{(1)}, J^{(2)}$ (note anyway that the model is invariant under the permutation of $J^{(1)}, J^{(2)}, J^{(3)}$). When the difference between $J^{(1)}$ and $J^{(2)}$ is large the two critical points are quite separated, hence when one Ising model is critical the other is not, and one expects that the system is equivalent to a perturbed Ising model, and that the presence of $J^{(3)}$ at most change the location of the critical point. On the contrary when the difference between $J^{(1)}$ and $J^{(2)}$ (or equivalently the anysotropy $u$) is small the two Ising models become critical almost at the same point, and one expects that the interaction between the two systems has a much more dramatic effect.

As in the previous chapter, the partition function of the model can be written as a sum of sixteen partition functions labeled by $\gamma_1, \gamma_2 = (\varepsilon_1, \varepsilon'_1), (\varepsilon_2, \varepsilon'_2)$ (corresponding to choosing each $\varepsilon_j$ and $\varepsilon'_j$ as $\pm$):

$$Z_{AT} = \frac{1}{4}(\cosh \beta J^{(3)})^{2L^2} \sum_{\gamma_1, \gamma_2} (-1)^{\delta_{\gamma_1}+\delta_{\gamma_2}} Z_{AT}^{\gamma_1, \gamma_2}, \qquad (11.5)$$

each of which is given by a functional integral

$$Z_{AT}^{\gamma_1, \gamma_2} = [4(1+\widehat{\lambda} t^{(1)} t^{(2)})]^{L^2} \prod_{j=1}^{2} (\cosh \beta J^{(j)})^{L^2} (-1)^{L^2}$$

$$\int \prod_{\mathbf{x}\in\Lambda}^{j=1,2} dH_{\mathbf{x}}^{(j)} d\overline{H}_{\mathbf{x}}^{(j)} dV_{\mathbf{x}}^{(j)} d\overline{V}_{\mathbf{x}}^{(j)} \, e^{S_{\gamma_1}^{(1)}(t_\lambda^{(1)})+S_{\gamma_2}^{(2)}(t_\lambda^{(2)})+V_\lambda} \qquad (11.6)$$

where $S_{\gamma_1}^{(1)}$ is given by (9.10) and, if we define:

$$\lambda^{(1)} = \frac{\widehat{\lambda}[1-(t^{(1)})^2]t^{(2)}}{1+\widehat{\lambda} t^{(1)} t^{(2)}}, \qquad \lambda^{(2)} = \frac{\widehat{\lambda}(1-(t^{(2)})^2)t^{(1)}}{1+\widehat{\lambda} t^{(1)} t^{(2)}}$$

$$\lambda = \frac{\widehat{\lambda}(1-(t^{(1)})^2)(1-(t^{(2)})^2)}{(1+\widehat{\lambda} t^{(1)} t^{(2)})^2}, \qquad (11.7)$$

where $\widehat{\lambda} = \tanh \beta \lambda$ and $t_\lambda^{(j)}$ is given by $t_\lambda^{(j)} = t^{(j)} + \lambda^{(j)}$ and $V_\lambda$ by:

$$V_\lambda = \sum_{\mathbf{x}\in\Lambda_M} \lambda \left( \overline{H}_{\mathbf{x}}^{(1)} H_{\mathbf{x}+\widehat{e}_1}^{(1)} \overline{H}_{\mathbf{x}}^{(2)} H_{\mathbf{x}+\widehat{e}_1}^{(2)} + \overline{V}_{\mathbf{x}}^{(1)} V_{\mathbf{x}+\widehat{e}_0}^{(1)} \overline{V}_{\mathbf{x}}^{(2)} V_{\mathbf{x}+\widehat{e}_0}^{(2)} \right). \qquad (11.8)$$

We shall study in detail only the partition function $Z_{AT}^{-} = Z_{AT}^{(-,-),(-,-)}$, i.e. the partition function in which all Grassmannian variables verify antiperiodic boundary conditions, as if $(\lambda, t, u)$ does not belong to the *critical surface*, the partition function $Z_{AT}^{\gamma_1, \gamma_2}$ divided by $Z_I^{(1)\gamma_1} Z_I^{(2)\gamma_2}$ is exponentially insensitive to boundary conditions as $L \to \infty$. Proceeding as in chapt.10

$$Z_{AT}^{-} = e^{-EL^2} \int P(d\psi) P(d\chi) e^{Q(\psi,\chi)+V(\psi,\chi)}, \qquad (11.9)$$

where: $E$ is a suitable constant; $Q(\psi, \chi)$ collects the quadratic terms of the form $\psi^{\alpha_1}_{\omega_1,\mathbf{x}_1}\chi^{\alpha_2}_{\omega_2,\mathbf{x}_2}$; $V(\psi,\chi)$ is the quartic interaction (it is equal to $V_\lambda$, see (11.8), in terms of the $\psi^\pm_\omega$, $\chi^\pm_\omega$ variables); $P(d\phi)$, $\phi = \psi, \chi$, is:

$$P(d\phi) = \mathcal{N}_\phi^{-1} \prod_{\mathbf{k}\in\mathcal{D}}\prod_{\omega=\pm 1} d\phi^+_{\mathbf{k},\omega}d\phi^-_{\mathbf{k},\omega}\exp\left\{-\frac{t_\lambda}{4L^2}\sum_{\mathbf{k}\in D_{-,-}}\mathbf{\Phi}^{+,\mathbf{T}}_{\mathbf{k}}A_\phi(\mathbf{k})\mathbf{\Phi}_{\mathbf{k}}\right\}$$
(11.10)

where $A_\phi(\mathbf{k}) =$

$$\begin{pmatrix} i\sin k + \sin k_0 & -i\sigma_\phi(\mathbf{k}) & -\frac{\mu}{2}(i\sin k + \sin k_0) & i\mu(\mathbf{k}) \\ i\sigma_\phi(\mathbf{k}) & i\sin k - \sin k_0 & -i\mu(\mathbf{k}) & -\frac{\mu}{2}(i\sin k - \sin k_0) \\ -\frac{\mu}{2}(i\sin k + \sin k_0) & i\mu(\mathbf{k}) & i\sin k + \sin k_0 & -i\sigma_\phi(\mathbf{k}) \\ -i\mu(\mathbf{k}) & -\frac{\mu}{2}(i\sin k - \sin k_0) & i\sigma_\phi(\mathbf{k}) & i\sin k - \sin k_0 \end{pmatrix}$$

where

$$\mathbf{\Phi}^{+,\mathbf{T}}_{\mathbf{k}} = (\widehat{\phi}^+_{1,\mathbf{k}}, \widehat{\phi}^+_{-1,\mathbf{k}}, \widehat{\phi}^-_{1,-\mathbf{k}}, \widehat{\phi}^-_{-1,-\mathbf{k}}) \quad, \quad \mathbf{\Phi}^{\mathbf{T}}_{\mathbf{k}} = (\widehat{\phi}^-_{1,\mathbf{k}}, \widehat{\phi}^-_{-1,\mathbf{k}}, \widehat{\phi}^+_{1,-\mathbf{k}}, \widehat{\phi}^+_{-1,-\mathbf{k}}),$$
(11.11)

$\mathcal{N}_\phi$ is chosen in such a way that $\int P(d\phi) = 1$ and, if we define:

$$t_\lambda = \frac{t^{(1)}_\lambda + t^{(2)}_\lambda}{2} \quad , \quad u_\lambda = \frac{t^{(1)}_\lambda - t^{(2)}_\lambda}{2}$$
(11.12)
$$c(\mathbf{k}) = \cos k_0 + \cos k - 2 \quad , \quad d(\mathbf{k}) = (u_\lambda/t_\lambda)(2 - \cos k - \cos k_0)$$

we have

$$\sigma_\psi(\mathbf{k}) = 2\left(1 - \frac{\sqrt{2}-1}{t_\lambda}\right) + c(\mathbf{k}) \quad , \quad \sigma_\chi(\mathbf{k}) = 2\left(1 + \frac{\sqrt{2}+1}{t_\lambda}\right) + c(\mathbf{k})$$
$$\mu = -2\frac{u_\lambda}{t_\lambda} \quad , \quad \mu(\mathbf{k}) = \mu + d(\mathbf{k})$$
(11.13)

Note that there are, contrary to the cases treated previously, two mass terms, namely $\widehat{\phi}^+_{\omega,\mathbf{k}}\widehat{\phi}^-_{-\omega,\mathbf{k}}$ and $\widehat{\phi}^\varepsilon_{\omega,\mathbf{k}}\widehat{\phi}^\varepsilon_{\omega,-\mathbf{k}}$; this second mass term is absent in the isotropic case.

## 11.2 Anomalous universality

As we are interested in the critical behaviour, we exclude the extremely high and low temperature region *i.e.* $J^{(1)}$ or $J^{(2)} \gg O(1)$ and $J^{(1)}$ or $J^{(2)} \ll O(1)$; this assumption is done for semplicity and we could easily remove it. We will require then that, given $m \in \mathbb{R}^+$, $(t,u) \in D_m$ where

$$D_m = \{(t,u) \in \mathbb{R}^2 : 10^{-m} \leq t - u \leq 10^m, \quad 10^{-m} \leq t + u \leq 10^m\}$$
(11.14)

In this chapter we will prove that following result (originally proved in Refs.[49; 50], to which we refer for details).

**Theorem 11.1.** *The AT model admits two critical points of the form:*

$$t_c^\pm(\lambda, u) = \sqrt{2} - 1 + \nu(\lambda) \pm |u|^{1+\eta}(1 + \delta(\lambda, u)) . \quad (11.15)$$

*Here $\nu$ and $\delta$ are $O(\lambda)$ corrections and $\eta = -b\lambda + O(\lambda^2)$ with $b > 0$. If $t \neq t_c^\pm$ the free energy of the model is analytic in $\lambda, t, u$ and the specific heat $C_v$ is equal to:*

$$-F_1 \Delta^{2\eta_c} \log \frac{|t - t_c^-| \cdot |t - t_c^+|}{\Delta^2} + F_2 \frac{1 - \Delta^{2\eta_c}}{\eta_c} + F_3 , \quad (11.16)$$

*where: $2\Delta^2 = (t - t_c^-)^2 + (t - t_c^+)^2$; $\eta_c = a\lambda + O(\lambda^2)$, $a \neq 0$; and $F_1$, $F_2$, $F_3$ are functions of $t, u, \lambda$, bounded above and below by $O(1)$ constants.*

First note that the location of the critical points is dramatically changed by the interaction. The difference of the interacting critical temperatures normalized with the free one $G(\lambda, u) \equiv (t_c^+(\lambda, u) - t_c^-(\lambda, u))/(t_c^+(0, u) - t_c^-(0, u))$ rescales with the anisotropy parameter as a power law $\sim |u|^\eta$, and in the limit $u \to 0$ it vanishes or diverges, depending on the sign of $\lambda$ (this is because $\eta = -b\lambda + O(\lambda^2)$, with $b > 0$).

There is universality for the specific heat, in the sense that it diverges logarithmically at the critical points, as in the Ising model. However the coefficient of the log is *anomalous*: in fact if $t$ is near to one of the critical temperatures $\Delta \simeq \sqrt{2}|u|^{1+\eta}$ so that the coefficient in front of the logarithm behaves like $\sim |u|^{2(1+\eta)\eta_c}$, with $\eta_c$ a new anomalous exponent $O(\lambda)$; in particular it is vanishing or diverging as $u \to 0$ depending on the sign of $\lambda$. We can say that the system shows an *anomalous universality* which is a sort a new paradigmatic behaviour: the singularity at the critical points is described in terms of universal critical indexes but nevertheless in the isotropic limit $u \to 0$, some quantities, like the difference of the critical temperatures and the constant in front of the logarithm in the specific heat, scale with anomalous critical indexes, and they vanish or diverge, depending on the sign of $\lambda$.

Note also that (11.16) clarifies how the universality–nonuniversality crossover is realized as $u \to 0$. When $u \neq 0$ only the first term in (11.16) can be log-singular in correspondence of the two critical points; however the logarithmic term dominates on the second one only if $t$ varies inside an extremely small region $O(|u|^{1+\eta}e^{-c/|\lambda|})$ around the critical points (here $c$ is a positive $O(1)$ constant). Outside such region the power law behaviour

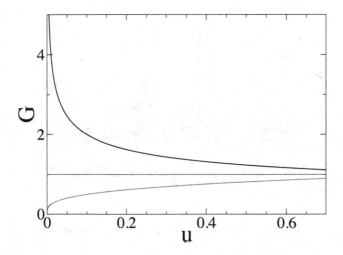

Fig. 11.1 Graphical representation of $G(\lambda, u)$ for $\lambda > 0$ and $\lambda < 0$

corresponding to the second addend dominates. When $u \to 0$ one recovers the power law decay found in the isotropic case

$$C_v \simeq F_2 \frac{1 - |t - t_c|^{2\eta_c}}{\eta_c} \qquad (11.17)$$

In Fig. 11.2 we plot the qualitative behaviour of $C_v$ as a function of $t$.

## 11.3 Integration of the $\chi$ variables

The propagators $<\phi^{\sigma}_{\mathbf{x},\omega}\phi^{\sigma'}_{\mathbf{y},\omega'}>$ of the fermionic integration $P(d\phi)$ verify the following bound, for some $A, \kappa > 0$:

$$|<\phi^{\sigma}_{\mathbf{x},\omega}\phi^{\sigma'}_{\mathbf{y},\omega'}>| \leq A e^{-\kappa \bar{m}_{\phi}|\mathbf{x}-\mathbf{y}|}, \qquad (11.18)$$

where $\bar{m}_{\phi}$ is the minimum between $|m^{(1)}_{\phi}|$ and $|m^{(2)}_{\phi}|$ and

$$m^{(1)}_{\phi} = \frac{2}{t_\lambda}\left(t^{(1)}_\lambda - t_\phi\right) \quad , \quad m^{(2)}_{\phi} = \frac{2}{t_\lambda}\left(t^{(2)}_\lambda - t_\phi\right). \qquad (11.19)$$

Note that both $m^{(1)}_\chi$ and $m^{(2)}_\chi$ are $O(1)$. After the integration of the $\chi$ variables we get

$$Z^{-}_{AT} = e^{-M^2 E_1} \int \overline{P}(d\psi) e^{-\mathcal{V}^{(1)}(\sqrt{Z_1}\psi)}, \qquad \mathcal{V}^{(1)}(0) = 0, \qquad (11.20)$$

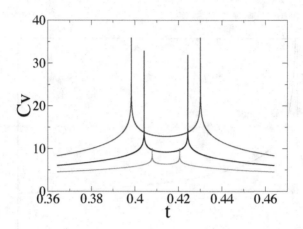

Fig. 11.2  Graphical representation of $C_v$ for $\lambda > 0$, $\lambda = 0$ and $\lambda < 0$

where, if $\sigma = \sigma_\psi(0)$ and $t_\psi = \sqrt{2} - 1$

$$Z_1 = t_\psi \quad , \quad \sigma_1 = \frac{\sigma}{1 - \frac{\sigma}{2}} \quad , \quad \mu_1 = \frac{\mu}{1 - \frac{\sigma}{2}} \quad (11.21)$$

and

$$\overline{P}(d\psi) = \mathcal{N}^{-1} \prod_{\mathbf{k} \in \mathcal{D}}^{\omega = \pm 1} d\psi^+_{\omega,\mathbf{k}} d\psi^-_{\omega,\mathbf{k}} \exp\left[-\frac{1}{4L^2} \sum_{\mathbf{k} \in \mathcal{D}} Z_1 \Psi^{+,T}_{\mathbf{k}} A^{(1)}_\psi(\mathbf{k}) \Psi_{\mathbf{k}}\right] \quad (11.22)$$

with

$$A^{(1)}_\psi(\mathbf{k}) = \begin{pmatrix} M^{(1)}(\mathbf{k}) & N^{(1)}(\mathbf{k}) \\ N^{(1)}(\mathbf{k}) & M^{(1)}(\mathbf{k}) \end{pmatrix}$$

and

$$M^{(1)}(\mathbf{k}) = \begin{pmatrix} i \sin k + \sin k_0 + a_1^+(\mathbf{k}) & -i(\sigma_1 + c_1(\mathbf{k})) \\ i(\sigma_1 + c_1(\mathbf{k})) & i \sin k - \sin k_0 + a_1^-(\mathbf{k}) \end{pmatrix}$$

$$N^{(1)}(\mathbf{k}) = \begin{pmatrix} b_1^+(\mathbf{k}) & i(\mu_1 + d_1(\mathbf{k})) \\ -i(\mu_1 + d_1(\mathbf{k})) & b_1^-(\mathbf{k}) \end{pmatrix}$$

where $a_1^\pm(\mathbf{k}), b_1^\pm(\mathbf{k})$ are analytic odd functions of $\mathbf{k}$ and $c_1(\mathbf{k}), d_1(\mathbf{k})$ are real analytic even functions of $\mathbf{k}$; moreover, in a neighborhood of $\mathbf{k} = \mathbf{0}$, $a_1^\pm(\mathbf{k}) = O(\sigma_1 \mathbf{k}) + O(\mathbf{k}^3)$, $b_1^\pm(\mathbf{k}) = O(\mu_1 \mathbf{k}) + O(\mathbf{k}^3)$, $c_1(\mathbf{k}) = O(\mathbf{k}^2)$ and $d_1(\mathbf{k}) = O(\mu_1 \mathbf{k}^2)$; the determinant $|\det A_\psi(\mathbf{k})|$ can be bounded above and below by some constant times $[(\sigma_1 - \mu_1)^2 + |c(\mathbf{k})|][(\sigma_1 + \mu_1)^2 + |c(\mathbf{k})|]$.

The symmetry properties are the same as in §10.5; again the local part of the kernels of the terms with four quartic in the fields is real; regarding the quadratic terms the following properties holds.

*a)* Let us consider the term with $2n = 2$, $\alpha_1 = -\alpha_2 = +$ and $\omega_1 = -\omega_2 = \omega$; let us denote it with $\sum_{\omega,\mathbf{k}} W_\omega(\mathbf{k};\mu_1)\widehat{\psi}^+_{\omega,\mathbf{k}}\widehat{\psi}^-_{-\omega,\mathbf{k}}$. Under parity it becomes

$$\sum_{\omega,\mathbf{k}} W_\omega(\mathbf{k};\mu_1)(i\omega)\widehat{\psi}^+_{\omega,-\mathbf{k}}(-i\omega)\widehat{\psi}^-_{-\omega,-\mathbf{k}} = \sum_{\omega,\mathbf{k}} W_\omega(-\mathbf{k};\mu_1)\widehat{\psi}^+_{\omega,\mathbf{k}}\widehat{\psi}^-_{-\omega,\mathbf{k}}, \tag{11.23}$$

so that $W_\omega(\mathbf{k};\mu_1)$ is even in $\mathbf{k}$.
Under complex conjugation it becomes

$$\sum_{\omega,\mathbf{k}} W_\omega(\mathbf{k};\mu_1)^*\widehat{\psi}^-_{-\omega,\mathbf{k}}\widehat{\psi}^+_{\omega,\mathbf{k}} = -\sum_{\omega,\mathbf{k}} W_\omega(\mathbf{k};\mu_1)^*\widehat{\psi}^+_{\omega,\mathbf{k}}\widehat{\psi}^-_{-\omega,\mathbf{k}}, \tag{11.24}$$

so that $W_\omega(\mathbf{k};\mu_1)$ is purely imaginary.
Under hole-particle it becomes

$$\sum_{\omega,\mathbf{k}} W_\omega(\mathbf{k};\mu_1)\widetilde{\psi}^-_{\omega,-\mathbf{k}}\widetilde{\psi}^+_{-\omega,-\mathbf{k}} = -\sum_{\omega,\mathbf{k}} W_{-\omega}(\mathbf{k};\mu_1)\widetilde{\psi}^+_{\omega,\mathbf{k}}\widetilde{\psi}^-_{-\omega,\mathbf{k}}, \tag{11.25}$$

so that $W_\omega(\mathbf{k};\mu_1)$ is odd in $\omega$.
Under $(1)\longleftrightarrow(2)$ it becomes:

$$\sum_{\omega,\mathbf{k}} W_\omega(\mathbf{k};-\mu_1)(-i)\widehat{\psi}^-_{-\omega,-\mathbf{k}}(i)\widehat{\psi}^+_{\omega,-\mathbf{k}} = \sum_{\omega,\mathbf{k}} W_\omega(\mathbf{k};-\mu_1)\widehat{\psi}^+_{\omega,\mathbf{k}}\widehat{\psi}^-_{-\omega,\mathbf{k}}, \tag{11.26}$$

so that $W_\omega(\mathbf{k};\mu_1)$ is even in $\mu_1$.

*b)* Let us consider the term with $2n = 2$, $\alpha_1 = \alpha_2 = \alpha$ and $\omega_1 = -\omega_2 = \omega$ and let us denote it with $\sum_{\omega,\alpha,\mathbf{k}} W^\alpha_\omega(\mathbf{k};\mu_1)\widehat{\psi}^\alpha_{\omega,\mathbf{k}}\widehat{\psi}^\alpha_{-\omega,-\mathbf{k}}$. We proceed as in item (a) and, by using parity, we see that $W^\alpha_\omega(\mathbf{k};\mu_1)$ is even in $\mathbf{k}$ and odd in $\omega$.
By using complex conjugation, we see that $W^\alpha_\omega(\mathbf{k};\mu_1) = -W^{-\alpha}_\omega(\mathbf{k};\mu_1)^*$.
By using hole-particle, we see that $W^\alpha_\omega(\mathbf{k};\mu_1)$ is even in $\alpha$ and $W^\alpha_\omega(\mathbf{k};\mu_1) = -W^{-\alpha}_\omega(\mathbf{k};\mu_1)^*$ implies that $W^\alpha_\omega(\mathbf{k};\mu_1)$ is purely imaginary.
By using $(1)\longleftrightarrow(2)$ we see that $W^\alpha_\omega(\mathbf{k};\mu_1)$ is odd in $\mu_1$.

*c)* Let us consider the term with $2n = 2$, $\alpha_1 = -\alpha_2 = +$, $\omega_1 = \omega_2 = \omega$ and let us denote it with $\sum_{\omega,\mathbf{k}} W_\omega(\mathbf{k};\mu_1)\widehat{\psi}^+_{\omega,\mathbf{k}}\widehat{\psi}^-_{\omega,\mathbf{k}}$. By using parity we see that $W_\omega(\mathbf{k};\mu_1)$ is odd in $\mathbf{k}$.
By using reflection we see that $W_\omega(k, k_0;\mu_1) = W_{-\omega}(k, -k_0;\mu_1)$.
By using complex conjugation we see that $W_\omega(k, k_0;\mu_1) = W^*_\omega(-k, k_0;\mu_1)$.
By using rotation we find $W_\omega(k, k_0;\mu_1) = -i\omega W_\omega(k_0, -k;\mu_1)$.

By using (1)⟷(2) we see that $W_\omega(\mathbf{k}; -\mu_1)$ is even in $\mu_1$.

We now define, if $\mathbf{k}_{\eta,\eta'} = (\frac{\eta\pi}{L}, \frac{\eta'\pi}{L})$

$$G_1(\mathbf{k}) = \frac{1}{4}\sum_{\eta,\eta'=\pm} W_\omega(\bar{\mathbf{k}}_{\eta\eta'}; \mu_1)(\eta\frac{\sin k}{\sin\pi/L} + \eta'\frac{\sin k_0}{\sin\pi/L}). \quad (11.27)$$

We can rewrite $G_1(\mathbf{k}) = a_\omega \sin k + b_\omega \sin k_0$, with

$$a_\omega = a_{-\omega} = -a_\omega^* = i\omega b_\omega = ia$$
$$b_\omega = -b_{-\omega} = b_\omega^* = -i\omega a_\omega = \omega b = -i\omega ia \quad (11.28)$$

with $a = b$ real and independent of $\omega$. As a consequence, $G_1(\mathbf{k}) = G_1(i\sin k + \omega\sin k_0)$ for some real constant $G_1$.

d) Let us consider the term with $2n = 2$, $\alpha_1 = \alpha_2 = \alpha$, $\omega_1 = \omega_2 = \omega$ and let us denote it with $\sum_{\alpha,\omega,\mathbf{k}} W_\omega^\alpha(\mathbf{k};\mu_1)\widehat{\psi}_{\omega,\mathbf{k}}^\alpha\widehat{\psi}_{\omega,-\mathbf{k}}^\alpha$. Repeating the proof in item c) we see that $W_\omega^\alpha(\mathbf{k};\mu_1)$ is odd in $\mathbf{k}$ and in $\mu_1$ and, if we define

$$F_1(\mathbf{k}) = \frac{1}{4}\sum_{\eta,\eta'} W_\omega^\alpha(\bar{\mathbf{k}}_{\eta\eta'};\mu_1)(\eta\frac{\sin k}{\sin\pi/L} + \eta'\frac{\sin k_0}{\sin\pi/L}), \quad (11.29)$$

we can rewrite $F_1(\mathbf{k}) = F_1(i\sin k + \omega\sin k_0)$. Since $W_\omega^\alpha(\mathbf{k};\mu_1)$ is odd in $\mu_1$, we find $F_1 = O(\lambda\mu_1)$.

This concludes the study of the properties of the kernels of $\mathcal{V}^{(1)}$ we shall need in the following. Repeating the proof above it can also seen that the corrections $a_1^\pm(\mathbf{k})$, $b_1^\pm(\mathbf{k})$, appearing in (11.31), are analytic odd functions of $\mathbf{k}$, while $c_1(\mathbf{k})$ and $d_1(\mathbf{k})$ are real and even; the explicit computation of the lower order terms in the Taylor expansion in $\mathbf{k}$ shows that, in a neighborhood of $\mathbf{k} = 0$, $a_1^\pm(\mathbf{k}) = O(\sigma_1\mathbf{k}) + O(\mathbf{k}^3)$, $b_1^\pm(\mathbf{k}) = O(\mu_1\mathbf{k}) + O(\mathbf{k}^3)$, $c_1(\mathbf{k}) = O(\mathbf{k}^2)$ and $d_1(\mathbf{k}) = O(\mu_1\mathbf{k}^2)$.

It is convenient to write $P(d\psi)$ as

$$\overline{P}(d\psi) = P_{Z_1,\bar{\sigma}_1,\mu_1}(d\psi)e^{-\nu F_\sigma(\psi)}, \quad (11.30)$$

where $F_\sigma(\psi) = (1/2L^2)\sum_{\mathbf{k},\omega}(-i\omega)\widehat{\psi}_{\omega,\mathbf{k}}^+\widehat{\psi}_{-\omega,\mathbf{k}}^-$ and $P_{Z_1,\bar{\sigma}_1,\mu_1}(d\psi)$ equal to $\overline{P}(d\psi)$ with $\sigma_1$ replaced by $\bar{\sigma}_1 = \sigma_1 - \nu$.

## 11.4 Integration of the $\psi$ variables: first regime

In the integration of the $\psi$-variables one has to identify two separate regimes of momentum scales in the integration, in which different variables must be used to describe the system. It is important to stress that the scale separating the two regimes is not *a priori* fixed, but it is dynamically determined. In the first regime (of larger momentum scales), it is convenient to use Dirac Grassman variables $\psi^+_{\omega,\mathbf{x}}, \psi^-_{\omega,\mathbf{x}}$, $\omega = \pm 1$; in this regime the two Ising systems lose their individuality as the natural variables which we use are the Dirac variables which are a combination of the original Majorana fermions $\psi^{(1)}, \psi^{(2)}$ associated to the spin variables $\sigma^{(1)}_\mathbf{x}, \sigma^{(2)}_\mathbf{x}$.

In the effective action describing the system in this range of momenta, there are *two* relevant quadratic effective interaction, corresponding to $\psi^+_\omega \psi^-_{-\omega}$ and $\psi^\alpha_\omega \psi^\alpha_{-\omega}$, two marginal quadratic interaction and one marginal quartic interaction. The presence of two relevant terms is strongly related to the fact that the sytem has two critical points. The analyis in this regime is very similar to the one performed in the isotropic case (up to the presence of an extra mass term); in particular the flow of the effective coupling is controlled thanks to the vanishing of the Beta function proved in chapt. 5.

In the second regime of scales, where it is convenient to describe the system in terms of the original Majorana fermions; it turns out that on scale $h_1^*$ one of the two fermionic variables can be integrated out without any further multiscale decomposition ($h_1^*$ is chosen exactly by Ashking that one of the two variables is massive on scale $h_1^*$); and the remaining variable can again be integrated by a multiscale perturbative expansion based on Renormalization Group; in this second regime the quartic terms are irrelevant and the analysis is very similar to the one in chapt. 9.

In the first regime, after the integration of the scale $0, -1, .., h$ we get

$$Z^-_{AT} = \int P_{Z_h, \sigma_h, \mu_h}(d\psi^{(\leq h)}) e^{-\mathcal{V}^{(h)}(\sqrt{Z_h}\psi^{(\leq h)}) - L^2 E_h}, \quad \mathcal{V}^{(h)}(0) = 0, \tag{11.31}$$

where the quantities $Z_h$, $\sigma_h$, $\mu_h$, $C_h$, $P_{Z_h,\sigma_h,\mu_h}(d\psi^{(\leq h)})$, $\mathcal{V}^{(h)}$ and $E_h$ have to be defined recursively. $P_{Z_h,\sigma_h,\mu_h}(d\psi^{(\leq h)})$ is defined by (11.22) in which we replace $Z_1, \bar\sigma_1, \mu_1, a^\omega_1, b^\omega_1, c_1, d_1$ with $Z_h, \sigma_h, \mu_h, a^\omega_h, b^\omega_h, c_h, d_h$. Moreover

$$\mathcal{V}^{(h)}(\psi) = \sum_{n=1}^{\infty} \frac{1}{L^{2n}} \sum_{\substack{\mathbf{k}_1,\ldots,\mathbf{k}_{2n-1}, \\ \underline{\alpha},\underline{\omega}}} \prod_{i=1}^{2n} \widehat{\psi}^{\alpha_i(\leq h)}_{\omega_i,\mathbf{k}_i} \widehat{W}^{(h)}_{2n,\underline{\alpha},\underline{\omega}}(\mathbf{k}_1,\ldots,\mathbf{k}_{2n-1}) \delta(\sum_{i=1}^{2n} \alpha_i \mathbf{k}_i)$$

$$= \sum_{n=1}^{\infty} \sum_{\substack{\mathbf{x}_1,\ldots,\mathbf{x}_{2n}, \\ \underline{\sigma},\underline{j},\underline{\omega},\underline{\alpha}}} \prod_{i=1}^{2n} \partial^{\sigma_i}_{j_i} \psi^{\alpha_i(\leq h)}_{\omega_i,\mathbf{x}_i} W^{(h)}_{2n,\underline{\sigma},\underline{j},\underline{\alpha},\underline{\omega}}(\mathbf{x}_1,\ldots,\mathbf{x}_{2n}) \tag{11.32}$$

where in the last line $j_i = 0, 1$, $\sigma_i \geq 0$ and $\partial_j$ is the forward discrete derivative in the $\widehat{e}_j$ direction.

$\mathcal{L}$ will be non zero only if acting on a kernel $\widehat{W}_{2n,\underline{\alpha},\underline{\omega}}^{(h)}$ with $n = 1, 2$. In this case $\mathcal{L}$ will be the combination of four different operators: $\mathcal{L}_j$, $j = 0, 1$, whose effect on a function of $\mathbf{k}$ will be essentially to extract the term of order $j$ from its Taylor series in $\mathbf{k}$; and $\mathcal{P}_j$, $j = 0, 1$, whose effect on a functional of the sequence $\sigma_h(\mathbf{k}), \mu_h(\mathbf{k}), \ldots, \sigma_1, \mu_1$ will be essentially to extract the term of order $j$ from its power series in $\sigma_h(\mathbf{k}), \mu_h(\mathbf{k}), \ldots, \sigma_1, \mu_1$.

The action of $\mathcal{L}_j$, $j = 0, 1$, on the kernels $\widehat{W}_{2n,\underline{\alpha},\underline{\omega}}^{(h)}(\mathbf{k}_1, \ldots, \mathbf{k}_{2n})$ is defined as in chapt.7 and and the action of $\mathcal{P}_j$, $j = 0, 1$, on the kernels $\widehat{W}_{2n,\underline{\alpha},\underline{\omega}}$, thought as functionals of the sequence $\sigma_h(\mathbf{k}), \mu_h(\mathbf{k}), \ldots, \sigma_1, \mu_1$ is defined as follows.

$$\mathcal{P}_0 \widehat{W}_{2n,\underline{\alpha},\underline{\omega}} = \widehat{W}_{2n,\underline{\alpha},\underline{\omega}} \Big|_{\underline{\sigma}^{(h)} = \underline{\mu}^{(h)} = 0}$$

$$\mathcal{P}_1 \widehat{W}_{2n,\underline{\alpha},\underline{\omega}} = \sum_{k \geq h, \mathbf{k}} \left[ \sigma_k(\mathbf{k}) \frac{\partial \widehat{W}_{2n,\underline{\alpha},\underline{\omega}}}{\partial \sigma_k(\mathbf{k})} \Big|_{\underline{\sigma}^{(h)} = \underline{\mu}^{(h)} = 0} + \mu_k(\mathbf{k}) \frac{\partial \widehat{W}_{2n,\underline{\alpha},\underline{\omega}}}{\partial \mu_k(\mathbf{k})} \Big|_{\underline{\sigma}^{(h)} = \underline{\mu}^{(h)} = 0} \right]$$

(11.33)

Given $\mathcal{L}_j, \mathcal{P}_j$, $j = 0, 1$ as above, we define the action of $\mathcal{L}$ on the kernels $\widehat{W}_{2n,\underline{\alpha},\underline{\omega}}$ as follows.

1) If $n = 1$, then $\mathcal{L}\widehat{W}_{2,\underline{\alpha},\underline{\omega}} = \mathcal{L}_0(\mathcal{P}_0 + \mathcal{P}_1)\widehat{W}_{2,\underline{\alpha},\underline{\omega}}$ if $\omega_1 + \omega_2 = 0$ and $\alpha_1 + \alpha_2 = 0$; $\mathcal{L}\widehat{W}_{2,\underline{\alpha},\underline{\omega}} = \mathcal{L}_0 \mathcal{P}_1 \widehat{W}_{2,\underline{\alpha},\underline{\omega}}$ if $\omega_1 + \omega_2 = 0$ and $\alpha_1 + \alpha_2 \neq 0$; $\mathcal{L}\widehat{W}_{2,\underline{\alpha},\underline{\omega}} = \mathcal{L}_1 \mathcal{P}_0 \widehat{W}_{2,\underline{\alpha},\underline{\omega}}$ if $\omega_1 + \omega_2 \neq 0$ and $\alpha_1 + \alpha_2 = 0$; finally $\mathcal{L}\widehat{W}_{2,\underline{\alpha},\underline{\omega}} = 0$ if $\omega_1 + \omega_2 \neq 0$ and $\alpha_1 + \alpha_2 \neq 0$.

2) If $n = 2$, then $\mathcal{L}\widehat{W}_{4,\underline{\alpha},\underline{\omega}} = \mathcal{L}_0 \mathcal{P}_0 \widehat{W}_{4,\underline{\alpha},\underline{\omega}}$.

3) If $n > 2$, then $\mathcal{L}\widehat{W}_{2n,\underline{\alpha},\underline{\omega}} = 0$.

Finally, the effect of $\mathcal{L}$ on $\mathcal{V}^{(h)}$ is, by definition, to replace on the r.h.s. of (11.32) $\widehat{W}_{2n,\underline{\alpha},\underline{\omega}}$ with $\mathcal{L}\widehat{W}_{2n,\underline{\alpha},\underline{\omega}}$.
Then

$$\mathcal{L}\mathcal{V}^{(h)}(\psi^{(\leq h)}) = (s_h + \gamma^h n_h) F_\sigma^{(\leq h)} + m_h F_\mu^{(\leq h)} + l_h F_\lambda^{(\leq h)} + z_h F_\zeta^{(\leq h)} ,\quad (11.34)$$

where $s_h, n_h, m_h, l_h$ and $z_h$ are real constants and: $s_h$ is linear in $\underline{\sigma}^{(h)}$ and independent of $\underline{\mu}^{(h)}$; $m_h$ is linear in $\underline{\mu}^{(h)}$ and independent of $\underline{\sigma}^{(h)}$; $n_h, l_h, z_h$

are independent of $\underline{\sigma}^{(h)}, \underline{\mu}^{(h)}$; moreover

$$F_\sigma^{(\leq h)}(\psi^{(\leq h)}) = \frac{1}{2L^2} \sum_{\mathbf{k}} \sum_{\omega=\pm 1} (-i\omega) \widehat{\psi}_{\omega,\mathbf{k}}^{+(\leq h)} \widehat{\psi}_{-\omega,\mathbf{k}}^{-(\leq h)} = \frac{1}{L^2} \sum_{\mathbf{k}} \widehat{F}_\sigma^{(\leq h)}(\mathbf{k})$$

$$F_\mu^{(\leq h)}(\psi^{(\leq h)}) = \frac{1}{4L^2} \sum_{\mathbf{k}} \sum_{\alpha,\omega=\pm 1} i\omega \widehat{\psi}_{\omega,\mathbf{k}}^{\alpha(\leq h)} \widehat{\psi}_{-\omega,-\mathbf{k}}^{\alpha(\leq h)} = \frac{1}{L^2} \sum_{\mathbf{k}} \widehat{F}_\mu^{(\leq h)}(\mathbf{k}) ,$$

$$F_\lambda^{(\leq h)}(\psi^{(\leq h)}) = \frac{1}{L^8} \sum_{\mathbf{k}_1,\ldots,\mathbf{k}_4} \widehat{\psi}_{1,\mathbf{k}_1}^{+(\leq h)} \widehat{\psi}_{-1,\mathbf{k}_2}^{+(\leq h)} \widehat{\psi}_{-1,\mathbf{k}_3}^{-(\leq h)} \widehat{\psi}_{1,\mathbf{k}_4}^{-(\leq h)} \delta(\mathbf{k}_1+\mathbf{k}_2-\mathbf{k}_3-\mathbf{k}_4)$$

$$F_\zeta^{(\leq h)}(\psi^{(\leq h)}) = \frac{1}{2L^2} \sum_{\mathbf{k}} \sum_{\omega=\pm 1} (i\sin k + \omega \sin k_0) \widehat{\psi}_{\omega,\mathbf{k}}^{+(\leq h)} \widehat{\psi}_{\omega,\mathbf{k}}^{-(\leq h)} \quad (11.35)$$

where $\delta(\mathbf{k}) = L^2 \sum_{\mathbf{n} \in \mathbb{Z}^2} \delta_{\mathbf{k},2\pi\mathbf{n}}$.

The effect of $\mathcal{R}$ is such that $\mathcal{R}\widehat{W}_{2n,\underline{\alpha},\underline{\omega}}$ is at least quadratic in $\mathbf{k}, \underline{\sigma}^{(h)}, \underline{\mu}^{(h)}$ if $n = 1$ and at least linear in $\mathbf{k}, \underline{\sigma}^{(h)}, \underline{\mu}^{(h)}$ when $n = 2$. This will give dimensional gain factors in the bounds for $\mathcal{R}\widehat{W}_{2n,\underline{\alpha},\underline{\omega}}^{(h)}$ w.r.t. the bounds for $\widehat{W}_{2n,\underline{\alpha},\underline{\omega}}^{(h)}$.

Once that the above definitions are given we can describe our integration procedure for $h \leq 0$. We start from (11.31) and we rewrite it as

$$\int P_{Z_h,\sigma_h,\mu_h}(d\psi^{(\leq h)}) e^{-\mathcal{L}\mathcal{V}^{(h)}(\sqrt{Z_h}\psi^{(\leq h)}) - \mathcal{R}\mathcal{V}^{(h)}(\sqrt{Z_h}\psi^{(\leq h)}) - L^2 E_h} , \quad (11.36)$$

with $\mathcal{L}\mathcal{V}^{(h)}$ as in (11.34). Then we include the quadratic part of $\mathcal{L}\mathcal{V}^{(h)}$ (unless for the $\sim n_h$ term) in the fermionic integration and we rescale the fields, obtaining

$$\int P_{Z_{h-1},\sigma_{h-1},\mu_{h-1}}(d\psi^{(\leq h-1)}) \int P_{Z_{h-1},\sigma_{h-1},\mu_{h-1}}(d\psi^{(h)}) e^{-\widehat{\mathcal{V}}^{(h)}(\sqrt{Z_{h-1}}\psi^{(\leq h)})} \quad (11.37)$$

The flow equations are given by

$$\frac{Z_{h-1}}{Z_h} = 1 + z_h \quad , \quad \frac{\sigma_{h-1}}{\sigma_h} = 1 + \frac{s_h/\sigma_h - z_h}{1 + z_h} \quad , \quad \frac{\mu_{h-1}}{\mu_h} = 1 + \frac{m_h/\mu_h - z_h}{1 + z_h} ,$$
$$(11.38)$$

together with those for the running coupling constants:

$$\lambda_{h-1} = \lambda_h + \beta_\lambda^h(\lambda_h, \nu_h; \ldots; \lambda_1, \nu_1)$$
$$\nu_{h-1} = \gamma \nu_h + \beta_\nu^h(\lambda_h, \nu_h; \ldots; \lambda_1, \nu_1) \quad (11.39)$$

Proceeding as before, if $\lambda$ is small enough, there exists an analytic function $\nu^*(\lambda)$ independent of $t, u$ such that the running coupling constants $\{\lambda_h, \nu_h\}_{h \leq 1}$ with $\nu_1 = \nu^*(\lambda)$ verify $|\nu_h| \leq c|\lambda|\gamma^{(\vartheta/2)h}$ and $|\lambda_h| \leq c|\lambda|$.

If $\lambda$ is small enough and $\nu_1$ is chosen as above, the solution of (11.38) can be written as:

$$Z_h = \gamma^{\eta_z(h-1)+F_\zeta^h} \quad , \quad \mu_h = \mu_1 \gamma^{\eta_\mu(h-1)+F_\mu^h} \quad , \quad \sigma_h = \sigma_1 \gamma^{\eta_\sigma(h-1)+F_\sigma^h}$$
(11.40)

where $\eta_z, \eta_\mu, \eta_z$ and $F_\zeta^h, F_\mu^h, F_\sigma^h$ are $O(\lambda)$ functions, independent of $\sigma_1, \mu_1$; moreover $\lambda_h = \lambda + O(\lambda^2)$ from the key property of vanishing of the beta function.

The integration is iterated until a scale $h_1^*$ defined $h_1^* = \left[\log_\gamma |\sigma_1|^{\frac{1}{1-\eta_\sigma}}\right]$ if $|\sigma_1|^{\frac{1}{1-\eta_\sigma}} > 2|\mu_1|^{\frac{1}{1-\eta_\mu}}$, and $h_1^* = \left[\log_\gamma |u|^{\frac{1}{1-\eta_\mu}}\right]$ if $|\sigma_1|^{\frac{1}{1-\eta_\sigma}} \leq 2|\mu_1|^{\frac{1}{1-\eta_\mu}}$. From the above definition it follows that

$$C_2 \gamma^{h_1^*} \leq |\sigma_{h_1^*}| + |\mu_{h_1^*}| \leq C_1 \gamma^{h_1^*} \,,$$
(11.41)

with $C_1, C_2$ independent of $\lambda, \mu_1, \sigma_1$.

In fact, if $|\sigma_1|^{\frac{1}{1-\eta_\sigma}} > 2|\mu_1|^{\frac{1}{1-\eta_\mu}}$, then $\gamma^{h_1^*-1} = c_\sigma |\sigma_1|^{\frac{1}{1-\eta_\sigma}}$, with $1 \leq c_\sigma < \gamma$, so that $C_2 \gamma^{h_1^*} \leq |\sigma_{h_1^*}| \leq C_1/10 \gamma^{h_1^*}$, for some $C_1, C_2 = O(1)$. Furthermore we find

$$\frac{|\mu_{h_1^*}|}{|\sigma_{h_1^*}|} = c_\sigma^{\eta_\mu - \eta_\sigma} |\mu_1||\sigma_1|^{-\frac{1-\eta_\mu}{1-\eta_\sigma}} \gamma^{F_\mu^{h_1^*} - F_\sigma^{h_1^*}} < 1$$
(11.42)

and (11.41) follows.

If $|\sigma_1|^{\frac{1}{1-\eta_\sigma}} \leq 2|\mu_1|^{\frac{1}{1-\eta_\mu}}$, then $\gamma^{h_1^*-1} = c_u |u|^{\frac{1}{1-\eta_\mu}}$, with $1 \leq c_u < \gamma$, so that $C_1 \gamma^{h_1^*} \leq |\mu_{h_1^*}| \leq C_2/10 \gamma^{h_1^*}$. Furthermore

$$\frac{|\sigma_{h_1^*}|}{|\mu_{h_1^*}|} = c_u^{\eta_\sigma - \eta_\mu} |\sigma_1||u|^{-\frac{1-\eta_\sigma}{1-\eta_\mu}} \gamma^{F_\sigma^{h_1^*} - F_\mu^{h_1^*}} < 10$$
(11.43)

and (11.41) again follows.

The integration of the scales $\leq h_1^*$ must be performed in a different way, as discussed below.

## 11.5 Integration of the $\psi$ variables: second regime

Once that all the scales $> h_1^*$ are integrated out, it is more convenient to describe the system in terms of the fields $\psi_\omega^{(1)}, \psi_\omega^{(2)}$, $\omega = \pm 1$, defined through the following change of variables, $\alpha = \pm$:

$$\widehat{\psi}_{\omega,\mathbf{k}}^{\alpha(\leq h_1^*)} = \frac{1}{\sqrt{2}} (\widehat{\psi}_{\omega,-\alpha\mathbf{k}}^{(1,\leq h_1^*)} - i\alpha \widehat{\psi}_{\omega,-\alpha\mathbf{k}}^{(2,\leq h_1^*)})$$
(11.44)

If we perform this change of variables, we find $P_{Z_{h_1^*},\sigma_{h_1^*},\mu_{h_1^*}} = \prod_{j=1}^2 P^{(j)}_{Z_{h_1^*},m^{(j)}_{h_1^*}}$ where, if $\Psi_\mathbf{k}^{(j,\le h_1^*),T} = (\psi^{(j,\le h_1^*)}_{1,\mathbf{k}}, \psi^{(j,\le h_1^*)}_{-1,\mathbf{k}})$,

$$P^{(j)}_{Z_{h_1^*},m^{(j)}_{h_1^*}}(d\psi^{(j,\le h_1^*)}) = \qquad (11.45)$$

$$= \frac{1}{N^{(j)}_{h_1^*}} \prod_{\mathbf{k},\omega} d\psi^{(j,\le h_1^*)}_{\omega,\mathbf{k}} \exp\left\{-\frac{Z_{h_1^*}}{4L^2} \sum_\mathbf{k} C_{h_1^*}(\mathbf{k}) \Psi^{(j,\le h_1^*),T}_\mathbf{k} A_j^{(h_1^*)}(\mathbf{k}) \Psi^{(j,\le h_1^*)}_{-\mathbf{k}}\right\}$$

with

$$A_j^{(h_1^*)}(\mathbf{k}) = \begin{pmatrix} (-i\sin k - \sin k_0) + a^{+(j)}_{h_1^*}(\mathbf{k}) & -i\big(m^{(j)}_{h_1^*}(\mathbf{k}) + c^{(j)}_{h_1^*}(\mathbf{k})\big) \\ i\big(m^{(j)}_{h_1^*}(\mathbf{k}) + c^{(j)}_{h_1^*}(\mathbf{k})\big) & (-i\sin k + \sin k_0) + a^{-(j)}_{h_1^*}(\mathbf{k}) \end{pmatrix}$$

and

$$m^{(1)}_{h_1^*} = \mu_{h_1^*} + \sigma_{h_1^*} \qquad m^{(2)}_{h_1^*} = \mu_{h_1^*} - \sigma_{h_1^*} \qquad (11.46)$$

Note that, by (11.41), $\max\{|m^{(1)}_{h_1^*}|, |m^{(2)}_{h_1^*}|\} = |\sigma_{h_1^*}| + |\mu_{h_1^*}| = O(\gamma^{h_1^*})$. From now on, for definiteness we shall suppose that $\max\{|m^{(1)}_{h_1^*}|, |m^{(2)}_{h_1^*}|\} \equiv |m^{(1)}_{h_1^*}|$. Then, it is easy to realize that the propagator $g^{(1,\le h_1^*)}_{\omega_1,\omega_2}$ is bounded as follows.

$$|\partial^{n_0}_{x_0} \partial^{n_1}_x g^{(1,\le h_1^*)}_{\omega_1,\omega_2}(\mathbf{x})| \le C_{N,n} \frac{\gamma^{(1+n)h_1^*}}{1 + (\gamma^{h_1^*}|\mathbf{x}|)^N} \quad , \quad n = n_0 + n_1 \;, \qquad (11.47)$$

namely $g^{(1,\le h_1^*)}_{\omega_1,\omega_2}$ satisfies the same bound as the single scale propagator on scale $h = h_1^*$. This suggests to integrate out $\psi^{(1,\le h_1^*)}$, without any other scale decomposition, so obtaining

$$Z^-_{AT} = \int P^{(2)}_{Z_{h_1^*},\widehat{m}^{(2)}_{h_1^*}}(d\psi^{(2,\le h_1^*)}) e^{-\overline{\mathcal{V}}^{(h_1^*)}(\sqrt{Z_{h_1^*}}\psi^{(2,\le h_1^*)}) - L^2 \overline{E}_{h_1^*}} \;, \qquad (11.48)$$

where: $\widehat{m}^{(2)}_{h_1^*}(\mathbf{k}) = m^{(2)}_{h_1^*}(\mathbf{k}) - \gamma^{h_1^*} \pi_{h_1^*} C^{-1}_{h_1^*}(\mathbf{k})$, with $\pi_{h_1^*}$ a free parameter.

Now we shall perform an iterative integration of the field $\psi^{(2)}$ essentially identical to the one in chapt. 9. If $h = h_1^*, h_1^* - 1, \ldots$, we shall write:

$$Z^-_{AT} = \int P^{(2)}_{Z_h,\widehat{m}^{(2)}_h}(d\psi^{(2,\le h)}) e^{-\overline{\mathcal{V}}^{(h)}(\sqrt{Z_h}\psi^{(2,\le h)}) - L^2 E_h} \;. \qquad (11.49)$$

We define $\mathcal{L}$ as a combination of four operators $\mathcal{L}_j$ and $\overline{\mathcal{P}}_j$, $j = 0, 1$. $\mathcal{L}_j$ are defined as before, while $\overline{\mathcal{P}}_0$ and $\overline{\mathcal{P}}_1$, in analogy with (11.33), are defined as the operators extracting from a functional of $\widehat{m}^{(2)}_h(\mathbf{k})$, $h \le h_1^*$, the contributions independent and linear in $\widehat{m}^{(2)}_h(\mathbf{k})$. We define the action of $\mathcal{L}$ on the kernels $\overline{W}^{(h)}_{2n,\underline{\omega}}$ as follows.

1) If $n = 1$, then $\mathcal{L}\overline{W}_{2,\underline{\omega}}^{(h)} = \mathcal{L}_0(\overline{\mathcal{P}}_0 + \overline{\mathcal{P}}_1)\overline{W}_{2,\underline{\omega}}^{(h)}$ if $\omega_1 + \omega_2 = 0$ and $\mathcal{L}\overline{W}_{2,\underline{\omega}}^{(h)} = \mathcal{L}_1\overline{\mathcal{P}}_0\overline{W}_{2,\underline{\omega}}^{(h)}$ if $\omega_1 + \omega_2 \neq 0$

2) If $n > 2$, then $\mathcal{L}\overline{W}_{2n,\underline{\omega}}^{(h)} = 0$.

Then
$$\mathcal{L}\overline{\mathcal{V}}^{(h)} = (s_h + \gamma^h p_h)F_\sigma^{(2,\leq h)} + z_h F_\zeta^{(2,\leq h)}, \qquad (11.50)$$

where $s_h, p_h$ and $z_h$ are real constants and: $s_h$ is linear in $\widehat{m}_k^{(2)}(\mathbf{k})$, $h \leq k \leq h_1^*$; $p_h$ and $z_h$ are independent of $\widehat{m}_k^{(2)}(\mathbf{k})$. Furthermore $F_\sigma^{(2,\leq h)}$ and $F_\zeta^{(2,\leq h)}$ are given by the first and the last of (11.35) with $\widehat{\psi}_{\omega,\mathbf{k}}^{(2,\leq h)} \widehat{\psi}_{\omega',-\mathbf{k}}^{(2,\leq h)}$ replacing $\widehat{\psi}_{\omega,\mathbf{k}}^{+(\leq h)} \widehat{\psi}_{\omega',\mathbf{k}}^{-(\leq h)}$.

If $\mathcal{L}$ and $\mathcal{R} = 1 - \mathcal{L}$ are defined as in previous subsection, we can rewrite (11.49) as:

$$\int P_{Z_h, \widehat{m}_h^{(2)}}^{(2)} (d\psi^{(2,\leq h)}) e^{-\mathcal{L}\overline{\mathcal{V}}^{(h)}(\sqrt{Z_h}\psi^{(2,\leq h)}) - \mathcal{R}\overline{\mathcal{V}}^{(h)}(\sqrt{Z_h}\psi^{(2,\leq h)}) - L^2 E_h}. \qquad (11.51)$$

Furthermore, using (11.50) and defining:

$$\widehat{Z}_{h-1}(\mathbf{k}) = Z_h(1 + C_h^{-1}(\mathbf{k})z_h), \quad \widehat{m}_{h-1}^{(2)}(\mathbf{k}) = \frac{Z_h}{\widehat{Z}_{h-1}(\mathbf{k})}\left(\widehat{m}_h^{(2)}(\mathbf{k}) + C_h^{-1}(\mathbf{k})s_h\right), \qquad (11.52)$$

we see that (11.51) is equal to

$$\int P_{\widehat{Z}_{h-1}, \widehat{m}_{h-1}^{(2)}}^{(2)} (d\psi^{(2,\leq h)}) e^{-\gamma^h p_h F_\sigma^{(2,\leq h)}(\sqrt{Z_h}\psi^{(2),\leq h}) - \mathcal{R}\overline{\mathcal{V}}^{(h)}(\sqrt{Z_h}\psi^{(2),\leq h}) - L^2(E_h + t_h)} \qquad (11.53)$$

Again, we rescale the potential:

$$\widetilde{\mathcal{V}}^{(h)}(\sqrt{Z_{h-1}}\psi^{(\leq h)}) = \gamma^h \pi_h F_\sigma^{(2,\leq h)}(\sqrt{Z_{h-1}}\psi^{(2,\leq h)}) + \mathcal{R}\overline{\mathcal{V}}^h(\sqrt{Z_h}\psi^{(2,\leq h)}), \qquad (11.54)$$

where $Z_{h-1} = \widehat{Z}_{h-1}(\mathbf{0})$ and $\pi_h = (Z_h/Z_{h-1})p_h$; we perform the single scale integration and we define the new effective potential as

$$e^{-\overline{\mathcal{V}}^{(h-1)}(\sqrt{Z_{h-1}}\psi^{(2,\leq h-1)}) - M^2 \widetilde{E}_h} = \int P_{Z_{h-1}, \widehat{m}_{h-1}^{(2)}}^{(2)} (d\psi^{(2,h)}) e^{-\widetilde{\mathcal{V}}^h(\sqrt{Z_h}\psi^{(2,\leq h)})}. \qquad (11.55)$$

For the same reasons as in chapt 9 we get for $h \leq h_1^*$ and some constants $c, \vartheta > 0$

$$e^{-c|\lambda|} \leq \frac{\widehat{m}_h^{(2)}}{\widehat{m}_{h-1}^{(2)}} \leq e^{c|\lambda|}, \quad e^{-c|\lambda|^2} \leq \frac{Z_h}{Z_{h-1}} \leq e^{c|\lambda|^2}, \quad |\pi_h| \leq c|\lambda| \qquad (11.56)$$

Moreover if $\lambda$, $\sigma_1$ and $\mu_1$ are small enough there exists $\pi_{h_1^*}^*(\lambda, \sigma_1, \mu_1)$ such that, if we fix $\pi_{h_1^*} = \pi_{h_1^*}^*(\lambda, \sigma_1, \mu_1)$, for $h \leq h_1^*$ we have:

$$|\pi_h| \leq c|\lambda|\gamma^{(\vartheta/2)(h-h_1^*)} \quad , \quad \widehat{m}_h^{(2)} = \widehat{m}_{h_1^*}^{(2)}\gamma^{F_m^h} \quad , \quad Z_h = Z_{h_1^*}\gamma^{\overline{F}_\zeta^h} \,, \quad (11.57)$$

where $F_m^h$ and $\overline{F}_\zeta^h$ are $O(\lambda)$. Moreover:

$$\left|\pi_{h_1^*}^*(\lambda, \sigma_1, \mu_1) - \pi_{h_1^*}^*(\lambda, \sigma_1', \mu_1')\right|$$
$$\leq c|\lambda|\left(\gamma^{(\eta_\sigma - 1)h_1^*}|\sigma_1 - \sigma_1'| + \gamma^{(\eta_\mu - 1)h_1^*}|\mu_1 - \mu_1'|\right). \quad (11.58)$$

We iterate the preceding construction up to the scale $h_2^*$ defined as the scale s.t. $|\widehat{m}_k^{(2)}| \leq \gamma^{k-1}$ for any $h_2^* \leq k \leq h_1^*$ and $|\widehat{m}_{h_2^*-1}^{(2)}| > \gamma^{h_2^*-2}$.

Once we have integrated all the fields $\psi^{(2,>h_2^*)}$, we can integrate $\psi^{(2,\leq h_2^*)}$ without any further multiscale decomposition. Note in fact that by definition the propagator satisfies the same bound (11.47) with $h_2^*$ replacing $h_1^*$.

## 11.6 Critical behaviour

In the present subsection we prove that, if $u$ and $t - \sqrt{2} + 1$ are small, there are precisely two critical points, of the form (14.4). More precisely, keeping in mind that the equation for the critical point is simply $\gamma^{h_2^*} = 0$ (see the end of previous subsection), we prove that $\gamma^{h_2^*} = 0$ only if $(\lambda, t, u) = (\lambda, t_c^\pm(\lambda, u), u)$, where $t_c^\pm(\lambda, u)$ is given by (14.4).

From the definition of $h_2^*$ given above, it follows that $h_2^*$ satisfies the following equation:

$$\gamma^{h_2^*-1} = c_m \gamma^{F_m^{h_2^*}}\left||\sigma_{h_1^*}| - |\mu_{h_1^*}| - \alpha_\sigma \gamma^{h_1^*+1}\pi_{h_1^*+1}\right|, \quad (11.59)$$

for some $1 \leq c_m < \gamma$ and $\alpha_\sigma = \operatorname{sign}\sigma_1$. Then, the equation $\gamma^{h_2^*} = 0$ can be rewritten as:

$$|\sigma_{h_1^*}| - |\mu_{h_1^*}| - \alpha_\sigma \gamma^{h_1^*+1}\pi_{h_1^*+1} = 0 \,. \quad (11.60)$$

We first show that (11.60) cannot be solved when $|\sigma_1|^{\frac{1}{1-\eta_\sigma}} > 2|\mu_1|^{\frac{1}{1-\eta_\mu}}$. In fact,

$$|\sigma_1|\gamma^{\eta_\sigma(h_1^*-1)+F_\sigma^{h_1^*}} - |\mu_1|\gamma^{\eta_\mu(h_1^*-1)+F_\mu^{h_1^*}} - \alpha_\sigma \gamma^{h_1^*+1}\pi_{h_1^*+1} = \quad (11.61)$$

$$|\sigma_1|^{1+\frac{\eta_\sigma}{1-\eta_\sigma}}c_1 - \left(|\mu_1||\sigma_1|^{-\frac{1-\eta_\mu}{1-\eta_\sigma}}\right)|\sigma_1|^{\frac{1-\eta_\mu}{1-\eta_\sigma}-\frac{\eta_\mu}{1-\eta_\sigma}}c_1' - \alpha_\sigma \gamma^{h_1^*+1}\pi_{h_1^*+1} \geq \frac{\gamma^{h_1^*-1}}{3\gamma}$$

where $c_1, c'_1$ are constants $= 1 + O(\lambda)$, $\pi_{h_1^*+1} = O(\lambda)$ and $\gamma^{h_1^*-1} = c_\sigma |\sigma_1|^{\frac{1}{1-\eta_\sigma}}$, with $1 \leq c_\sigma < \gamma$. Now, if $|\mu_1| > 0$, the r.h.s. of (11.62) equation is strictly positive.

So, let us consider the case $|\sigma_1|^{\frac{1}{1-\eta_\sigma}} \leq 2|\mu_1|^{\frac{1}{1-\eta_\mu}}$ (s.t. $h_1^* = [\log_\gamma |u|^{\frac{1}{1-\eta_\mu}}]$). In this case (11.60) can be easily solved to find:

$$|\sigma_1| = |\mu_1||u|^{\frac{\eta_\mu-\eta_\sigma}{1-\eta_\mu}} c_u^{\eta_\mu-\eta_\sigma} \gamma^{F_\mu^{h_1^*}-F_\sigma^{h_1^*}} + |u|^{\frac{1-\eta_\sigma}{1-\eta_\mu}} c_u^{1-\eta_\sigma} \alpha_\sigma \gamma^{2-F_\sigma^{h_1^*}} \pi_{h_1^*+1} \,. \tag{11.62}$$

Note that $c_u^{\eta_\mu-\eta_\sigma} \gamma^{F_\mu^{h_1^*}-F_\sigma^{h_1^*}} = 1 + O(\lambda)$ is just a function of $u$, (it does not depend on $t$), because of our definition of $h_1^*$. Moreover $\pi_{h_1^*+1}$ is a smooth function of $t$: if we call $\pi_{h_1^*+1}(t,u)$ resp. $\pi_{h_1^*+1}(t',u)$ the correction corresponding to the initial data $\sigma_1(t,u), \mu_1(t,u)$ resp. $\sigma_1(t',u), \mu_1(t',u)$, we have

$$|\pi_{h_1^*+1}(t,u) - \pi_{h_1^*+1}(t',u)| \leq c|\lambda||u|^{\frac{\eta_\sigma-1}{1-\eta_\mu}}|t-t'| \,, \tag{11.63}$$

where we used (11.58) and the bounds $|\sigma_1 - \sigma'_1| \leq c|t-t'|$ and $|\mu_1 - \mu'_1| \leq c|u||t-t'|$, following from the definitions of $(\sigma_1, \mu_1)$ in terms of $(\sigma, \mu)$ and of $(t,u)$, see (11.7), (11.21) and $\sigma = 2(1 - t_\psi/t_\lambda)$.

Using the same definitions we also realize that (11.62) can be rewritten as

$$t = \left[ \sqrt{2} - 1 + \frac{\nu(\lambda)}{2} \pm |u|^{1+\eta}\left(1 + \lambda f(t,u)\right) \right] \frac{1 + \widehat{\lambda}(t^2 - u^2)}{1 + \widehat{\lambda}} \,, \tag{11.64}$$

where

$$1 + \eta = \frac{1 - \eta_\sigma}{1 - \eta_\mu} \,, \tag{11.65}$$

and the crucial property is that $\eta = -b\lambda + O(\lambda^2)$, $b > 0$. We also recall that both $\eta$ and $\nu$ are functions of $\lambda$ and are independent of $t, u$. Moreover $f(t,u)$ is a suitable bounded function s.t. $|f(t,u) - f(t',u)| \leq c|u|^{-(1+\eta)}|t-t'|$, as it follows from the Lipshitz property of $\pi_{h_1^*+1}$ (11.63). The r.h.s. of (11.64) is Lipshitz in $t$ with constant $O(\lambda)$, so that (11.64) can be inverted w.r.t. $t$ by contractions and, for both choices of the sign, we find a unique solution

$$t = t_c^\pm(\lambda, u) = \sqrt{2} - 1 + \nu^*(\lambda) \pm |u|^{1+\eta}\left(1 + f_\pm^*(\lambda, u)\right) , \tag{11.66}$$

with $|f_\pm^*(\lambda, u)| \leq c|\lambda|$, for some $c$.

Finally the specific heat can be written as

$$C_v = \frac{1}{|\Lambda|} \sum_{\mathbf{x},\mathbf{y}\in\Lambda_M} \sum_{\omega_1,\omega_2=\pm 1} \sum_{h,h'=h_2^*}^{1} \frac{(Z_{h\vee h'}^{(1)})^2}{Z_{h-1}Z_{h'-1}}$$

$$\left[ G^{(h)}_{(+,\omega_1),(+,\omega_2)}(\mathbf{x}-\mathbf{y}) G^{(h')}_{(-,-\omega_2),(-,-\omega_1)}(\mathbf{y}-\mathbf{x}) + \right.$$

$$\left. + G^{(h)}_{(+,\omega_1),(-,-\omega_2)}(\mathbf{x}-\mathbf{y}) G^{(h')}_{(-,-\omega_1),(+,\omega_2)}(\mathbf{x}-\mathbf{y}) \right] +$$

$$\frac{1}{|\Lambda|} \sum_{\mathbf{x},\mathbf{y}\in\Lambda_M} \sum_{h_2^*}^{1} \left(\frac{\overline{Z}_h}{Z_h}\right)^2 \Omega^{(h)}_{\Lambda_M}(\mathbf{x}-\mathbf{y}) \qquad (11.67)$$

where $h \vee h' = \max\{h, h'\}$ and $G^{(h)}_{(\alpha_1,\omega_1),(\alpha_2,\omega_2)}(\mathbf{x})$ must be interpreted as

$$G^{(h)}_{(\alpha_1\omega_1),(\alpha_2,\omega_2)}(\mathbf{x}) = g^{(h)}_{(\alpha_1\omega_1),(\alpha_2,\omega_2)}(\mathbf{x}) \qquad (11.68)$$

if $h > h_1^*$, as $g^{(1,\leq h_1^*)}_{\omega_1,\omega_2}(\mathbf{x}) + g^{(2,h_1^*)}_{\omega_1,\omega_2}(\mathbf{x})$ if $h = h_1^*$ as $g^{(2,h)}_{\omega_1,\omega_2}(\mathbf{x})$ if $h_2^* < h < h_1^*$, and as $g^{(2,\leq h_2^*)}_{\omega_1,\omega_2}(\mathbf{x})$ if $h = h_2^*$. Moreover, if $N, n_0, n_1 \geq 0$ and $n = n_0 + n_1$, $|\partial_x^{n_0}\partial_{x_0}\Omega^{(h)}_{\Lambda_M}(\mathbf{x})| \leq C_{N,n}|\lambda|\frac{\gamma^{(2+n)h}}{1+(\gamma^h|\mathbf{d}(\mathbf{x})|)^N}$. Now, calling $\eta_0$ the exponent associated to $\overline{Z}_h/Z_h$, from (11.67) we find:

$$C_v = -C_1 \gamma^{2\eta_0 h_1^*} \log_\gamma \gamma^{h_1^* - h_2^*} (1 + \Omega^{(1)}_{h_1^*,h_2^*}(\lambda))$$

$$+ C_2 \frac{1 - \gamma^{2\eta_0(h_1^*-1)}}{2\eta_0} (1 + \Omega^{(2)}_{h_1^*}(\lambda)) \qquad (11.69)$$

where $|\Omega^{(1)}_{h_1^*,h_2^*}(\lambda)|, |\Omega^{(2)}_{h_1^*}(\lambda)| \leq c|\lambda|$, for some $c$, from which Theorem 11.1 follows.

# PART 4
# Quantum Liquids

# Chapter 12

# Spinless Luttinger Liquids

## 12.1 Fermions on a chain

In the last part of the book we consider the low temperature properties of system of interacting fermions, describing conduction electrons in a metal.

We start considering a system of interacting spinless fermions on a $1d$ lattice, with hamiltonian

$$H = -\frac{1}{2}\sum_{x\in\Lambda}(a_x^+ a_{x+1}^- + a_{x+1}^+ a_x^-) + U\sum_{x,y\in\Lambda} v(x-y)a_x^+ a_x^- a_y^+ a_y^- - \mu\sum_{x\in\Lambda} a_{x,\sigma}^+ a_{x,\sigma}^-$$
(12.1)

where $\Lambda$ is an interval of $L$ points on the one dimensional lattice of step 1, which will be chosen equal to $(-[L/2], [(L-1)]/2)$ and $a_x^\pm$ is a set of fermionic creation or annihilation operators on the Fock space satisfying periodic boundary conditions; $U$ is the coupling, $t$ is the hopping parameter, $v(\mathbf{x}-\mathbf{y})$ is a short range potential and $\mu$ the chemical potential.

As we discussed in chapt. 2, the physical properties can be obtained by the knowledge of the Schwinger functions. In the non-interacting case $U=0$ the 2-point Schwinger function $g(\mathbf{x}-\mathbf{y})$ is given by

$$g(\mathbf{x}-\mathbf{y}) = \frac{1}{\beta L}\sum_{\mathbf{k}\in\mathcal{D}} \frac{e^{-i\mathbf{k}(\mathbf{x}-\mathbf{y})}}{-ik_0 + \mu - t(1-\cos k)}$$
(12.2)

with $\mathbf{k} = (k_0, k)$, $\mathcal{D} \equiv \mathcal{D}_L \times \mathcal{D}_\beta$, with $\mathcal{D}_L \equiv \{k = 2\pi n/L, n \in Z, -[L/2] \le n \le [(L-1)/2]\}$ and $\mathcal{D}_\beta \equiv \{k_0 = 2(n+1/2)\pi/\beta, n \in Z\}$.

At zero temperature, $\hat{g}(\mathbf{k})$ is singular at $k_0 = 0$ and $k = \pm p_F^0$, where $p_F^0 = \cos^{-1}(1-\mu)/t$, $v_0 = \sin p_F^0$ and

$$\hat{g}(\mathbf{k}) \simeq_{k_0 \simeq 0, k \simeq \pm p_F^0} \frac{1}{-ik_0 + v_0(|k| - p_F^0)}$$
(12.3)

The presence of singularities for non-vanishing momenta is reflected in the presence of oscillating factors of the Fourier transform, that is

$$g(x-y,0) \simeq_{|x-y|\to\infty} \frac{\cos p_F^0(x-y)}{|x-y|} \qquad (12.4)$$

The occupation number $n_k^0$, by (1.39) and (12.2), is equal to $n_k^0 = 1$ if $|k| \leq p_F^0$ and 0 otherwise.

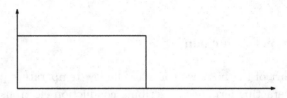

Fig. 12.1  Graphical representation of $n_k^0$

The *density correlation function* (1.40) is given by $\Omega_0(\mathbf{x}) = g(\mathbf{x})g(-\mathbf{x})$, and the static density correlations for $x \neq 0$, $G_0(x) = \Omega_0(x,0)$ can be written as

$$G_0(x) = \frac{1}{2\pi^2 x^2}(1 + \cos 2p_F^0 x)[1 + O(\frac{1}{|x|})] \qquad (12.5)$$

where $\mu = 1 - \cos p_F^0$. Note that the dominant part of $G_0(x)$ has an oscillating and a non oscillating part, both decaying as $O(x^{-2})$ for large $x$.

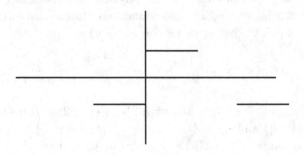

Fig. 12.2  Graphical representation of $G^0(k)$

Denoting by $\widehat{G}_0(k)$ the Fourier transform of $G_0(x)$, we get, if $\varepsilon(k) =$

$1 - \cos k - \mu$

$$\widehat{G}_0(k) = \int_{-\pi}^{\pi} dp \, \chi(\varepsilon(p) < 0) \chi(\varepsilon(p+k) > 0) \qquad (12.6)$$

that is, if $p_F^0 \leq \frac{\pi}{2}$ for definiteness, $\widehat{G}_0(k) = |k|$ if $|k| \leq 2p_F^0$, and $= 2p_F^0$ if $\pi \geq |k| \geq 2p_F^0$. The first derivative is then discontinuous, that is $\partial_k \widehat{G}_0(k) = 1$ for $0 \leq k \leq 2p_F^0$, $\partial_k \widehat{G}_0(k) = -1$ for $-2p_F^0 \leq k \leq 0$ and 0 otherwise.

## 12.2 Grassman representation

Note first we cannot expect that the value of the Fermi momentum and the Fermi velocity are left unchanged from the interaction; in order to take this into account we can write the factor 1 of the hopping term as $(1 - \delta) + \delta \equiv t + \delta$

$$\mu = t \cos p_F - \nu \qquad (12.7)$$

with $\delta, \nu$ suitable counterterms.

It is well know, see for instance Ref.[16], that the Schwinger functions of the model (12.1) can be equivalently rewritten in terms of Grassmann functional integrals, which can be analyzed by the methods used in the previous chapters. Let us temporarily introduce an ultraviolet cutoff on $k_0$ (our lattice model has an intrinsic ultraviolet cut-off in the $k$ variables) in the following way. Consider the smooth support function $H_0(t)$, $t \in \mathbb{R}$, such that $H_0(t) = H_0(-t)$ and $= 1$ if $|t| < 1/\gamma$, and $\in (0, 1)$ if $1/\gamma < |t| < 1$ and $= 0$ if $|t| > e_0$. Then, given a (large) integer $N$, we restrict the set $\mathcal{D}_\beta$ of $k_0$ possible values to the set $\{k_0 = \frac{2\pi}{\beta}(n_0 + \frac{1}{2}) : H_0(\gamma^{-N} k_0) > 0\}$, which we shall denote with the same symbol, as well as the set $\mathcal{D}_{\beta,L}$, which is a finite set for each $N$. Given $N$, we consider the Grassmann algebra generated by the Grassmannian variables $\{\widehat{\psi}_{\mathbf{k},\sigma}^\pm\}_{\mathbf{k} \in \mathcal{D}_{\beta,L}}$ and a Grassmann integration $\int \left[ \prod_{\mathbf{k} \in \mathcal{D}_{\beta,L}} d\widehat{\psi}_\mathbf{k}^+ d\widehat{\psi}_\mathbf{k}^- \right]$. If we define the free propagator $\widehat{g}_\mathbf{k}$ as $\widehat{g}_\mathbf{k} = H_0(\gamma^{-N} k_0) \left[ -ik_0 + t \cos k - t \cos p_F \right]^{-1}$ and the "Gaussian integration" $P(d\psi)$ as

$$P(d\psi) = \left[ \prod_{\mathbf{k} \in \mathcal{D}_{\beta,L}} (L\beta \widehat{g}_\mathbf{k}) d\widehat{\psi}_\mathbf{k}^+ d\widehat{\psi}_\mathbf{k}^- \right] \cdot \exp\left\{ -\sum_{\mathbf{k} \in \mathcal{D}_{\beta,L}} (L\beta \widehat{g}_\mathbf{k})^{-1} \widehat{\psi}_\mathbf{k}^+ \widehat{\psi}_\mathbf{k}^- \right\},$$
$$(12.8)$$

it holds that

$$\int P(d\psi) \widehat{\psi}_{\mathbf{k}_1}^- \widehat{\psi}_{\mathbf{k}_2}^+ = L\beta \delta_{\mathbf{k}_1, \mathbf{k}_2} \widehat{g}_{\mathbf{k}_1}, \qquad (12.9)$$

so that
$$\lim_{N\to\infty} \frac{1}{L\beta} \sum_{\mathbf{k}\in\mathcal{D}_{\beta,L}} e^{-i\mathbf{k}(\mathbf{x}-\mathbf{y})} \widehat{g}_{\mathbf{k}} = \lim_{N\to\infty} \int P(d\psi) \psi^-_{\mathbf{x}} \psi^+_{\mathbf{y}} = g(\mathbf{x}-\mathbf{y}) \quad (12.10)$$

The partition function can be written as
$$e^{-L\beta F_{L,\beta}} = \frac{\text{Tr}[e^{-\beta H}]}{\text{Tr}[e^{-\beta H_0}]} = \lim_{N\to\infty} \int P(d\psi) e^{-V(\psi)} \quad (12.11)$$
where
$$\mathcal{V} = U \int_{-\beta/2}^{\beta/2} dx_0 \sum_{x\in\Lambda} \psi^+_{\mathbf{x},+} \psi^-_{\mathbf{x},+} \psi^+_{\mathbf{x},-} \psi^-_{\mathbf{x},-} + \quad (12.12)$$
$$\nu \int_{-\beta/2}^{\beta/2} dx_0 \sum_{x\in\Lambda,\sigma} \psi^+_{\mathbf{x},\sigma} \psi^-_{\mathbf{x},\sigma} + \delta \int_{-\beta/2}^{\beta/2} dx_0 \sum_{x,y\in\Lambda,\sigma} t_{x,y} \psi^+_{\mathbf{x},\sigma} \psi^-_{\mathbf{x},\sigma} \quad (12.13)$$
where $t_{x,y} = \frac{1}{2}\delta_{y,x+1} + \frac{1}{2}\delta_{x,y+1}$, where the symbol $\int d\mathbf{x}$ must be interpreted as
$$\int d\mathbf{x} = \int_{-\beta/2}^{+\beta/2} dx_0 \sum_{x\in\Lambda} \quad (12.14)$$
Similarly, the Schwinger functions defined can be computed as
$$S(\mathbf{x}_1,\sigma_1,\varepsilon_1;\ldots;\mathbf{x}_n,\sigma_n,\varepsilon_n) = \lim_{N\to\infty} \frac{\int P(d\psi) e^{-V(\psi)} \psi^{\varepsilon_1}_{\mathbf{x}_1,\sigma_1} \cdots \psi^{\varepsilon_n}_{\mathbf{x}_n,\sigma_n}}{\int P(d\psi) e^{-V(\psi)}}.$$
$$(12.15)$$

## 12.3 Luttinger liquid behavior

The methods of non-perturbative renormalizations developed previously allow to prove that a system of spinless fermions on a $1d$ lattice has generically a *Luttinger liquid behavior*, in agreement with the definition given in §2.3. The following theorem was originally proved in Ref. [28], to which we refer for more details.

**Theorem 12.1.** *Assume $-1 < \mu < 1$ and $\mu \neq 0$; there exists an $\varepsilon > 0$ such that*

*a) the two point Schwinger function is given by, in the limit $L,\beta \to \infty$ and $|x-y| \geq 1$*

$$S(\mathbf{x},\mathbf{y}) = \sum_{\omega=\pm} \frac{e^{i\omega p_F(x-y)}}{v(x_0-y_0) + i\omega(x-y)} \frac{1+A_\omega(\mathbf{x},\mathbf{y})}{|\mathbf{x}-\mathbf{y}|^\eta} + \bar{S}(\mathbf{x},\mathbf{y}) \quad (12.16)$$

with

$$\eta = aU^2 + U^2 f_0(U) \quad p_F = \cos^{-1}\mu + f_1(U) \quad v = v_0 + f_2(U) \quad (12.17)$$

where $a > 0$, $|f_0(U)|, |f_1(U)|, |f_2(U)| \leq CU$ and

$$|\bar{\partial}_x^{n_1} \partial_{x_0}^{n_0} A_\omega(\mathbf{x},\mathbf{y})| \leq CU \frac{1}{|\mathbf{x}-\mathbf{y}|^{n_0+n_1}} \quad |\bar{S}(\mathbf{x},\mathbf{y})| \leq \frac{C}{1+|\mathbf{x}-\mathbf{y}|^{1+\vartheta}} \quad (12.18)$$

for suitable positive constants $C, \vartheta$, if $\bar{\partial}$ denotes the discrete derivative. Moreover the occupation number $n_k$ is continuous at $k = \pm p_F$ but its first derivative diverges at $k = \pm p_F$ as $|k - (\pm p_F)|^{-1+\eta}$.

b) The density-density correlation function can be written as, if $|x| \geq 1$

$$\Omega_{\mathbf{x},0} = \cos(2p_F x)\Omega^a(\mathbf{x}) + \Omega^b(\mathbf{x}) + \Omega^c(\mathbf{x}), \quad (12.19)$$

with

$$\Omega^a(\mathbf{x}) = \frac{1 + A_1(\mathbf{x})}{2\pi^2[x^2 + (vx_0)^2]^{1+\eta_1}}$$

$$\Omega^b(\mathbf{x}) = \frac{1}{2\pi^2[x^2 + (vx_0)^2]} \left\{ \frac{x_0^2 - (x/v_0)^2}{x^2 + (v_0 x_0)^2} + A_2(\mathbf{x}) \right\} \quad (12.20)$$

with $|A_i(\mathbf{x})| \leq C|U|$ and

$$|\Omega^c(\mathbf{x})| \leq \frac{C}{1+|\mathbf{x}-\mathbf{y}|^{2+\vartheta}} \quad |\bar{\partial}_x^{n_1} \partial_{x_0}^{n_0} \Omega^{a,b}(\mathbf{x})| \leq \frac{C}{|\mathbf{x}-\mathbf{y}|^{2+n_0+n_1}}, \quad (12.21)$$

for some constant $C$, where $\eta_1 = -bU + Uf_4(U)$ with $b > 0$ and $|f_4(U)| \leq CU$.

c) Let $\widehat{\Omega}(\mathbf{k})$, $\mathbf{k} = (k, k_0) \in [-\pi, \pi] \times \mathbb{R}^1$, the Fourier transform of $\Omega_{\mathbf{x},0}$. Let $G(x) = \Omega_{\mathbf{x},0}|_{x_0=0}$ and $\widehat{G}(k)$ its Fourier transform. Then $\widehat{G}(k)$ is bounded and $\partial_k \widehat{G}(k)$ has a first order discontinuity at $k = 0$, with a jump equal to $1 + O(U)$, and, at $k = \pm 2p_F$, it behaves as $|k - (\pm 2p_F)|^{2\eta_1}$; for $k \neq 0, \pm 2p_F$ $\partial_k \widehat{G}(k)$ is bounded.

From (12.16) we see that the 2-point Schwinger function have the same asymptotic behaviour as in the Luttinger model (1.51), with the appearance of a positive critical index functions of the interaction. More exactly the 2-point Schwinger function can be written as sum of two terms; one which is very similar to the corresponding quantities in the Luttinger model, and in which the dependence from $p_F$ is quite simple (they can be written as oscillating terms times terms which are free of oscillations, in the sense that each derivative increases the decay by a unit, see (12.21)) and another (non Luttinger like) in which the dependence on $p_F$ and the lattice steps is very complicate; this last term decays faster than the Luttinger like terms but the derivatives do not increase the decay for the presence of oscillating terms. The non Luttinger like terms have Fourier transform which is bounded; however sufficiently high derivatives of the Fourier transform can be singular for values different from $k = 0, \pm p_F, \pm 2p_F$. The interaction radically modify the occupation number, in which the discontinuity is eliminated; this is one of the most dramatic signature of the Luttinger liquid behaviour.

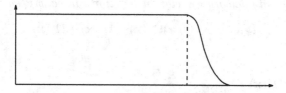

Fig. 12.3 Graphical representation of $n_k$

On the other hand, the interaction leaves invariant the singularity of the first derivative of the one dimensional Fourier transform of the density correlation in $k = 0$ (a first order discontinuity) while the singularity in $k = \pm 2p_F$ is changed by the interaction from a discontinuity to a power law singularity in the repulsive case, or the discontinuity is removed in the attractive case. This enhancement of the singularity at $2p_F$ in the repulsive case is considered a signal of the tendency of the system to develop density wave excitations with period $\pi/p_F$, generically incommensurate with the lattice.

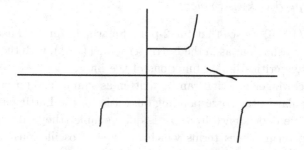

Fig. 12.4 Graphical representation of $\partial \widehat{G}(k)$ in the repulsive case

It is somewhat surprising that the interaction affects only the singulariry at $\pm 2p_F$ in the density correlation, leaving unchanged the singularity at $k=0$; as we will see, the reason is in some hidden symmetries in the model impling the validity of approximate Ward Identities.

## 12.4 The ultraviolet integration

In order to perform a multiscale analysis we note first that the ultraviolet problem is really weak; the lattice has the effect that the momenta are bounded $|k| \leq \pi$ and the ultraviolet problem related to the fact that $k_0$ is unbounded is of the same nature, but much easier, than the ones seen before. On the other hand, the infrared problem has a new feature; when $\beta = \infty$ $g(k)$ is singular in correspondence of two points, namely $(0, p_F)$ and $(0, -p_F)$; this is in contrast with the cases seen above, in which the singularity was at $(0, 0)$.

We start from the generating function

$$e^{\mathcal{W}(J,\phi)} = \int P(d\psi) e^{-\mathcal{V}(\psi) + \int d\mathbf{x}[J_\mathbf{x}, \psi_\mathbf{x}^+ \psi_\mathbf{x}^- + \phi_\mathbf{x}^+ \psi_\mathbf{x}^- + \psi_\mathbf{x}^+ \phi_\mathbf{x}^-]} \quad (12.22)$$

Let $T^1$ be the one dimensional torus, $||k - k'||_{T^1}$ the usual distance between $k$ and $k'$ in $T^1$. We introduce a *scaling parameter* $\gamma > 1$ and a positive smooth function $\chi(\mathbf{k}')$, $\mathbf{k}' = (k', k_0)$, such that $\chi(\mathbf{k}') = 1$ if $|\mathbf{k}'| < t_0 \equiv a_0 v_0/\gamma$ and $\chi(\mathbf{k}') = 0$ if $|\mathbf{k}'| > a_0 v_0$, where $|\mathbf{k}'| = \sqrt{k_0^2 + (v_0 ||k'||_{T^1})^2}$ and $a_0 = \min\{p_F/2, (\pi - p_F)/2\}$, $v_0 = \sin p_F$. The above choice is such that the supports of $\chi(k - p_F, k_0)$ and $\chi(k + p_F, k_0)$ are disjoint and the $C^\infty$ function on $T^1 \times R$

$$\widehat{f}_{u.v.}(\mathbf{k}) \equiv 1 - \chi(k - p_F, k_0) - \chi(k + p_F, k_0) \quad (12.23)$$

is equal to 0, if $[v_0||(|k| - p_F)||_{T^1}]^2 + k_0^2 < t_0^2$. We define

$$g(\mathbf{x} - \mathbf{y}) = g^{(u.v.)}(\mathbf{x} - \mathbf{y}) + g^{(i.r.)}(\mathbf{x} - \mathbf{y}) \quad (12.24)$$

with ($t = 1$ for notational semplicity)

$$g^{(u.v.)}(\mathbf{x} - \mathbf{y}) = \frac{1}{L\beta} \sum_{\mathbf{k} \in \mathcal{D}} e^{-i\mathbf{k}(\mathbf{x}-\mathbf{y})} \frac{\widehat{f}_{u.v.}(\mathbf{k})}{-ik_0 - \cos k + \cos p_F} \quad (12.25)$$

and

$$g^{(i.r.)}(\mathbf{x} - \mathbf{y}) = \frac{1}{L\beta} \sum_{\mathbf{k} \in \mathcal{D}} e^{-i\mathbf{k}(\mathbf{x}-\mathbf{y})} \frac{\prod_{\omega=\pm 1} \chi(k - \omega p_F, k_0)}{-ik_0 - \cos k + \cos p_F} \quad (12.26)$$

$g^{(u.v.)}(\mathbf{x}-\mathbf{y})$ is the *ultraviolet* part of the propagator while $g^{(i.r.)}(\mathbf{x}-\mathbf{y})$ is the infrared part; $\widehat{f}_{u.v.}(\mathbf{k})$ has support far from the points $(0, \pm p_F)$ in which the free propagator is singular while $1 - \widehat{f}_{u.v.}(\mathbf{k})$ has support around the two singularities $(0, \pm p_F)$.

Let us consider the $J = \phi^{\pm} = 0$ case; we write

$$\int P(d\psi) e^{-\mathcal{V}(\psi)} = \int P(d\psi^{(i.r.)}) e^{-\mathcal{V}^{(i.r.)}(\psi^{(i.r.)})} \qquad (12.27)$$

with

$$e^{-\mathcal{V}^{(i.r.)}(\psi^{(i.r.)})} = \lim_{N \to \infty} \int P(d\psi^{[1,N]}) e^{-VV(\psi^{(i.r.)}+\psi^{[1,N]})} \qquad (12.28)$$

where $P(d\psi^{[1,N]})$ is the fermionic integration with propagator

$$g^{[1,N]}(\mathbf{x}-\mathbf{y}) = \lim_{N \to \infty} \sum_{k=1}^{N} g^{(k)}(\mathbf{x}-\mathbf{y}) \qquad (12.29)$$

where

$$g^{(k)}(\mathbf{x}-\mathbf{y}) = \frac{1}{L\beta} \sum_{\mathbf{k} \in \mathcal{D}} h_k(k_0) e^{-i\mathbf{k}(\mathbf{x}-\mathbf{y})} g(\mathbf{k}) \qquad (12.30)$$

with $h_k(k_0) = H_0(\gamma^{-k}|k_0|) - H_0(\gamma^{-k+1}|k_0|)$ and $N$ and $M$ are proportional by the compact support properties of $h_k$. Note that for any integer $K \geq 0$, $g^{(k)}(\mathbf{x}-\mathbf{y})$ satisfies the bound

$$|g^{(k)}(\mathbf{x}-\mathbf{y})| \leq \frac{C_K}{1 + (\gamma^k(x_0-y_0)| + |x-y|)^K} \qquad (12.31)$$

We associate to any propagator $g^{(k)}(\mathbf{x}, \mathbf{y})$ a Grassmann field $\psi^{(k)}$ and an integration $P(d\psi^{(k)})$ with propagator $g^{(k)}(\mathbf{x}-\mathbf{y})$. We can integrate iteratively the fields on scale $N, N-1, \ldots, h+1$ and after each integration we can rewrite the r.h.s. of (12.27) in terms of a new effective potential $\mathcal{V}^{(h)}$ of the form

$$\mathcal{V}^{(h)}(\psi^{(\leq h)}) = \sum_{n=1}^{\infty} \int d\mathbf{x}_1 \ldots d\mathbf{x}_n \prod_{i=1}^{2n} \widehat{\psi}_{\mathbf{x}_i}^{(\leq h)\varepsilon_i} W_{2n}^{(h)}(\mathbf{x}_1, \ldots, \mathbf{x}_{2n}) \qquad (12.32)$$

with $W_{2n}^{(h)}$ admitting a representation in terms of *trees*, with value $W_\tau^{(h)}$ bounded by

$$\|W_\tau^{(h)}\| \leq C^n |\lambda|^n \gamma^{-h(n-1)} \prod_{v\, note.p.} \gamma^{-(h_v - h_{v'})(n_v - 1 + z_v)} \qquad (12.33)$$

where $n_v$ is the number of endpoints of type $\lambda$ following $v$ and $z_v = 1$ if $n_v = 1$ and $0$ otherwise. In deriving (12.33) we have used that there is improvment, with respect to the naive dimensional bound, in the contribution to $W_2$ obtained by the self-contraction of $\int dxdy v(\mathbf{x}-\mathbf{y})\psi_\mathbf{x}^+\psi_\mathbf{x}^-\psi_\mathbf{y}^+\psi_\mathbf{y}^-$. Indeed there are two possible graphs; the first is

$$\widehat{v}(0)\int d\mathbf{k}\frac{f_h(\mathbf{k})}{-ik_0+\varepsilon(k)-\mu} = \widehat{v}(0)\int d\mathbf{k}\frac{f_h(\mathbf{k})(\varepsilon(k)-\mu)}{(-ik_0+\varepsilon(k)-\mu)(ik_0+\varepsilon(k)-\mu)} \tag{12.34}$$

and the second is

$$\int d\mathbf{k}\widehat{v}(\mathbf{p})\frac{f_h(\mathbf{k}-\mathbf{p})}{-i(k_0-p_0)+\varepsilon(k-p)-\mu} \tag{12.35}$$

which are $O(\gamma^{-h_v})$ instead as $O(1)$, as in the naive dimensional bound.

## 12.5 Quasi-particle fields

We consider now the integration of the r.h.s. of (12.27). Contrary to the cases treated up to now, the infrared singularity of the propagator is given by two points $\pm p_F$; this suggests to decompose the fermionic field in two fields each of them with propagator singular in a single point.

We can rewrite $g^{(i.r.)}(\mathbf{x}-\mathbf{y})$ as

$$g^{(i.r.)}(\mathbf{x}-\mathbf{y}) = \sum_{\omega=\pm 1} g_\omega^{(\leq 0)}(\mathbf{x}-\mathbf{y}) \tag{12.36}$$

with

$$g_\omega^{(i.r.)}(\mathbf{x}-\mathbf{y}) = \frac{1}{L\beta}\sum_{k\in\mathcal{D}}e^{-i\mathbf{k}(\mathbf{x}-\mathbf{y})}\frac{\chi(k-\omega p_F,k_0)}{-ik_0-\cos k+\cos p_F} \tag{12.37}$$

and performing the change of variable $\mathbf{k}=\mathbf{k}'+\omega p_F$ we obtain

$$g_\omega^{(i.r.)}(\mathbf{x}-\mathbf{y}) \equiv e^{i\omega p_F(x-y)}\bar{g}_\omega^{(\leq 0)}(\mathbf{x}-\mathbf{y}) = \tag{12.38}$$

$$e^{i\omega p_F(x-y)}\frac{1}{L\beta}\sum_{k\in\mathcal{D}}e^{-i\mathbf{k}(\mathbf{x}-\mathbf{y})}\frac{\chi(k,k_0)}{-ik_0-\omega\sin p_F\sin k'+\cos p_F(\cos k'-1)}$$

By using the addition principle, we can write

$$\psi_\mathbf{x}^{\pm(i.r.)} = \sum_{\omega=\pm}e^{\pm i\omega p_F \mathbf{x}}\psi_{\mathbf{x},\omega}^{(\leq 0)} \tag{12.39}$$

where $\psi_{\mathbf{x},\omega}^{(\leq 0)}$ have propagator $\bar{g}_\omega^{(\leq 0)}(\mathbf{x}-\mathbf{y})$ and are called *quasi-particle fields*; hence we can express the r.h.s. of (12.27) in terms of quasi-particles fields as

$$\int P_{Z_0,C_0}(d\psi^{(\leq 0)})e^{\mathcal{V}^{(0)}} \tag{12.40}$$

where $Z_0 = 1$,
$$P_{Z_0,C_0}(d\psi^{(\leq 0)}) = \mathcal{N}_0^{-1} \prod_{\mathbf{k}' \in \mathcal{D}} \prod_{\omega=\pm 1} d\psi_{\mathbf{k}',\omega}^{+(\leq 0)} d\psi_{\mathbf{k}',\omega}^{-(\leq 0)} \qquad (12.41)$$

$$\exp -\frac{1}{\beta L} \sum_{\substack{\mathbf{k}' \in \mathcal{D} \\ C_0^{-1}(\mathbf{k}) > 0}} Z_0 C_0(\mathbf{k}) \psi_{\mathbf{k}',\omega}^{+(\leq 0)}(-ik_0 + \omega v_0 \sin k' + \cos p_F (\cos k' - 1)) \psi_{\mathbf{k}',\omega}^{-(\leq h)}$$

$$(12.42)$$

and $\mathcal{V}^0$ is obtained from (12.32) by (12.39) and it is given by
$$\mathcal{V}^{(0)}(\psi^{(\leq 0)}) = \qquad (12.43)$$

$$\sum_{n=1}^{\infty} \frac{1}{(L\beta)^{2n}} \sum_{\underline{\sigma}} \sum_{\mathbf{k}_1,\ldots,\mathbf{k}_{2n}} \prod_{i=1}^{2n} \hat{\psi}_{\mathbf{k}_i,\omega_i}^{(\leq 0)\varepsilon_i} \widehat{W}_{2n,\underline{\sigma},\underline{\omega}}^{(0)}(\mathbf{k}_1,\ldots,\mathbf{k}_{2n-1}) \delta(\sum_{i=1}^{2n} \varepsilon_i(\mathbf{k}_i + \omega_i p_F))$$

where $\underline{\omega} = (\omega_1, \ldots, \omega_{2n})$ and we used the notation
$$\delta(\mathbf{k}) = \delta(k)\delta(k_0), \quad \delta(k) = L \sum_{n \in \mathbb{Z}} \delta_{k,2\pi n}, \quad \delta(k_0) = \beta \delta_{k_0,0}. \qquad (12.44)$$

We can now perform the following decomposition
$$g_\omega^{(\leq 0)}(\mathbf{x} - \mathbf{y}) = \sum_{h=-\infty}^{0} g_\omega^{(h)}(\mathbf{x} - \mathbf{y}) \qquad (12.45)$$

and
$$g_\omega^{(h)}(\mathbf{x} - \mathbf{y}) = \frac{1}{L\beta} \sum_{\mathbf{k}' \in \mathcal{D}} e^{-i\mathbf{k}'(\mathbf{x}-\mathbf{y})} \frac{f_h(\mathbf{k}')}{-ik_0 - \cos(k' + \omega p_F) + \cos p_F} \qquad (12.46)$$

and $f_h(\mathbf{k}') = \chi(\gamma^{-h}\mathbf{k}') - \chi(\gamma^{-h+1}\mathbf{k}')$ and $\chi(\mathbf{k}') = \sum_{h=-\infty}^{0} f_h(\mathbf{k}')$; finally we define $C_h^{-1}(\mathbf{k}') = \sum_{k=-\infty}^{h} f_k(\mathbf{k}')$. Note that in the support of $f_h(\mathbf{k}')$ the denominator of $\hat{g}_\omega^{(h)}(\mathbf{k})$ is $O(\gamma^h)$; it holds

$$|g_\omega^{(h)}(\mathbf{x} - \mathbf{y})| \leq \gamma^h \frac{C_N}{1 + (\gamma^h |\mathbf{x} - \mathbf{y}|)^N} \qquad (12.47)$$

It is convenient to decompose the propagator in the following way
$$g_\omega^{(h)}(\mathbf{x} - \mathbf{y}) = g_{\omega,L}^{(h)}(\mathbf{x} - \mathbf{y}) + r_\omega^{(h)}(\mathbf{x} - \mathbf{y}) \qquad (12.48)$$

with
$$g_{\omega,L}^{(h)}(\mathbf{x} - \mathbf{y}) = \frac{1}{\beta L} \sum_{\mathbf{k}' \in \mathcal{D}} e^{-i\mathbf{k}(\mathbf{x}-\mathbf{y})} \frac{f_h(\mathbf{k}')}{-ik_0 + \omega k'} \qquad (12.49)$$

and
$$|r_\omega^{(h)}(\mathbf{x} - \mathbf{y})| \leq \gamma^{2h} \frac{C_N}{1 + (\gamma^h |\mathbf{x} - \mathbf{y}|)^N} \qquad (12.50)$$

that is $r_\omega^{(h)}(\mathbf{x}-\mathbf{y})$ has an extra small factor $\gamma^h$ in the bound with respect to (12.47).

The multiscale integration is very similar to the one in chapt. 4.; assume that we have integrated the fields $\psi^{(h+1)}, \psi^{(h+1)}, ...$ obtaining an expression of the form

$$\int P_{Z_h,C_h}(d\psi^{(\leq h)})\, e^{-\mathcal{V}^{(h)}(\sqrt{Z_h}\psi^{(\leq h)})-L\beta E_h}, \quad \mathcal{V}^{(h)}(0)=0, \quad (12.51)$$

where

$$P_{Z_h,C_h}(d\psi^{(\leq h)}) = \mathcal{N}_h^{-1} \prod_{\mathbf{k}'\in D}\prod_{\omega=\pm 1} d\psi_{\mathbf{k}',\omega}^{+(\leq h)} d\psi_{\mathbf{k}',\omega}^{-(\leq h)} \exp[-\frac{1}{\beta L}\sum_{\substack{\mathbf{k}'\in D \\ C_h^{-1}(\mathbf{k})>0}}$$

$$Z_h C_h(\mathbf{k})\psi_{\mathbf{k}',\omega,\sigma}^{+(\leq h)}(-ik_0 + \omega v_0 \sin k' + \cos p_F(\cos k'-1))\psi_{\mathbf{k}',\omega,\sigma}^{-(\leq h)}] \quad (12.52)$$

and

$$\mathcal{V}^{(h)}(\psi^{(\leq h)}) = \sum_{n=1}^{\infty} \frac{1}{(L\beta)^{2n}} \sum_{\underline{\sigma}} \sum_{\mathbf{k}'_1,...,\mathbf{k}'_{2n}} \prod_{i=1}^{2n} \widehat{\psi}_{\mathbf{k}'_i,\omega_i,\sigma_i}^{(\leq 0)\varepsilon_i}$$

$$\cdot \widehat{W}_{2n,\underline{\sigma},\underline{\omega}}^{(h)}(\mathbf{k}'_1,...,\mathbf{k}'_{2n-1})\,\delta(\sum_{i=1}^{2n}\varepsilon_i \mathbf{k}'_i + \sum_{i=1}^{2n}\varepsilon_i\omega_i p_F) \quad (12.53)$$

We split the effective potential $\mathcal{V}^{(h)}$ as $\mathcal{L}\mathcal{V}^{(h)}+\mathcal{R}\mathcal{V}^{(h)}$, where $\mathcal{L}$ is defined in the following way

1) If $2n = 4$, then

$$\mathcal{L}\widehat{W}_{4,\underline{\sigma},\underline{\omega}}^{(h)}(\mathbf{k}'_1,\mathbf{k}'_2,\mathbf{k}'_3) = \widehat{W}_{4,\underline{\sigma},\underline{\omega}}^{(h)}(\bar{\mathbf{k}}_{++},\bar{\mathbf{k}}_{++},\bar{\mathbf{k}}_{++}), \quad (12.54)$$

where

$$\bar{\mathbf{k}}_{\eta\eta'} = \left(\eta\frac{\pi}{L}, \eta'\frac{\pi}{\beta}\right). \quad (12.55)$$

2) If $2n = 2$ and, possibly after a suitable permutation of the fields, $\underline{\sigma} = (+,-)$, then

$$\mathcal{L}\widehat{W}_{2,\underline{\sigma},\underline{\omega}}^{(h)}(\mathbf{k}') = \frac{1}{4}\sum_{\eta,\eta'=\pm 1}\widehat{W}_{2,\underline{\sigma},\underline{\omega}}^{(h)}(\bar{\mathbf{k}}_{\eta\eta'})\cdot$$

$$\cdot\left\{1 + \left[\eta\frac{L}{\pi}\left(b_L + a_L\frac{E(k')}{v_0^*}\right) + \eta'\frac{\beta}{\pi}k_0\right]\right\} \quad (12.56)$$

where $E(k') = v_0 \sin k' + \cos p_F(\cos k'-1)$,

$$a_L\frac{L}{\pi}\sin\frac{\pi}{L} = 1, \quad \frac{\cos p_F}{v_0}(1-\cos\frac{\pi}{L}) + b_L\frac{L}{\pi}\sin\frac{\pi}{L} = 0. \quad (12.57)$$

In order to better understand this definition, note that, if $L = \beta = \infty$,

$$\mathcal{L}\widehat{W}_{2,\underline{\sigma},\underline{\omega}}^{(h)}(\mathbf{k}') = \widehat{W}_{2,\underline{\sigma},\underline{\omega}}^{(h)}(0) + \delta_{\omega_1,\omega_2}\left[\frac{E(k')}{v_0^*}\frac{\partial \widehat{W}_{2,\underline{\sigma},\underline{\omega}}^{(h)}}{\partial k'}(0) + k_0 \frac{\partial \widehat{W}_{2,\underline{\sigma},\underline{\omega}}^{(h)}}{\partial k_0}(0)\right].$$
(12.58)

3) In all the other cases

$$\mathcal{L}\widehat{W}_{2n,\underline{\sigma},\underline{\omega}}^{h}(\mathbf{k}'_1,\ldots,\mathbf{k}'_{2n-1}) = 0.$$
(12.59)

The definition of $\mathcal{L}$ is essentially identical to the one in chapt.4 up to the fact that the lattice and of the finite volume and temperature is takn into account.

We get

$$\mathcal{L}\mathcal{V}^{(h)}(\psi) = \gamma^h n_h F_\nu^{(h)}(\psi) + z_h F_z^{(h)}(\psi) + a_h F_a^{(h)}(\psi) + l_h F^{(h)}(\psi) \quad (12.60)$$

where

$$F_\nu = \frac{1}{\beta L}\sum_{\mathbf{k}'}\sum_{\omega,\sigma}\psi^+_{\mathbf{k}',\omega,\sigma}\psi^-_{\mathbf{k}',\omega,\sigma}$$

$$F_z = \frac{1}{\beta L}\sum_{\mathbf{k}'}(-ik_0)\sum_{\omega,\sigma}\psi^+_{\mathbf{k}',\omega,\sigma}\psi^-_{\mathbf{k}',\omega,\sigma}$$

$$F_a = \frac{1}{\beta L}\sum_{\mathbf{k}'}[\omega \sin p_F \sin k' + \cos p_F(\cos k' - 1)]\sum_{\omega,\sigma}\psi^+_{\mathbf{k}',\omega,\sigma}\psi^-_{\mathbf{k}',\omega,\sigma} \quad (12.61)$$

$$F_\lambda = \frac{1}{(\beta L)^4}\sum_{\mathbf{k}'_1,\mathbf{k}'_2,\mathbf{k}'_3,\mathbf{k}'_4}\sum_\omega \delta(\sum_i \varepsilon_i \mathbf{k}'_i) \quad \psi^+_{\mathbf{k}'_1,\omega}\psi^-_{\mathbf{k}'_2,\omega}\psi^+_{\mathbf{k}'_3,-\omega}\psi^-_{\mathbf{k}'_4,-\omega}$$

We change the free integration in the following way We write (12.62) as

$$\int P_{Z_h,C_h}(d\psi^{(\leq h)})e^{-\mathcal{L}\mathcal{V}^{(h)}(\sqrt{Z_h}\psi^{(\leq h)}) - \mathcal{R}\mathcal{V}^{(h)}(\sqrt{Z_h}\psi^{(\leq h)}) - L\beta E_h}, \quad (12.62)$$

and we include the quadratic part of $\mathcal{L}\mathcal{V}^{(h)}$ given by $z_h \int d\mathbf{k}' \sum_{\omega,\sigma}\psi^+_{\mathbf{k}',\omega,\sigma}(-ik_0 + \omega \sin k' + \cos p_F(\cos k' - 1))\psi^-_{\mathbf{k}',\omega,\sigma}$ in the free integration; we call

$$\mathcal{L}\bar{\mathcal{V}}^h = \mathcal{L}\mathcal{V}^{(h)} - z_h \int d\mathbf{k}'\sum_{\omega,\sigma}\psi^+_{\mathbf{k}',\omega,\sigma}(-ik_0 + \omega \sin k' + \cos p_F(\cos k' - 1))\psi^-_{\mathbf{k}',\omega,\sigma}$$
(12.63)

so that we obtain

$$\int P_{\widetilde{Z}_{h-1},C_h}(d\psi^{(\leq h)})e^{-\mathcal{L}\bar{\mathcal{V}}^h(\sqrt{Z_h}\psi^{(\leq h)}) - \mathcal{R}\mathcal{V}^{(h)}(\sqrt{Z_h}\psi^{(\leq h)}) - L\beta E_h}, \quad (12.64)$$

where

$$\widetilde{Z}_{h-1}(\mathbf{k}) = Z_h(1 + z_h C_h^{-1}(\mathbf{k})) \quad (12.65)$$

It is convenient to rescale the fields:

$$\widehat{\mathcal{V}}^{(h)}(\sqrt{Z_{h-1}}\psi^{(\leq h)}) = \lambda_h F_\lambda(\sqrt{Z_{h-1}}\psi^{(\leq h)}) + \quad (12.66)$$

$$+\delta_h F_a(\sqrt{Z_{h-1}}\psi^{(\leq h)}) + \gamma^h \nu_h F_\nu(\sqrt{Z_{h-1}}\psi^{(\leq h)}) + \mathcal{R}\mathcal{V}^{(h)}(\sqrt{Z_h}\psi^{(\leq h)}),$$
$$(12.67)$$

and we integrate $\psi^{(h)}$; the procedure can be then iterated.

Again we get that the kernels of the effective potential admit an expansion in terms of trees which is well defined provided that the running coupling constants are small enough, by a sligth adaptation of Lemma 4.2.

## 12.6 The flow of the running coupling constants

After the decomposition (12.39), we see that the analysis of a system of interacting spinless fermions is remarkably similar to the infrared problem of massless QED2; the propagators coincides up to small corrections which is $O(\gamma^h)$ smaller and the effective interactions differs only for the presence of the $\nu_h, \delta_h$ terms which (as we will see) can be chosen as $O(\gamma^h)$ by properly choosing the counterterms $\nu, \delta$. The fact that the effective action of the model (13.1) becomes closer and closer to the effective action of infared QED2 will be crucial in order to control the flow of the running coupling constants, as we can use the cancellations due to the relativistic symmetry we have discussed in chapt. 6, even if the model (13.1) describe non relativistic fermions. This remarkable analogy between non relativistic fermions in $d=1$ and QFT in $d=1+1$ was discovered by Tomonaga Ref.[51] starting from the simple observation that a non-relativistic quadratic dispersion relation become linear close to the Fermi surface, up to small corrections.

In order to control the flow of the running coupling constants we choose $\nu, \delta$ so that $|\nu_h| \leq cU\gamma^{\vartheta h}$, $|\delta_h| \leq cU\gamma^{\vartheta h}$ for any $h$. We can write the beta function for $\delta_h$ as

$$\beta_\delta^{(h)} = \beta_{\delta,a}^{(h)} + \beta_{\delta,b}^{(h)} \quad (12.68)$$

where $\beta_{\delta,a}^{(h)}$ us given by a sum of trees with no end-points $\nu_k, \delta_k$ and only propagators $g_{L,\omega}^{(k)}$ (4.77); by the symmetry in the exchange $x, x_0$ of $g_{L,\omega}^{(k)}$, and remembering that $\beta_\delta^{(h)} = \sum_\tau [z(\tau) - a(\tau)]$ it holds that

$$|\beta_{\delta,a}^{(h)}| \leq C|U|\gamma^{2\vartheta h} \quad (12.69)$$

A similar decomposition can be done also for

$$\beta_\nu^{(h)} = \beta_{\nu,a}^{(h)} + \beta_{\nu,b}^{(h)} \quad (12.70)$$

again with
$$|\beta_{\nu,a}^{(h)}| \leq C|U|\gamma^{2\vartheta h} \tag{12.71}$$

by the parity property $g_{L,\omega}^{(h)}(\mathbf{x},\mathbf{y}) = -g_{L,\omega}^{(h)}(\mathbf{y},\mathbf{x})$. If we want to fix $\nu,\delta$ in such a way that $\nu_{-\infty} = \delta_{-\infty} = 0$, we must have, if $(\nu_1,\delta_1) = (\nu,\delta)$:

$$\nu = -\sum_{k=-\infty}^{1} \gamma^{k-2}\beta_\nu^{(k)}(g_k,\delta_k,\nu_k;\ldots;g_1,\delta_1,\nu_1). \tag{12.72}$$

$$\delta = -\sum_{k=-\infty}^{1} \beta_\delta^{(k)}(g_k,\delta_k,\nu_k;\ldots;g_1,\delta_1,\nu_1). \tag{12.73}$$

If we manage to fix $\nu,\delta$ as in (12.72), (12.73) we also get:

$$\nu_h = -\sum_{k\leq h} \gamma^{k-h-1}\beta_\nu^{(k)}(g_k,\delta_k,\nu_k;\ldots;g_1,\delta_1,\nu_1). \tag{12.74}$$

$$\delta_h = -\sum_{k\leq h} \beta_\delta^{(k)}(g_k,\delta_k,\nu_k;\ldots;g_1,\delta_1,\nu_1). \tag{12.75}$$

and by a fixed point argument similar to the previous ones it follows that $|\nu_h| \leq cU\gamma^{\vartheta h}$, $|\delta_h| \leq cU\gamma^{\vartheta h}$ for any $h$.

We have finally to prove that effective coupling $\lambda_h$ remain close to its initial value. The flow equation for $\lambda_h$ is

$$\lambda_{h-1} = \lambda_h + \beta_\lambda^h(\lambda_h,\nu_h;\ldots;\lambda_1,\nu_1) \tag{12.76}$$

and we can proceed as in chapt. 10 writing the beta function as

$$\beta_\lambda^h(\lambda_h,\nu_h;\ldots;\lambda_1,\nu_1) = \beta_{\lambda,L,h}(\lambda_h,\ldots,\lambda_h) +$$
$$\sum_{k=h+1}^{1} D_\lambda^{h,k} + r_\lambda^h(\lambda_h,\ldots,\lambda_1) + \sum_{k\geq h} \nu_k \widetilde{\beta}_\lambda^{h,k} + \sum_{k\geq h} \delta_k \widehat{\beta}_\lambda^{h,k} \tag{12.77}$$

where $\beta_{\lambda,L,h}$ is the same appearing in QED2 (with $K = 0$), for which Theorem 6.1 holds, so that

$$|\beta_{\lambda,L,h}| \leq c|\lambda|^2 \gamma^{\vartheta h}, \quad |D_\lambda^{h,k}| \leq c|\lambda|\gamma^{\vartheta(h-k)}|\lambda_k - \lambda_h|,$$
$$|r_\lambda^h| \leq c|\lambda|^2 \gamma^{(\vartheta/2)h}, \quad |\widetilde{\beta}_\lambda^{h,k}| \leq c|\lambda|\gamma^{\vartheta(h-k)} \tag{12.78}$$

We get then

$$\lambda_h = U + O(U^2), |\nu_h|,|\delta_h| \leq C|U|\gamma^h \quad \frac{Z_h}{Z_{h-1}} = 1 + O(U^2) \tag{12.79}$$

## 12.7 Density correlations

The density-density correlation can be written in terms of a Grassmann integral in the following way

$$< \rho(\mathbf{x})\rho(\mathbf{y}) >_T = \frac{\partial^2 \mathcal{S}}{\partial \phi(\mathbf{x}) \partial \phi(\mathbf{y})} \qquad (12.80)$$

where

$$\mathcal{S}(\phi) = \log \int P(d\psi) e^{-\mathcal{V} - \int d\mathbf{x} J(\mathbf{x}) \psi_\mathbf{x}^+ \psi_\mathbf{x}^-} \qquad (12.81)$$

We shall evaluate $\mathcal{S}$ in a way which is very close to that used for the integration of the partition function. We introduce the scale decomposition described above and we perform iteratively the integration of the single scale fields, starting from the field of scale 1.

After integrating the fields $\psi^{(1)}, \ldots \psi^{(h+1)}$ we find

$$e^{\mathcal{S}(\phi)} = e^{-L\beta E_h + S^{(h+1)}(\phi)} \int P_{Z_h, C_h}(d\psi^{\leq h}) e^{-\mathcal{V}^{(h)}(\sqrt{Z_h}\psi^{(\leq h)}) + \mathcal{B}^{(h)}(\sqrt{Z_h}\psi^{(\leq h)}, \phi)}, \qquad (12.82)$$

where $P_{Z_h, \sigma_h, C_h}(d\psi^{(\leq h)})$ and $\mathcal{V}^h$ are given by (12.52) and (12.53), respectively, while $S^{(h+1)}(\phi)$, which denotes the sum over all the terms dependent on $\phi$ but independent of the $\psi$ field, and $\mathcal{B}^{(h)}(\psi^{(\leq h)}, \phi)$, which denotes the sum over all the terms containing at least one $\phi$ field and two $\psi$ fields, can be represented in the form, if $\int d\mathbf{x} = \int_{-\frac{\beta}{2}}^{\frac{\beta}{2}} dx_0 \sum_{x \in \Lambda}$

$$S^{(h+1)}(\phi) = \sum_{m=1}^{\infty} \int d\mathbf{x}_1 \cdots d\mathbf{x}_m S_m^{(h+1)}(\mathbf{x}_1, \ldots, \mathbf{x}_m) \left[\prod_{i=1}^{m} \phi(\mathbf{x}_i)\right] \qquad (12.83)$$

$$\mathcal{B}^{(h)}(\psi^{(\leq h)}, \phi) = \sum_{m=1}^{\infty} \sum_{n=1}^{\infty} \sum_{\underline{\sigma},\underline{\omega}} \int d\mathbf{x}_1 \cdots d\mathbf{x}_m d\mathbf{y}_1 \cdots d\mathbf{y}_{2n} \cdot$$

$$\cdot B_{m,2n,\underline{\sigma},\underline{\omega}}^{(h)}(\mathbf{x}_1, \ldots, \mathbf{x}_m; \mathbf{y}_1, \ldots, \mathbf{y}_{2n}) \left[\prod_{i=1}^{m} \phi(\mathbf{x}_i)\right] \left[\prod_{i=1}^{2n} \psi_{\mathbf{y}_i, \omega_i}^{(\leq h)\sigma_i}\right]$$

Since the field $\phi$ is equivalent, from the point of view of dimensional considerations, to two $\psi$ fields, the only terms in the r.h.s. of (12.84) which are not irrelevant are those with $m = 1$ and $n = 1$, which are marginal. Hence we extend the definition of the localization operator $\mathcal{L}$, so that its action on $\mathcal{B}^{(h)}(\psi^{(\leq h)}, \phi)$ in described in the following way, by its action on the kernels $B_{m,2n,\sigma,\omega}^{(h)}(\mathbf{p}, \mathbf{k}_1, \ldots, \mathbf{k}_n)$: If $m = 1$, $n = 1$ then

$$\mathcal{L} B_{1,2,\sigma,\omega}^{(h)}(\mathbf{p}; \mathbf{k}_1, \mathbf{k}_2) = B_{1,2,\sigma,\omega}^{(h)}(0; 0, 0) \qquad (12.84)$$

and $\mathcal{L} = 0$ in all the other cases.

It follows that

$$\mathcal{L}\mathcal{B}^{(h)}(\psi^{(\le h)}, \phi) = \frac{Z_h^{(1)}}{Z_h} F_1^{(\le h)} + \frac{Z_h^{(2)}}{Z_h} F_2^{(\le h)},  \quad (12.85)$$

where $Z_h^{(1)}$ and $Z_h^{(2)}$ are real numbers, such that $Z_1^{(1)} = Z_1^{(2)} = 1$ and

$$F_1^{(\le h)} = \sum_\omega \int d\mathbf{x}\phi(\mathbf{x}) e^{2i\omega \mathbf{p}_F x} \psi_{\mathbf{x},\omega}^{(\le h)+} \psi_{\mathbf{x},-\omega}^{(\le h)-}, \quad (12.86)$$

$$F_2^{(\le h)} = \sum_{\sigma=\pm 1} \int d\mathbf{x}\phi(\mathbf{x}) \psi_{\mathbf{x},\omega}^{(\le h)\sigma} \psi_{\mathbf{x},\omega}^{(\le h)-}. \quad (12.87)$$

Of course also the new renormalization constants related to the density-density correlation function obey to a Beta function equation of the form

$$\frac{Z_{h-1}^{(i)}}{Z_h^{(i)}} = 1 + z_h^{(i)}, \quad i = 1, 2. \quad (12.88)$$

When $i = 2$, the flow equation can be written as

$$\frac{Z_{h-1}^{(2)}}{Z_h^{(i)}} = 1 + \beta_z^h(\lambda_h, .., \lambda_1) + r_z^h \quad (12.89)$$

where $\beta_z^h(\lambda_h, .., \lambda_1)$ is given by a sum of trees with all end-points with $h_v \le -1$ and propagators with $g_L^h$, and $r_z^h$ is the rest; by the short memory property and (12.50) we get $|r_z^h| \le C\lambda_h^2 g^{\vartheta h}$. On the other hand $\beta_z^h(\lambda_h, .., \lambda_h)$ is identical to the one in Theorem 6.1 and it is bounded by $|\beta_z^h| \le \lambda_h^2 g^{\vartheta h}$ so that

$$|\frac{Z_{h-1}^{(2)}}{Z_h^{(i)}} - 1| \le C|U| \quad (12.90)$$

On the other hand

$$Z_h^{(1)} = \gamma^{\eta_1 h}(1 + O(U)) \quad (12.91)$$

with $\eta_1 = aU^2 + O(U^3)$.

The density correlations can be then written as

$$N(\mathbf{x}, \mathbf{y}) = \cos 2p_F(x-y) H^a(\mathbf{x}, \mathbf{y}) + H^b(\mathbf{x}, \mathbf{y}) + H^c(\mathbf{x}, \mathbf{y}) \quad (12.92)$$

where

$$H^a(\mathbf{x}, \mathbf{y}) = \sum_{h=-\infty}^{0} [\frac{Z_h^{(1)}}{Z_h}]^2 [\sum_{\omega=\pm} g_\omega^{(h)}(\mathbf{x}, \mathbf{y}) g_{-\omega}^{(h)}(\mathbf{y}, \mathbf{x}) + \bar{\Omega}_a^{(h)}(\mathbf{x}, \mathbf{y})] \quad (12.93)$$

$$H^b(\mathbf{x},\mathbf{y}) = \sum_{h=-\infty}^{1} [\frac{Z_h^{(2)}}{Z_h}]^2 [\sum_{\omega=\pm} g_\omega^{(h)}(\mathbf{x},\mathbf{y})g_\omega^{(h)}(\mathbf{y},\mathbf{x}) + \Omega_b^{(h)}(\mathbf{x},\mathbf{y})] \quad (12.94)$$

$$H^c = \sum_{h=-\infty}^{0} \bar{\Omega}_c^{(h)}(\mathbf{x},\mathbf{y}) \quad (12.95)$$

where we include in $\bar{\Omega}_{a,\omega}^{(h)}(\mathbf{x},\mathbf{y}), \bar{\Omega}_{b,\omega}^{(h)}(\mathbf{x},\mathbf{y})$ only the terms obtained contracting $\mathcal{L}_4\mathcal{V}^{(k)}$, $k \leq 0$ with propagators $g_L^{(h)}(\mathbf{r})$ and with two vertices $Z_k^{(1)}$ or $Z_k^{(2)}$ respectively; it holds that for $i = a, b$, as a consequence of the fact that all the oscillating factor $e^{\pm i p_F x_i}$ cancel out as $\sum_i \varepsilon_i \omega_i$ in the monomials in $\mathcal{L}_4\mathcal{V}^{(k)}$

$$|\partial^n \bar{\Omega}_{a,b}^{(h)}(\mathbf{x},\mathbf{y})| \leq \gamma^{(2+n)h}|U|\frac{C_N}{1+(\gamma^h|\mathbf{x}-\mathbf{y}|)^N} \quad (12.96)$$

Moreover

$$|\bar{\Omega}_c^{(h)}(\mathbf{x},\mathbf{y})| \leq \gamma^{\frac{5}{2}h}|U|\frac{C_N}{1+(\gamma^h|\mathbf{x}-\mathbf{y}|)^N} \quad (12.97)$$

and the extra $\gamma^{\frac{h}{2}}$ in the above bound is due to short memory property,togheter with the fact that $\bar{\Omega}_c^{(h)}(\mathbf{x},\mathbf{y})$ is sum of terms containing or a $\nu_k, \delta_h$ (remember that $|\nu_k| \leq CU\gamma^{\frac{k}{2}}, |\delta_k| \leq C|U|\gamma^{\frac{k}{2}}$), or $r_\omega^{(k)}(\mathbf{x}-\mathbf{y})$ or $g_\omega^{(u.v.)}(\mathbf{x}-\mathbf{y})$. We can write

$$|H^c(\mathbf{x},\mathbf{y})| \leq \frac{C|U|}{1+|\mathbf{x}-\mathbf{y}|^{\frac{5}{2}}} \quad (12.98)$$

Its bidimensional Fourier transform, or the one dimensional Fourier transform and its derivative, are then trivially bounded. Moreover it holds that

$$|\partial^n H^a(\mathbf{x},\mathbf{y})| \leq \frac{C}{1+|\mathbf{x}-\mathbf{y}|^{2-\eta+n}} \quad |\partial^n H^b(\mathbf{x},\mathbf{y})| \leq \frac{C}{1+|\mathbf{x}-\mathbf{y}|^{2+n}} \quad (12.99)$$

with $\eta = O(U)$ and positive and $H^a(\mathbf{x},\mathbf{y}) = H^a(-\mathbf{x},-\mathbf{y})$, $H^b(\mathbf{x},\mathbf{y}) = H^b(-\mathbf{x},-\mathbf{y})$; this is due to the fact that $H^a(\mathbf{x},\mathbf{y})$ and $H^b(\mathbf{x},\mathbf{y})$ are sum over an even number of odd propagators $g_L^h(\mathbf{r}) = -g_L^h(-\mathbf{r})$. We see that $H^a(\mathbf{x},\mathbf{y})$ and $H^b(\mathbf{x},\mathbf{y})$ are free of oscillations, and the only oscillating factor is the prefactor $\cos 2p_F(x-y)$ in the first term; on the other hand $H^c(\mathbf{x},\mathbf{y})$ has oscillating factors with period $\frac{2\pi}{2np_F}$ with any $n$ but it has a much faster decay for $|\mathbf{x}-\mathbf{y}| \to \infty$.

We discuss now the properties of the Fourier transform of $N_\varepsilon(\mathbf{x},\mathbf{y})$. The bidimensional Fourier transform, or the one dimensional Fourier transform and its derivative, of $H^c_\varepsilon(\mathbf{x},\mathbf{y})$ are trivially bounded by (12.98); in particular there are no singularities at $2np_F$ for $|n|\geq 3$ in the weak coupling regime.

Let us consider, for $i=a,b$, $H^i(\mathbf{x},\mathbf{y})|_{\mathbf{y}=0}\equiv H^i_\varepsilon(x,x_0)$; the Fourier transform of $H^i(x,x_0)$ is of course bounded by (12.98) and its derivative is given by

$$|\int dx e^{ikx} ix H^i(x,0)| \leq \qquad (12.100)$$

$$|\frac{1}{k}\int dx[e^{ikx}-1]\partial_x[xH^i_\varepsilon(x,0)]| \leq |\frac{1}{k}\int_{|x|\geq|k|^{-1}} dx[e^{ikx}-1]\partial_x[xH^{i,1}(x,0)]|$$

$$|\frac{1}{k}\int_{|x|\leq|k|^{-1}} dx[e^{ikx}-1-ikx]\partial_x[xH^i(x,0)]|$$

where we used the fact that $\partial_x[xH^i_\varepsilon(x,0)]$ is an even function of $x$. Hence, if $|k|\geq 1$, $|\int dx e^{ikx}H^i_\varepsilon(x,0)|\leq C|k|^{-1}$, while, if $0<|k|\leq 1$

$$|\int dx e^{ikx}H^a_\varepsilon(x,0)|\leq C[1+|k|^{-\eta}] \qquad (12.101)$$

and

$$|\int dx e^{ikx}H^b_\varepsilon(x,0)|\leq C \qquad (12.102)$$

Finally note that the dominant contribution close to $k\simeq 2p_F$ comes from

$$\sum_{h=-\infty}^{0}[\frac{Z^{(1),\varepsilon}_h}{Z_h}]^2\sum_{\omega=\pm}g^{(h)}_\omega(\mathbf{x},\mathbf{y})g^{(h)}_{-\omega}(\mathbf{y},\mathbf{x}) \qquad (12.103)$$

and it is given by, up to a constant

$$\int_{p_F/2}^{\min(-p_F+k,p_F)} dp|p-p_F|^{-\eta} \qquad (12.104)$$

which is equal to $\eta^{-1}|\min(-p_F+k,p_F)-p_F|^{1-\eta}$. On the other hand the dominant contribution close to $k\simeq 0$ comes from

$$\sum_{h=-\infty}^{0}[\frac{Z^{(2),\varepsilon}_h}{Z_h}]^2\sum_{\omega=\pm}g^{(h)}_\omega(\mathbf{x},\mathbf{y})g^{(h)}_\omega(\mathbf{y},\mathbf{x}) \qquad (12.105)$$

and is given by, up to a constant, $\int_{p_F/2}^{\min(-p_F+k,p_F)} dp$.

## 12.8 Quantum spin chains

The model (12.1) is also interesting becouse it can be exactly mapped in the *Heisenberg-Ising XYZ chain*. If $(S_x^1, S_x^2, S_x^3) = \frac{1}{2}(\sigma_x^1, \sigma_x^2, \sigma_x^3)$, for $i = 1, 2, ..., L$, $\sigma_i^\alpha$, $\alpha = 1, 2, 3$, being the Pauli matrices, the Hamiltonian of the $XYZ$ chain is given by

$$H = -\sum_{x=1}^{L-1}[J_1 S_x^1 S_{x+1}^1 + J_2 S_x^2 S_{x+1}^2 + J_3 S_x^3 S_{x+1}^3 + h S_x^3] - h S_L^3 + U_L^1, \quad (12.106)$$

where the last term, to be fixed later, depends on the boundary conditions. The space-time *spin correlation function* at temperature $\beta^{-1}$ is given by

$$\Omega_{L,\beta}^\alpha(\mathbf{x}) = <S_{\mathbf{x}}^\alpha S_0^\alpha>_{L,\beta} - <S_{\mathbf{x}}^\alpha>_{L,\beta}<S_0^\alpha>_{L,\beta}, \quad (12.107)$$

where $\mathbf{x} = (x, x_0)$, $S_{\mathbf{x}}^\alpha = e^{Hx_0} S_x^\alpha e^{-Hx_0}$ and $<.>_{L,\beta} = Tr[e^{-\beta H}.]/Tr[e^{-\beta H}]$ denotes the expectation in the grand canonical ensemble. We shall use also the notation $\Omega^\alpha(\mathbf{x}) \equiv \lim_{L,\beta \to \infty} \Omega_{L,\beta}^\alpha(\mathbf{x})$.

The Hamiltonian (12.106) can be written as a *fermionic interacting spinless Hamiltonian*. In fact, it is easy to check that the operators

$$a_x^\pm \equiv \left[\prod_{y=1}^{x-1}(-\sigma_y^3)\right] \sigma_x^\pm \quad (12.108)$$

are a set of anticommuting operators and that, if $\sigma_x^\pm = (\sigma_x^1 \pm i\sigma_x^2)/2$, we can write

$$\sigma_x^- = e^{-i\pi \sum_{y=1}^{x-1} a_y^+ a_y^-} a_x^-, \quad \sigma_x^+ = a_x^+ e^{i\pi \sum_{y=1}^{x-1} a_y^+ a_y^-}, \quad \sigma_x^3 = 2a_x^+ a_x^- - 1. \quad (12.109)$$

Hence, if we fix the units so that $J_1 + J_2 = 2$ and we introduce the *anisotropy* $u = (J_1 - J_2)/(J_1 + J_2)$, we get

$$H = \sum_{x=1}^{L-1} \left\{ -\frac{1}{2}[a_x^+ a_{x+1}^- + a_{x+1}^+ a_x^-] - \frac{u}{2}[a_x^+ a_{x+1}^+ + a_{x+1}^- a_x^-] \right.$$
$$\left. - J_3(a_x^+ a_x^- - \frac{1}{2})(a_{x+1}^+ a_{x+1}^- - \frac{1}{2}) \right\} - h \sum_{x=1}^{L}(a_x^+ a_x^- - \frac{1}{2}) + U_L^2 \quad (12.110)$$

where $U_L^2$ is the boundary term in the new variables. We choose it so that the fermionic Hamiltonian (14.4) coincides with the Hamiltonian of a fermion system on the lattice with periodic boundary conditions, that is we put $U_L^2$ equal to the term in the first sum in the r.h.s. of (14.4) with

$x = L$ and $a_{L+1}^{\pm} = a_1^{\pm}$. It is easy to see that this choice corresponds to fix the boundary conditions for the spin variables so that

$$U_L^1 = -\frac{1}{2}[\sigma_L^+ e^{i\pi\mathcal{N}}\sigma_1^- + \sigma_L^- e^{i\pi\mathcal{N}}\sigma_1^+] - \frac{u}{2}[\sigma_L^+ e^{i\pi\mathcal{N}}\sigma_1^+ + \sigma_L^- e^{i\pi\mathcal{N}}\sigma_1^-] - \frac{J_3}{4}\sigma_L^3\sigma_1^3, \tag{12.111}$$

where $\mathcal{N} = \sum_{x=1}^{L} a_x^+ a_x$. Strictly speaking, with this choice $U_L^1$ does not look really like a boundary term, because $\mathcal{N}$ depends on all the spins of the chain. However $[(-1)^{\mathcal{N}}, H] = 0$; hence the Hilbert space splits up in two subspaces on which $(-1)^{\mathcal{N}}$ is equal to 1 or to $-1$ and on each of these subspaces $U_L^1$ really depends only on the boundary spins. One expects that, in the $L \to \infty$ limit, the correlation functions are independent on the boundary term.

In this fermionic variables, $\Omega^3(\mathbf{x})$ aquires a particularly simple form

$$\Omega_{L,\beta}^3(\mathbf{x}) = <\rho_{\mathbf{x}}\rho_{\mathbf{0}}>_{L,\beta} - <\rho_{\mathbf{x}}>_{L,\beta}<\rho_{\mathbf{0}}>_{L,\beta}, \tag{12.112}$$

with $\rho_x = a_x^+ a_x^-$.

If $J_3 = 0$ it turns out that, if $|u| < 1$, $\Omega^3(x,0)$ is of the following form:

$$\Omega^3(x,0) = -\frac{\alpha^{|x|}}{\pi^2 x^2}\sin^2\left(\frac{\pi x}{2}\right)F(-|x|\log\alpha, |x|), \quad \alpha = (1-|u|)/(1+|u|), \tag{12.113}$$

where $F(\gamma, n)$ is a bounded function, such that, if $\gamma \leq 1$, $F(\gamma, n) = 1 + O(\gamma \log \gamma) + O(1/n)$, while, if $\gamma \geq 1$ and $n \geq 2\gamma$, $F(\gamma, n) = \pi/2 + O(1/\gamma)$.

For $|h| > 0$, it is not possible to get a so explicit expression for $\Omega^3(x,0)$. However, it is not difficult to prove that, if $|u| < \sin p_F$, $|\Omega^3(x,0)| \leq \alpha^{|x|}$ and, if $x \neq 0$ and $|ux| \leq 1$

$$\Omega^3(x,0) = -\frac{1}{\pi^2 x^2}\sin^2(p_F x)[1 + O(|ux|\log |ux|) + O(1/|x|)]. \tag{12.114}$$

Note that, if $u = 0$, a very easy calculation shows that $\Omega^3(x,0) = -(\pi^2 x^2)^{-2}\sin^2(p_F x)$.

If two parameters are equal (e.g. $J_1 = J_2$), but $J_3 \neq 0$, the model is called $XXZ$ model. In the case $h = 0$ the ground state energy has been computed and it has been proved that there is a gap in the spectrum, which, if $J_1 - J_2$ and $J_3$ are not too large, is given approximately by, see for instance [14]

$$\Delta = 8\pi \frac{\sin \mu}{\mu}|J_1|\left(\frac{|J_1^2 - J_2^2|}{16(J_1^2 - J_3^2)}\right)^{\frac{\pi}{2\mu}} \tag{12.115}$$

with $\cos \mu = -J_3/J_1$.

The fermionic representation of the $XYZ$ spin chain allow to repeat the previous analysis and to obtain the following theorem.

**Theorem 12.2.** *For $J_3$ small enough and $p_F^0 = \arccos(-h) \neq \pi$ the spin correlation function $\Omega_{L,\beta}^3(\mathbf{x})$ is a bounded function of $\mathbf{x} = (x, x_0)$, $x = 1, \ldots, L$, $x_0 \in [0, \beta]$, periodic in $x$ and $x_0$ of period $L$ and $\beta$ respectively, continuous as a function of $x_0$. We can write*

$$\Omega_{L,\beta}^3(\mathbf{x}) = \cos(2p_F x)\Omega_{L,\beta}^{3,a}(\mathbf{x}) + \Omega_{L,\beta}^{3,b}(\mathbf{x}) + \Omega_{L,\beta}^{3,c}(\mathbf{x}) , \qquad (12.116)$$

*with $p_F = p_F^0 + f(J_3)$, $|f| \leq C|J_3|$ and $\Omega_{L,\beta}^{3,i}(\mathbf{x})$, $i = a, b, c$, continuous bounded functions, which are infinitely times differentiable as functions of $x_0$, if $i = a, b$. Moreover, there exist two constants $\eta_1$ and $\eta_2$ of the form $\eta_1 = a_1 J_3 + O(J_3^2)$ and $\eta_2 = -a_2 J_3 + O(J_3^2)$, $a_1$ and $a_2$ being positive constants, uniformly bounded in $L$, $\beta$, $p_F$ such that the following is true.*

*If $|\mathbf{x}| \geq 1$, given any positive integers $n$ and $N$, there exist positive constants $\vartheta < 1$ and $C_{n,N}$, independent of $L$, $\beta$, $p_F$ and $(u, J_3) \in \mathcal{A}$, so that, for any integers $n_0, n_1 \geq 0$ and putting $n = n_0 + n_1$,*

$$|\partial_{x_0}^{n_0} \bar{\partial}_x^{n_1} \Omega_{L,\beta}^{3,a}(\mathbf{x})| \leq \frac{1}{|\mathbf{x}|^{2+2\eta_1+n}} \frac{C_{n,N}}{1+[\Delta|\mathbf{x}|]^N} , \qquad (12.117)$$

$$|\partial_{x_0}^{n_0} \bar{\partial}_x^{n_1} \Omega_{L,\beta}^{3,b}(\mathbf{x})| \leq \frac{1}{|\mathbf{x}|^{2+n}} \frac{C_{n,N}}{1+[\Delta|\mathbf{x}|]^N} , \qquad (12.118)$$

$$|\Omega_{L,\beta}^{3,c}(\mathbf{x})| \leq \frac{1}{|\mathbf{x}|^2}\left[\frac{1}{|\mathbf{x}|^\vartheta} + \frac{(\Delta|\mathbf{x}|)^\vartheta}{|\mathbf{x}|^{\min\{0, 2\eta_1\}}}\right]\frac{C_{0,N}}{1+[\Delta|\mathbf{x}|]^N} , \qquad (12.119)$$

*where $\bar{\partial}_x$ denotes the discrete derivative and*

$$\Delta = \max\{|u|^{1+\eta_2}, \sqrt{(v_0\beta)^{-2} + L^{-2}}\} . \qquad (12.120)$$

The only difference with respect to the previous case is in the presence, when $u \neq 0$, of an extra mass term; after the integration of the fields $\psi^{(u,v)}, \psi^{(0)}, \ldots, \psi^{(h)}$ we obtain

$$\int P_{Z_h, \sigma_h}(d\psi^{(\leq h)}) e^{-\mathcal{V}^{(h)}(\sqrt{Z_h}\psi^{(\leq h)}) - L\beta E_h} , \quad \mathcal{V}^{(h)}(0) = 0 , \qquad (12.121)$$

where

$$P_{Z_h, \sigma_h}(d\psi^{(\leq h)}) = \prod_{\mathbf{k}': C_h^{-1}(\mathbf{k}')>0} \prod_{\omega=\pm 1} \frac{d\widehat{\psi}_{\mathbf{k}',\omega}^{(\leq h)+} d\widehat{\psi}_{\mathbf{k}',\omega}^{(\leq h)-}}{\mathcal{N}(\mathbf{k}')} . \qquad (12.122)$$

$$\exp\left\{-\frac{1}{L\beta}\sum_{\mathbf{k}': C_h^{-1}(\mathbf{k}')>0} C_h(\mathbf{k}') Z_h \sum_{\omega, \omega'=\pm 1} \widehat{\psi}_{\mathbf{k}',\omega}^{(\leq h)+} T_{\omega,\omega'}^{(h+1)} \widehat{\psi}_{\mathbf{k}',\omega'}^{(\leq h)-}\right\}$$

and the $2 \times 2$ matrix $T_h(\mathbf{k}')$ is given by

$$T_h(\mathbf{k}') = \begin{pmatrix} -ik_0 + E(k') & i\sigma_{h-1}(\mathbf{k}') \\ -i\sigma_{h-1}(\mathbf{k}') & -ik_0 - E(k') \end{pmatrix}$$

The relevant part of the effective potential has the extra term

$$F_\sigma^{(\leq h)} = \sum_{\omega=\pm 1} \frac{i\omega}{(L\beta)} \sum_{\mathbf{k}' \in \mathcal{D}'_{L,\beta}} \widehat{\psi}^{(\leq h)+}_{\mathbf{k}',\omega} \widehat{\psi}^{(\leq h)-}_{\mathbf{k}',-\omega} \qquad (12.123)$$

and $s_0 = O(u\varepsilon)$. Hence the anisotropy term $u \neq 0$ is essentially a mass term and the integration proceeds in a way essentially identical to the one described in chapt. 5, and the flow verifies (5.102),(5.104); we refer to Ref. [28] for more details.

## 12.9 Crystals and quasi-crystals

System of interacting spinless fermions have been extensively investigated. In addition to fermions on a lattice, also continuum fermions can be investigated with hamiltonian

$$H = \int d\vec{x} a_x^+ \frac{-\partial_x^2}{2m} a_x^- + \lambda \int dx dy v(x-y) a_x^+ a_x^- a_y^+ a_y^- \qquad (12.124)$$

The analysis has been performed in Ref. [37], which has provided the first example of Luttinger liquid behaviour in a non solvable model. The fermionic integration has propagator

$$g(\mathbf{k}) = \frac{1}{-ik_0 + \frac{k^2}{2m} - \mu} \qquad (12.125)$$

and the ultraviolet part of the propagator can be written as

$$g^{(\text{u.v.})}(\mathbf{x} - \mathbf{y}) = \sum_{h=0}^{\infty} C^{(h)}(\mathbf{x} - \mathbf{y}), \qquad (12.126)$$

with

$$C^{(h)}(\mathbf{x} - \mathbf{y}) = \gamma^{\frac{h}{2}} \bar{C}_h \left( \gamma^h (x_0 - y_0), \gamma^{\frac{h}{2}} (x - y) \right), \qquad (12.127)$$

such that

$$|\bar{C}_h(x_0, x)| \leq \frac{C_N}{1 + |\mathbf{x}|^N}. \qquad (12.128)$$

Then $\mathcal{V}^{(0)}$ is written as a sum over trees $\tau \in \mathcal{T}_{n,0}$ which is bounded, if $|\nu| \leq C|\lambda|$, by

$$C^n |\lambda|^n \sum_{\tau \in \mathcal{T}_{n,0}} \prod_{v \in \tau} \gamma^{(h_v - h_{v'}) D_v / 4}, \qquad (12.129)$$

with
$$D_v = |P_v| + 2m_v^4 + 4m_v^2 - 6 , \qquad (12.130)$$

and $|P_v|$ is the number of external fields of the cluster $v$ while $m_v^4, m_v^2$ are the number of $\lambda$ or $\nu$ vertices inside the cluster $G_v$.

We have seen that the sum over trees can be done of $D_v > 0$; this is what happens in this case except in a finite number of cases, namely
(1) $|P_v| = 2$, $m_v^4 = 2$, $m_v^2 = 0$;
(2) $|P_v| = 2$, $m_v^4 = 1$, $m_v^2 = 0$.

However an explicit analysis of the above cases shows that the bounds can be improved and $D_v > 0$ also in that cases, see Ref. [37]. After the ultraviolet integration is performed, the integration of the infrared scales is sone as in §13.1, and the large distance behaviour of the 2-point Schwinger function is given by (12.16).

The Schwinger function with hamiltonian (1.31) with a periodic potential $uc(x)\psi^x\psi_x$ with $c(x) = c(x+a)$ has been analyzed in [52; 53; 54]. When $p_F \neq n\pi/a$, $n$ integer, one has Luttinger liquid behaviour (the 2-point Schwinger function is given by (12.16)) while for $p_F = n\pi/a$ there is a faster than any power decay

$$|S(\mathbf{x}-\mathbf{y})| \leq \frac{1}{|x-y|^{1+\eta}} \frac{C_M}{1 + [(uc_n)^{1+\tilde{\eta}}|\mathbf{x}-\mathbf{y}|]^M} \qquad (12.131)$$

where $\eta = a\lambda^2 + O(\lambda^3)$, $\tilde{\eta} = b\lambda + O(\lambda^2)$.

For a long time solid state systems were considered as either *crystalline* (*i.e.* lattice periodic) or *amorphous*. The lattice periodicity was then described in terms of interactions with periodic potentials. However in recent times several solid state systems with a *quasi-periodic* structure have been discovered. In some cases such materials have a basic structure and a periodic modulation superimposed on it, such that the periodicity of the modulation is *incommensurate* with the periodicity of the basic structure. Another possibility is that of structures composed by two periodic lattice subsystems, with mutually incommensurate periods.

In order to study the electronic properties of quasi-periodic systems, in case of lattice systems one can add to the Hamiltonian an interaction with an external field $\varphi(x)$ with $\varphi(x) = \varphi(x+T)$ with an *irrational* $T$, so that $T$ is incommensurate with the period of the lattice (which is 1 in the units we have chosen). The hamiltonian has the form

$$H = \sum_{x \in \Lambda} \left[ \frac{1}{2} \left( -\psi_x^+ \psi_{x+1}^- - \psi_x^+ \psi_{x-1}^- + 2\psi_x^+ \psi_x^- \right) \right] - \mu \sum_{x \in \Lambda} \psi_x^+ \psi_x^- \qquad (12.132)$$

$$+ u \sum_{x \in \Lambda} \varphi(x) \psi_x^+ \psi_x^- + \lambda \sum_{x,y \in \Lambda} v(x-y) \psi_x^+ \psi_y^+ \psi_y^- \psi_x^- + \nu \sum_{x \in \Lambda} \psi_x^+ \psi_x^-. \quad (12.133)$$

Let us fix $p_F = \bar{m} p$, with $\bar{m} \in \mathbb{N}$ and $p = \pi/T$, if $T$ is the period of the potential $\varphi$. By the definition of $p$ we can write $\varphi(x) = \bar{\varphi}(2px)$ with $\bar{\varphi}$ is a $2\pi$-periodic function and $p/\pi$ is an irrational number; moreover the Fourier transform of $\bar{\varphi}$ is exponentially decreasing (*i.e.* $\bar{\varphi}$ is supposed to be analytic in a strip around the real axis). In order to perform a rigorous analysis one cannot assume that $p/\pi$ is a generic irrational number, but it has to belong to a class of numbers called *Diophantine* characterized by the following arithmetic properties: there exist two constants $C_0$ and $\tau$ such that, for any integers $k, n$,

$$|2np + 2k\pi| \geq C_0 |n|^{-\tau} \qquad \forall (n,k) \in \mathbb{Z}^2 \setminus \{(0,0)\} ; \quad (12.134)$$

the Diophantine vectors $(p, \pi)$ are of full measure for $\tau > 1$. Under such assumptions it has been proved in Ref. [55; 56] (see also Ref.[57]) that, for $p_F^0 = np$, the 2-point Schwinger function verifies (12.131).

# Chapter 13

# The $1d$ Hubbard Model

## 13.1 Spinning fermions

In the previous chapter we have seen that interacting spinless fermions in $d=1$ show Luttinger liquid behavior. In this chapter we consider the presence of the spin analyzing the $d=1$ Hubbard model, describing electrons in a crystalline lattice, hopping from one site of a lattice to another and interacting by a repulsive (Coulomb) force with coupling $U>0$.

The Hubbard model Hamiltonian is given by

$$H = -\frac{1}{2} \sum_{x \in \Lambda} \sum_{\sigma = \pm} (a^+_{x,\sigma} a^-_{x+1,\sigma} + a^+_{x+1,\sigma} a^-_{x,\sigma}) + \qquad (13.1)$$

$$U \sum_{x \in \Lambda} a^+_{x,+} a^-_{x,+} a^+_{x,-} a^-_{x,-} - \mu \sum_{x \in \Lambda} \sum_{\sigma = \pm} a^+_{x,\sigma} a^-_{x,\sigma}$$

where $\Lambda$ is an interval of $L$ points on the one dimensional lattice of step 1, which will be chosen equal to $(-[L/2], [(L-1)]/2)$ and $a^{\pm}_{x,\sigma}$ is a set of fermionic creation or annihilation operators with spin $\sigma = \pm$ satisfying periodic boundary conditions. The hamiltonian vrifies an $SU(2)$ spin symmetry.

The charge and spin density *correlation functions* are given by, if $x_0 \geq y_0$

$$N^\varepsilon_{L,\beta}(\mathbf{x}-\mathbf{y}) = <\rho^\varepsilon_\mathbf{x} \rho^\varepsilon_\mathbf{y}> - <\rho^\varepsilon_\mathbf{x}><\rho^\varepsilon_\mathbf{y}> \qquad (13.2)$$

where $\varepsilon = 0, 1$, $\rho^0_\mathbf{x} = \frac{1}{\sqrt{2}} \sum_{\sigma = \pm} a^+_{\mathbf{x},\sigma} a^-_{\mathbf{x},\sigma}$ is the *charge density* and $\rho^1_\mathbf{x} = \frac{1}{\sqrt{2}} \sum_{\sigma = \pm} \sigma a^+_{\mathbf{x},\sigma} a^-_{\mathbf{x},\sigma}$ is the *spin density*. We define also the static correlation functions as

$$S^\varepsilon_{L,\beta}(x) = N^\varepsilon_{L,\beta}(\mathbf{x})|_{x_0=0^+} \qquad (13.3)$$

In the not half filled band case and in the weak coupling regime, the (repulsive) Hubbard model (13.2) is a Luttinger liquid. Indeed the following theorem is true (see Refs. [58; 59] for details).

**Theorem 13.1.** *Consider the hamiltonian (13.2) with $-1 < \mu < 1$ and $\mu \neq 0$ (not filled or half filled band case); there exists an $\varepsilon > 0$ such that, for $0 < U < \varepsilon$ the two point Schwinger function is given by, in the limit $L, \beta \to \infty$*

$$S(\mathbf{x},\mathbf{y}) = \sum_{\omega=\pm} \frac{e^{i\omega p_F(x-y)}}{v(x_0 - y_0) + i\omega(x-y)} \frac{1 + A_\omega(\mathbf{x},\mathbf{y})}{|\mathbf{x} - \mathbf{y}|^\eta} \qquad (13.4)$$

*with*

$$\eta = aU^2 + U^2 f_0(U) \quad p_F = \cos^{-1}\mu + f_1(U) \quad v = v_0 + f_2(U) \qquad (13.5)$$

*where $a > 0$, $|f_0(U)|, |f_1(U)|, |f_2(U)| \leq CU$ and $|A_\omega(\mathbf{x},\mathbf{y})| \leq CU$. Moreover the occupation number $n_k$ is continuous at $k = \pm p_F$ but its first derivative diverges at $k = \pm p_F$ as $|k - (\pm p_F)|^{-1+\eta}$.*

*The static spin and charge density correlations can be written as*

$$S^\varepsilon(x) = \cos(2p_F x) \frac{1 + A_{1,\varepsilon}(x)}{2\pi^2 x^{2-\eta_\varepsilon}} + \frac{1 + A_{2,\varepsilon}(x)}{2\pi^2 x^2} + O(\frac{1}{|x|^{2+\vartheta}}) \qquad (13.6)$$

*where $|A_{i,\varepsilon}(x)| \leq CU$, $C_1 U \leq \eta_\varepsilon \leq C_2 U$, $C, C_1, C_2, \vartheta$ positive constants.*

*The correlations $\widehat{S}^\varepsilon(k)$ are bounded for all $k \in [-\pi, \pi]$, while their first derivatives $\partial_k \widehat{S}^\varepsilon(k)$ are bounded for all $k \neq \pm 2p_F$. At $k = \pm 2p_F$, $\partial_k \widehat{S}^\varepsilon(k)$ diverges as $|k - (\pm 2p_F)|^{-\eta_\varepsilon}$ and close to $k = 0$ we can write*

$$\widehat{S}^\varepsilon(k) = \widehat{S}^\varepsilon_0(k) + U h^\varepsilon(k) \qquad (13.7)$$

*with $|h^\varepsilon(k)|, |\partial_k h^\varepsilon(k)| \leq C$.*

The presence of the spin has the effect that Luttinger liquid behaviour is not generic, but present only for repulsive interactions and for values of the Fermi momentum different from $\pi/2$ (not half filled band case).

## 13.2 The effective potential

The integration of the ultraviolet part is done exactly as in the previous chapter. Also the integration of the infrared scales is identical, with the only difference that (12.54) is replaced by

$$\mathcal{L}\widehat{W}^{(h)}_{4,\underline{\sigma},\underline{\omega}}(\mathbf{k}'_1, \mathbf{k}'_2, \mathbf{k}'_3) = \delta_{\sum_i \varepsilon_i \omega_i, 0} \widehat{W}^{(h)}_{4,\underline{\sigma},\underline{\omega}}(\bar{\mathbf{k}}_{++}, \bar{\mathbf{k}}_{++}, \bar{\mathbf{k}}_{++}), \qquad (13.8)$$

differing from (12.54) for the presence of $\delta_{\sum_i \varepsilon_i \omega_i, 0}$, saying that not all the quartic terms are relevant, but ony the ones such that the sum of the $\omega$ is vanishing. In the not half filled band case $p_F \neq \frac{\pi}{2}$ the condition $\delta(\sum_{i=1}^4 \varepsilon_i \omega_i p_F) \neq 0$ is equivalent to the condition $\sum_{i=1}^4 \varepsilon_i \omega_i \neq 0$. Then the action of $\mathcal{L}$ if $n = 2$ is non trivial only if $\sum_{i=1}^{4} \varepsilon_i \omega_i = 0$ and there are only the following possibilities for $\omega_1, \omega_2, \omega_3, \omega_4$:

$$(\omega, \omega, -\omega, -\omega); \quad (\omega, -\omega, -\omega, \omega); \quad (\omega, \omega, \omega, \omega) \qquad (13.9)$$

In the half filled band case $p_F = \frac{\pi}{2}$ the action of $\mathcal{L}$ is non trivial also if $\omega_1 = \omega_3 = -\omega_2 = -\omega_4$.

The multiscale analysis is identical to one performed in the previous chapter. The only difference is that there is not a a single quartic local monomial in the fermion, so that relevant part of the effective potential is now more complex. Indeed it is given by

$$\mathcal{L}\widehat{\mathcal{V}}^{(h)}(\psi) = \gamma^h \nu_h F_\nu + \delta_h F_\delta + \sum_{\sigma,\sigma'} [g_{1,h} F^{(h)}_{1,\sigma,\sigma'} + g_{2,h} F^{(h)}_{2,\sigma,\sigma'} + g_{4,h} F^{(h)}_{3,\sigma,\sigma'}] \qquad (13.10)$$

where, if $c_0 = \cos p_F, v_0 = \sin p_F$

$$F_\nu = \frac{1}{L\beta} \sum_{\mathbf{k}' \in \mathcal{D}} \sum_{\omega,\sigma} \widehat{\psi}^+_{\mathbf{k}',\omega,\sigma} \widehat{\psi}^-_{\mathbf{k}',\omega,\sigma}$$

$$F_\delta = \frac{1}{L\beta} \sum_{\mathbf{k}' \in \mathcal{D}} \sum_{\omega,\sigma} (\omega v_0 \sin k' + c_0(\cos k' - 1)) \widehat{\psi}^+_{\mathbf{k}',\omega,\sigma} \widehat{\psi}^-_{\mathbf{k}',\omega,\sigma}$$

$$F_{1,\sigma,\sigma'} = \sum_\omega \int d\mathbf{x} \psi^+_{\mathbf{x},\omega,\sigma} \psi^-_{\mathbf{x},-\omega,\sigma} \psi^+_{\mathbf{x},-\omega,\sigma'} \psi^-_{\mathbf{x},\omega,\sigma'}$$

$$F_{2,\sigma,\sigma'} = \sum_\omega \int d\mathbf{x} \psi^+_{\mathbf{x},\omega,\sigma} \psi^-_{\mathbf{x},\omega,\sigma} \psi^+_{\mathbf{x},-\omega,\sigma'} \psi^-_{\mathbf{x},-\omega,\sigma'}$$

$$F_{4,\sigma,\sigma'} = \sum_\omega \int d\mathbf{x} \psi^+_{\mathbf{x},\omega,\sigma} \psi^-_{\mathbf{x},\omega,\sigma} \psi^+_{\mathbf{x},\omega,\sigma'} \psi^-_{\mathbf{x},\omega,\sigma'} \qquad (13.11)$$

Note the absence of a term $\psi^+_{\mathbf{x},\omega,\sigma} \psi^-_{\mathbf{x},\omega,\sigma} \psi^+_{\mathbf{x},-\omega,\sigma'} \psi^-_{\mathbf{x}-,\omega,\sigma'}$ which is allowed by the definition of localization only in the half-filled case.

The kernels of the effective potential are bounded as chapt.6; the only important difference is that there exists a finite scale $\bar{h} = O(\log |p_F - \frac{\pi}{2}|)$ such that for $h \leq \bar{h}$ there are no contributions to the effective potential $\widehat{\mathcal{V}}^h$ (12.53) with $n = 2$ and a choice of $\omega, \varepsilon$ such that the condition $\sum_{i=1}^{4} \varepsilon_i \omega_i = 0$ is not verified. One gets, if $\vec{v}_k = (g_{1,h}, g_{2,h}, g_{4,h}, \delta_h, \nu_h)$, $\sup_{k \geq h} |\vec{v}_k| \leq e_h$

$$\int d\mathbf{x}_{v_0} |W^{(h)}_{\tau,\mathbf{P}}(\mathbf{x}_{v_0})| \leq C^n L\beta \varepsilon_h^n \gamma^{-hD_k(P_{v_0})} \qquad (13.12)$$

$$\cdot \prod_{v \text{ not e.p.}} \chi(P_v) \left\{ \frac{1}{s_v!} C^{\sum_{i=1}^{s_v} |P_{v_i}| - |P_v|} (Z_{h_v}/Z_{h_v-1})^{|P_v|/2} \gamma^{-[-2+\frac{|P_v|}{2}+z(P_v)]} \right\}$$

where $z(P_v) = 2$ if $|P_v| = 2$ and $z(P_v) = 1$ if $|P_v| = 1$ and $||\sum_{f \in P_v} \varepsilon(f)\omega(f)p_F||_{T^1} = 0$; moreover $\chi(P_v)$ are defined so that $\chi(P_v) = 0$ if $|P_v| = 4$, $h_v \leq \bar{h}$ and $||\sum_{f \in P_v} \varepsilon(f)\omega(f)p_F||_{T^1} \neq 0$, and $\chi(P_v) = 1$ otherwise.

The presence of the $\chi$-functions in (13.13) is easily understood by noting that one can insert freely such $\chi$ functions in momentum space, then one passes to coordinate space and make bounds using the Gram-Hadamard inequality. For any $v$ such that $h_v \leq \bar{h}$ it holds $-2 + \frac{|P_v|}{2} + z(P_v) \geq 1$, that is the dimension is negative, while if $h_v \geq \bar{h}$ it holds $-2 + \frac{|P_v|}{2} + z(P_v) \geq 0$; however the integration of a finite number of scales can be done without problems.

## 13.3 The flow of the running coupling constants

By the iterative integration procedure seen in the previous section it follows that the running coupling constants verify a recursive relation whose r.h.s. is called *Beta function*:

$$\frac{Z_{h-1}}{Z_h} = 1 + z_h(\vec{v}_h, .., \vec{v}_1) \quad \nu_{h-1} = \gamma \nu_h + \beta_\nu^{(h)}(\vec{v}_h, .., \vec{v}_1)$$

$$\delta_{h-1} = \delta_h + \beta_\delta^{(h)}(\vec{v}_h, .., \vec{v}_1) \quad g_{i,h-1} = g_{i,h} + \beta_{g,i}^{(h)}(\vec{v}_h, .., \vec{v}_1) \quad (13.13)$$

with $i = (1, 2, 3)$. By an explicit computation

$$g_{1,h-1} = g_{1,h} - ag_{1,h}^2 + O((\vec{v}_h)^2 \gamma^{\vartheta h}) + O((\vec{v}_h)^3)$$
$$g_{2,h-1} = g_{2,h} - \frac{a}{2}g_{1,h}^2 + O((\vec{v}_h)^2 \gamma^{\vartheta h}) + O((\vec{v}_h)^3) \quad (13.14)$$
$$g_{4,h-1} = g_{4,h} + O((\vec{v}_h)^2 \gamma^{\vartheta h}) + O((\vec{v}_h)^3) \quad (13.15)$$

with $a$ a positive constant.

If we neglect the cubic contributions $O((\vec{v}_h)^3)$ it is easy to see that the flow is bounded (in sense that the quartic running coupling remain smaller than $O(U)$ for any $h$) if $U > 0$; in the general case in which the interaction is non local the conditions is $g_{1,0} = Uv(2p_F) + O(U^2) > 0$. By taking into account all higher order terms could destroy such behavior; aim of the following sections is to prove that also taking into account the full Beta function the quartic running coupling remain smaller than $O(U)$.

We have two free parameters at our disposal, $\nu$ and $\delta$; we will show that we can fix them so that $\nu_h = O(U^2 \gamma^{\vartheta h})$ and $\delta_h = O(U^2 \gamma^{\vartheta h})$. We fix then our attention on the flow equation for $g_{1,h}, g_{2,h}, g_{4,h}$.

More explicitly (13.13) can be written as

$$g_{1,h-1} = g_{1,h} - ag_{1,h}^2 + G_h^1 + \sum_{k,k'} g_{1,k}g_{1,k'} H_{h,k,k'}^1 + R_h^1 \qquad (13.16)$$

$$g_{2,h-1} = g_{2,h} - \frac{1}{2}ag_{1,h}^2 + \beta_h^2 + G_h^2 + \sum_{k,k'} g_{1,k}g_{1,k'} H_{h,k,k'}^2 + R_h^2$$

$$g_{4,h-1} = g_{2,h} + \beta_h^4 + G_h^4 + \sum_{k,k'} g_{1,k}g_{1,k'} H_{h,k,k'}^4 + R_h^4 \qquad (13.17)$$

where the following definitions are used:

1) The functions $\beta_h^2, \beta_h^4, G_h^2, G_h^4, G_h^1, g_1 g_1 H^i$, with $i = 1, 2, 4$ are the sum of all the trees with only end-points at scale $\leq 0$ and with propagators $g_L^{(k)}$, see (4.77).

2) The terms contributing to $\beta_h^2, \beta_h^4$ are sum of trees with only end-points of type $g_2, g_4$.

3) The terms contributing to $G_h^1, G_h^2, G_h^4$ are sum of trees with only end-points of type $g_1, g_2, g_4$ and depend linearly from $g_{1,k}$. The terms at least quadratic in $g_1$ are included in $\sum_{k,k'} g_{1,k}g_{1,k'} H_{h,k,k'}^i$ and by the short memory property

$$|H_{h,k,k'}^i| \leq C\bar{v}_h \gamma^{\vartheta(h-k)} \gamma^{\vartheta(h-k')} \qquad (13.18)$$

4) In $R_h^{(i)}$ we include; terms depending from $\nu_h$ or $\delta_h$; terms with at least a propagator $r_1^h(\mathbf{x} - \mathbf{y})$; or terms with at least an endpoint at scale 1.

In writing (13.17) we have used that the beta function contributing to $g_1$ has at least a $g_1$; in fact consider a contribution to the antiparallel part of $g_1$; it is not invariant under the transformation $\psi_{1,\sigma}^\pm \to e^{\pm \sigma} \psi_{1,\sigma}^\pm$ and $\psi_{-1,\sigma}^\pm \to \psi_{-1,\sigma}^\pm$ while the terms corresponding to $g_2$ and $g_4$ are invariant.

The flow given by (13.17) is very difficult to study; luckily dramatic cancellations appear, given by, if $\bar{g}_h = \max_{k \geq h}(|g_k^1| + |g_k^2| + |g_k^4|)$ and $\bar{\mu}_h = \max_{k \geq h}(|g_k^2| + |g_k^2|)$, the following result.

**Lemma 13.1.** *The functions $\beta_h^2, \beta_h^4, G_h^2, G_h^4, G_h^1$, for $|\nu_h| \leq \varepsilon$ are such that, for a suitable constant $C$*

$$|\beta_h^2(\mu_h, ..., \mu_h)| \leq C\bar{\mu}_h^2 \gamma^{\vartheta h} \quad |\beta_h^4(\mu_h, ..., \mu_h)| \leq C\bar{\mu}_h^2 \gamma^{\vartheta h}$$
$$|G_h^2(g_h, ..., g_h)| \leq C\bar{g}_h^2 \gamma^{\vartheta h} \quad |G_h^4(g_h, ..., g_h)| \leq C\bar{g}_h^2 \gamma^{\vartheta h} \quad |G_h^1(g_h, ..., g_h)| \leq C\bar{g}_h^2 \gamma^{\vartheta h}$$
$$(13.19)$$

The above lemma says that a dramatic cancellation happens in the series for the above functions; each order is sum of many terms $O(1)$, but at the end the final sum is $O(\gamma^{\vartheta h})$, that is asymptotically vanishing. We call such property *partial vanishing of the Beta function* (partial becouse the $O(g_1^2)$ terms are not vanishing).

We proceed in the following way. We first *assign* a sequence $\underline{\nu} = \{\nu_h\}_{h\leq 1}$, $\underline{\delta} = \{\delta_h\}_{h\leq 1}$ not necessarily solving the flow equation for $\nu, \delta$, but such that $|\nu_h|, |\delta_h| \leq cU\gamma^{\vartheta h}$, for any $h \leq 1$. We then solve the flow equation for $g_{i,h}$, parametrically in $\nu, \delta$, and show that, *for any sequence $\underline{\nu}, \underline{\delta}$* with the supposed property, the solution $\underline{g}(\underline{\nu}, \underline{\delta}) = \{g_{1,h}(\underline{\nu}, \underline{\delta}), g_{2,h}(\underline{\nu}, \underline{\delta}), g_{4,h}(\underline{\nu}, \underline{\delta})\}_{h\leq 1}$ exists and has good decaying properties. We finally fix the sequence $\underline{\nu}$ via a convergent iterative procedure.

**Lemma 13.2.** *Assume that $|\nu_h|, |\delta_h| \leq cU\gamma^{\vartheta h}$ for any h. For $U > 0$ and small enough the flow is given by, for any h*

$$|g_{2,h} - g_{2,0} - g_{1,0}/2| \leq U^{3/2} \quad |g_{4,h} - g_{4,0}| \leq U^{3/2} \quad 0 < g_{1,h} \leq \frac{g_{1,0}}{1 - a/3 g_{1,0} h} \quad (13.20)$$

*Proof.* By using that $|\nu_h|, |\delta_h| \leq cU\gamma^{\vartheta h}$ it holds that

$$|R_h^i| \leq CU^2 \gamma^{\vartheta h} \qquad (13.21)$$

It is convenient to introduce $\widetilde{g}_{2,h} = 2g_{2,h} - g_{1,h}$; then using (13.19) and (13.21)

$$\widetilde{g}_{2,h-1} = \widetilde{g}_{2,h} + \sum_{k \geq h} D_{h,k} + \sum_{k \geq h}(2D_{h,k}^2 - D_{h,k}^1) + \sum_{k,k'} g_{1,k} g_{1,k'} \bar{H}_{h,k,k'} + \bar{R}_h \qquad (13.22)$$

with

$$D_{h,k} = \beta_h^2(\mu_h, ...\mu_h, \mu_k, \mu_{k+1}, .., \mu_0) - \beta_h^2(\mu_h, ...\mu_h, \mu_h, \mu_{k+1}, .., \mu_0) \quad (13.23)$$

$$D_{h,k}^i = G_h^i(g_h, ...g_h, g_k', g_{k'+1}, .., g_0) - G_h^i(g_h, ...g_h, g_h, g_{k'+1}, .., g_0) \quad i = 1, 2 \quad (13.24)$$

and a similar equation for $g_{4,h}$; $\bar{H}_{h,k,k'}$ verifies (13.18), $\bar{R}_h$ (13.21) and

$$|D_{h,k}| \leq C\gamma^{-2\vartheta(k-h)} U|g_h - g_k| \quad |D_{h,k}^i| \leq CU\gamma^{-2\vartheta(k-h)}|g_h - g_k| \quad (13.25)$$

Assume that for $k > h$

$$0 \leq g_{1,k-1} \leq \frac{g_{1,0}}{1 - a/3g_{1,0}(k-1)} \quad |g_{k-1} - g_k| \leq [U^{\frac{5}{4}} \gamma^{\vartheta k} + [\frac{g_{1,0}}{1 - a/3g_{1,0}k}]^2] \qquad (13.26)$$

We have then to prove that such inequalities hold for $k = h - 1$. Noting that

$$\sum_{k=h}^{-1} \gamma^{\vartheta(h-k)} \frac{1}{-k} = \frac{1}{-h} \sum_{k=h}^{-1} \gamma^{\vartheta(h-k)} + \sum_{k=h}^{-1} \gamma^{\vartheta(h-k)} \frac{(k-h)}{kh} \leq \frac{C_1}{-h} \qquad (13.27)$$

we obtain

$$\sum_{k,k'} g_{1,k} g_{1,k'} \bar{H}^2_{h,k,k'} \leq CU g^2_{1,h} \tag{13.28}$$

Then by (13.29) we get

$$|\tilde{g}_{2,h-1} - \tilde{g}_{2,h}| \leq C_3(U^2 \gamma^{\vartheta h} + U(\frac{g_{1,0}}{1 - \frac{a}{3} g_{1,0} h})^2) \tag{13.29}$$

and

$$|\tilde{g}_{2,h-1} - \tilde{g}_{2,0}| \leq C_3(U^2 \sum_{k=h}^{0} \gamma^{\vartheta k} + \sum_k U[\frac{g_{1,0}}{1 - a/3 g_{1,0} k}]^2 \leq U^{3/2} \tag{13.30}$$

In the same way in the flow for $g_4$ we use that there are no second order contributions quadratic in $g_{1,h}$. Finally we write, using (13.17) and the short memory property (namely that $\gamma^{\vartheta(h-k)} g_{1,k} \leq C g_{1,h}$)

$$g_{1,h-1} - g_{1,h} \leq -\frac{a}{3} g_{1,h} g_{1,h-1} \tag{13.31}$$

or

$$g_{1,h-1} \leq \frac{g_{1,h}}{1 + \frac{a}{3} g_{1,h}} \tag{13.32}$$

and as $\frac{x}{1+x}$ is an increasing function and by induction $0 < g_{1,h} \leq \frac{g_{1,0}}{1 - \frac{a}{3} g_{1,0} h}$ so that

$$g_{1,h-1} \leq \frac{g_{1,0}(1 - \frac{a}{3} h g_{1,0})^{-1}}{1 + \frac{a}{3} g_{1,0}(1 - \frac{a}{3} h g_{1,0})^{-1}} \leq \frac{g_{1,0}}{1 - \frac{a}{3} g_{1,0}(h-1)}. \tag{13.33}$$

Moreover $g_{1,h-1} = g_{1,h}(1 + O(U))$ and $g_{1,h} > 0$ so that $g_{1,h-1} > 0$. ∎

## 13.4 The auxiliary model

In order to prove Lemma 13.1 we follow the same strategy as in the spinless model trying to prove the cancellations in a model for which Ward Identities can be derived. There is however an important difference between spinless and spinning case.

In the spinless case, we decompose the beta function in the sum of two terms, one of which has a negligible effect on the flow and the other given by the sum of trees with only end-points associated with $\lambda_h$ and propagators $g_L^h$. This term coincides with the beta function of relativistic models like QED2 or the Thirring model. Such models are invariant, in the massless case, under the global phase symmetry

$$\psi^{\pm}_{\mathbf{x},\omega} \to e^{\pm i \alpha_\omega} \psi^{\pm}_{\mathbf{x},\omega} \tag{13.34}$$

Such invariance implies certain Ward Identities from which the cancellations necessary to prove the vanishing of the first term can be proved, so that at the end, as discussed in chapt.12, the effective coupling remains close to its initial value.

In the spinning case we could try to follow the same strategy introducing an auxiliary model with beta function asymptotically equal to the ones of the spinning Hubbard model. Such a model would describe fermions with linear dispersion relation, momentum cut-off, and local interaction given by the quartic part of $\mathcal{LV}^{(0)}$ (13.10). The problem is however that $\mathcal{LV}^{(0)}$ is *not invariant* under the generalization to the spinning case of (13.34), namely

$$\psi^{\pm}_{\mathbf{x},\omega,\sigma} \to e^{\pm i\alpha_{\omega,\sigma}} \psi^{\pm}_{\mathbf{x},\omega,\sigma} \qquad (13.35)$$

and its beta function is not vanishing.

On the other hand in the spinning case we do not need to prove a complete vanishing of the beta function, but the weaker property (13.19). We can then introduce an auxiliary model chosen so that it is invariant under (13.35) and from which (13.19) can be derived; the model has no a natural relativistic interpretation but it must seen as a technical device to deduce to properties we need.

The auxilary model we consider has generating function

$$\int P(d\psi^{[h,N]}) e^{\mathcal{V}_L + \int d\mathbf{x}[J_{\mathbf{x}} \psi^{+}_{\mathbf{x},\omega,\sigma} \psi^{-}_{\mathbf{x},\omega,\sigma} + \phi^{+}_{\mathbf{x},\omega,\sigma} \psi^{-}_{\mathbf{x},\omega,\sigma} + \psi^{+}_{\mathbf{x},\omega,\sigma} \phi^{-}_{\mathbf{x},\omega,\sigma}]} \qquad (13.36)$$

where $P(d\psi^{[h,N]})$ is the fermionic integration with propagator

$$g^{h}_{\omega,L}(\mathbf{x} - \mathbf{y}) = \frac{1}{\beta L} \sum_{\mathbf{k} \in \mathcal{D}_{L\beta}} C^{-1}_{h,N}(\mathbf{k}) \frac{e^{-i\mathbf{k}(\mathbf{x}-\mathbf{y})}}{-ik_0 + \omega k} \qquad (13.37)$$

with $C^{-1}_{h,0}(\mathbf{k}) = \sum_{k=-\infty}^{0} f^k(k_0, k)$ and

$$\mathcal{V}_L = \sum_{\omega} \int_{-\beta/2}^{\beta/2} dx_0 \int_{-L/2}^{L/2} dx v(\mathbf{x}-\mathbf{y}) [g_2^o \psi^{+}_{\mathbf{x},\omega,\sigma} \psi^{-}_{\mathbf{x},\omega,\sigma} \psi^{+}_{\mathbf{y},-\omega,\sigma} \psi^{-}_{\mathbf{y},-\omega,\sigma}$$

$$+ g_2^p \psi^{+}_{\mathbf{x},\omega,\sigma} \psi^{-}_{\mathbf{x},\omega,\sigma} \psi^{+}_{\mathbf{y},-\omega,-\sigma} \psi^{-}_{\mathbf{y},-\omega,-\sigma}$$

$$+ g_4 \psi^{+}_{\mathbf{x},\omega,\sigma} \psi^{-}_{\mathbf{x},\omega,\sigma} \psi^{+}_{\mathbf{y},\omega,-\sigma} \psi^{-}_{\mathbf{y},\omega,-\sigma}] \qquad (13.38)$$

with $v(\mathbf{x} - \mathbf{y})$ such that $||v||_1, ||v||_\infty$ is bounded. Note that the model is not $SU(2)$ invariant, as the interaction depends from the spin if $g_2^o \neq g_2^p$.

The Grassmann integration can be done by a multiscale analysis essentially identical to the one described in §3; however the symmetries of the interaction imply that the local part of the effective potential is given by

$$\mathcal{LV}^j_L = \sum_{\omega} \int_{-\beta/2}^{\beta/2} dx_0 \int_{-L/2}^{L/2} dx \tilde{g}^p_{2,j} \psi^{+}_{\mathbf{x},\omega,\sigma} \psi^{-}_{\mathbf{x},\omega,\sigma} \psi^{+}_{\mathbf{x},-\omega,\sigma} \psi^{-}_{\mathbf{x},-\omega,\sigma} + \qquad (13.39)$$

$$\widetilde{g}^o_{2,j}\psi^+_{\mathbf{x},\omega,\sigma}\psi^-_{\mathbf{x},\omega,\sigma}\psi^+_{\mathbf{x},-\omega,-\sigma}\psi^-_{\mathbf{x},-\omega,-\sigma} + \widetilde{g}_{4,j}\psi^+_{\mathbf{x},\omega,\sigma}\psi^-_{\mathbf{x}_2,\omega,\sigma}\psi^+_{\mathbf{x},\omega,-\sigma}\psi^-_{\mathbf{x},\omega,-\sigma}$$
(13.40)

Note in fact that the analogue of $\nu_h, \delta_h$ are vanishing by (in the limit $L, \beta \to \infty$) parity and invariance in the exchange $(x, x_0) \to (x_0, x)$; moreover the reference model is invariant under the total gauge transformation $\psi^\pm_{\mathbf{x},\omega,\sigma} \to e^{\pm i\alpha_{\omega,\sigma}}\psi^\pm_{\mathbf{x},\omega,\sigma}$ for any values of $\alpha_{\omega,\sigma}$, so that terms of the form $\psi^+_{\omega,\sigma}\psi^-_{-\omega,\sigma}\psi^+_{-\omega,-\sigma}\psi^-_{\omega,-\sigma}$ cannot be generated in the integration procedure as they violate such symmetry.

The Beta function is an analytic function of $\vec{v}^L_j$ and it can be written as, if $\alpha = (o,2), (p,2), 4$

$$\widetilde{\beta}^\alpha_j(\vec{v}^L_j, ..., \vec{v}^L_j) = \sum_{n_1,n_2,n_3} b^\alpha_{j,n_1,n_2,n_3}(\widetilde{g}^o_{2,j})^{n_1}(\widetilde{g}^p_{2,j})^{n_2}(\widetilde{g}_{4,j})^{n_3} \qquad (13.41)$$

We define $n \equiv n_1 + n_2 + n_3$ and $\vec{n} = (n_1, n_2, n_3)$. We will prove that

$$b^\alpha_{n_1,n_2,n_3} = 0 \qquad (13.42)$$

and this implies Lemma 13.1.

A) Let us start considering first the reference model in the spin symmetric case, that is if $g^o_{2,0} = g^p_{2,0}$. In such a case for any $k$ $g^o_{2,k} = g^p_{2,k}$ and $\beta^o_{2,k} = \beta^p_{2,k}$, so that the flow equation reduces to

$$\widetilde{g}_{2,h-1} = \widetilde{g}_{2,h-1} + \widetilde{\beta}^2_h(\widetilde{g}_{2,h}, \widetilde{g}_{4,h}; ...; \widetilde{g}_{2,0}, \widetilde{g}_{4,0}) \qquad (13.43)$$

$$\widetilde{g}_{4,h-1} = \widetilde{g}_{4,h-1} + \widetilde{\beta}^4_h(\widetilde{g}_{2,h}, \widetilde{g}_{4,h}; ...; \widetilde{g}_{2,0}, \widetilde{g}_{4,0}) \qquad (13.44)$$

It holds that the functions $\widetilde{\beta}^2_h$ and $\widetilde{\beta}^4_h$ essentially coincide with the functions $\beta^2_h, \beta^4_h$ of the Hubbard model defined in (13.17); that is, if $\mu_h = (g_{2,h}, g_{4,h})$, for a suitable constant $C$

$$|\widetilde{\beta}^2_h(\mu_h, ..., \mu_h) - \beta^2_h(\mu_h, ..\mu_h)| \le C\mu^2_h\gamma^{\vartheta h} \qquad (13.45)$$

$$|\widetilde{\beta}^4_h(\mu_h, ..., \mu_h) - \beta^4_h(\mu_h, ..\mu_h)| \le C\mu^2_h\gamma^{\vartheta h} \qquad (13.46)$$

The above equations prove (13.19).

B) We consider now the auxiliary model with $g^o_{2,0} \neq g^p_{2,0}$, so that there are three independent running coupling constants. We have seen that, for $\alpha = (2,p), (2,o), 4$

$$\widetilde{\beta}^\alpha_h(v^L_h, ..v^L_h) = \sum_{n_1,n_2,n_3} b^\alpha_{h,n_1,n_2,n_3}[g^o_{2,h}]^{n_1}[g^p_{2,h}]^{n_2}[g_{4,h}]^{n_3} \qquad (13.47)$$

On the other hand we can write the functions $G^\alpha_h$ in the Hubbard model, $\alpha = (2o), (2p), 4$, as

$$G^\alpha_h = \sum_{m_2,m_3} c^\alpha_{h,1,n_2,n_3}[g_{1,h}][g_{2,h}]^{m_2}[g_{4,h}]^{m_3} \qquad (13.48)$$

The coefficients $c^\alpha_{h,1,n_2,n_3}$ are given by sum of trees (or product of trees, for the presence of the $z^1_k$ terms) with (in total) one end-point $g_1$, $m_2$ end-points $g_2$ and $m_3$ end-points $g_4$; the SU(2) invariance of the Hubbard model implies that $G^{2o}_h = G^{2p}_h$. To $g_1$ and $g_2$ correspond two terms, the parallel or antiparallel part, and we can associate to the endpoints of the trees contributing to $c^\alpha_{h,1,m_2,m_3}$ an extra index distinguishing the parallel or antiparallel part; then we can write

$$c^\alpha_{h,1,m_2,m_3} = \sum_{m^o_1+m^p_1=1} \sum_{m^o_2+m^p_2=m_2} c^\alpha_{h,m^o_1,m^p_1,m^o_2,m^p_2,m_3} \qquad (13.49)$$

It holds that

$$c^\alpha_{h,1,m_2,m_3} = \sum_{m^o_2+m^p_2=m_2} c^\alpha_{h,0,1,m^o_2,m^p_2,m_3} \qquad (13.50)$$

that is only the spin parallel part of $g_1$ can contribute to $G^2_h$ or $G^4_h$; in fact making the the global gauge transformation $\psi^\pm_{1,\sigma} \to e^{i\sigma}\psi^\pm_{1,\sigma}$ and $\psi^\pm_{-1,\sigma} \to \psi^\pm_{-1,\sigma}$, the antiparallel part is not invariant, while the spin parallel (and the $g_2$, $g_4$ interactions) are invariant.

Finally note that the spin parallel $g_1$ interaction is equal (up to a sign) to the spin parallel $g_2$ interaction, so that, for $\alpha = (2o), (2p), 4$

$$c^\alpha_{0,1,m^o_2,m^p_2,m_3} = -b^\alpha_{m^o_2,m^p_2+1,m_3} = 0 \qquad (13.51)$$

C) It remains to consider $G^1_h$; we can consider equivalently the contribution to the spin parallel or the spin antiparallel, as they are equal by $SU(2)$ invariance of the Hubbard model, that is $G^{1o}_h = G^{1p}_h$. We consider the spin parallel part and we can write

$$G^{1p}_h = \sum_{m_2,m_3} c^{1p}_{h,1,m_2,m_3} [g_{1,h}][g_{2,h}]^{m_2}[g_{4,h}]^{m_3} \qquad (13.52)$$

with

$$c^{1p}_{1,m_2,m_3} = \sum_{m^o_1+m^p_1=1} \sum_{m^o_2+m^p_2=m_2} c^{1p}_{m^o_1,m^p_1,m^o_2,m^p_2,m_3} \qquad (13.53)$$

The single $g_1$ interaction cannot be antiparallel, again because making the global gauge transformation $\psi^\pm_{1,\sigma} \to e^{i\sigma}\psi^\pm_{1,\sigma}$ and $\psi^\pm_{-1,\sigma} \to \psi^\pm_{-1,\sigma}$, the antiparallel part is not invariant, while the spin parallel (and the $g_2$, $g_4$ interactions) are invariant. Hence

$$c^{1p}_{1,m_2,m_3} = \sum_{m^o_2+m^p_2=m_2} c^{1p}_{0,1,m^o_2,m^p_2,m_3} \qquad (13.54)$$

and
$$c^{1p}_{0,1,m_2^o,m_2^p,m_3} = b^{2p}_{m_2^o,m_2^p+1,m_3} = 0 \qquad (13.55)$$
as the contribution $(1p)$ and $(2p)$ are identical.

In order to prove (13.42) we proceed as in chapt.6 deriving WI for the auxialiry model. By the phase transformation (13.35) we obtain
$$D_\omega(\mathbf{p}) < \rho_{\mathbf{p},\omega,\sigma}; \psi^+_{\mathbf{k}_1,\omega',\sigma'}\psi^-_{\mathbf{k}_2,\omega',\sigma'} >_T = \delta_{\omega,\omega'}\delta_{\sigma,\sigma'}[< \psi^+_{\mathbf{k}_1,\omega',\sigma'}\psi^-_{\mathbf{k}_1,\omega',\sigma'} > \qquad (13.56)$$
$$- < \psi^+_{\mathbf{k}_2,\omega',\sigma'}\psi^-_{\mathbf{k}_2,\omega',\sigma'} >] + \Delta_{\omega,\sigma;\omega',\sigma'}(\mathbf{k},\mathbf{p}) \qquad (13.57)$$
where the same notations as in in (6.13) are used. The analogous of Lemma 6.11 holds (sligthly adapting the analysis in chapt.6)
$$\Delta_{\omega,\sigma;\omega',\sigma'}(\mathbf{k},\mathbf{p}) = \sum_{\sigma''}\nu_{\sigma,\sigma''}v_K(\mathbf{p})D_{-\omega}(\mathbf{p}) < \rho_{\mathbf{p},-\omega,\sigma''}; \psi^+_{\mathbf{k},\omega',\sigma'}\psi^-_{\mathbf{k}+\mathbf{p},\omega',\sigma'} > +$$
$$(13.58)$$
$$\sum_{\omega'',\sigma''}\mu_{\sigma''}v_K(\mathbf{p})D_\omega(\mathbf{p}) < \rho_{\mathbf{p},\omega,\sigma''}; \psi^+_{\mathbf{k},\omega',\sigma'}\psi^-_{\mathbf{k}+\mathbf{p},\omega',\sigma'} > +D_\omega(\mathbf{p})R^{(2,1)}(\mathbf{k},\mathbf{p})$$
$$(13.59)$$
with $R^{(2,1)}(\mathbf{k},\mathbf{p})$ verifying the bound (6.16) and
$$\nu_{\sigma,-\sigma} = \frac{g_2^o}{4\pi} \qquad \nu_{\sigma,\sigma} = \frac{g_2^p}{4\pi} \qquad \mu_{\sigma,-\sigma} = \frac{g_4^o}{4\pi} \qquad (13.60)$$
and 0 otherwise. Again a Schwinger-Dyson equation, analogous to (6.40), holds
$$< \psi^+_{\mathbf{k}_1,+,\sigma}\psi^-_{\mathbf{k}_2,+,\sigma}\psi^+_{\mathbf{k}_3,-,-\sigma}\psi^-_{\mathbf{k}_4,-,-\sigma} >_T = \qquad (13.61)$$
$$g_-(\mathbf{k}_4) < \psi^-_{\mathbf{k}_3,\omega,\sigma}\psi^+_{\mathbf{k}_3,\omega,\sigma} >_T [g_2^o < \psi^+_{\mathbf{k}_1,+,\sigma}\psi^-_{\mathbf{k}_2,+,\sigma}\rho_{\mathbf{k}_1-\mathbf{k}_2,+,\sigma} >_T$$
$$+g_2^p < \psi^+_{\mathbf{k}_1,+,\sigma}\psi^-_{\mathbf{k}_2,+,\sigma}\rho_{\mathbf{k}_1-\mathbf{k}_2,+,-\sigma} >_T +g_4 < \psi^+_{\mathbf{k}_1,+,\sigma}\psi^-_{\mathbf{k}_2,+,\sigma}\rho_{\mathbf{k}_1-\mathbf{k}_2,-,-\sigma} >_T]$$
$$+ \int d\mathbf{p}v_K(\mathbf{p})[g_2^o < \psi^+_{\mathbf{k}_1,+,\sigma}\psi^-_{\mathbf{k}_2,+,\sigma}\psi^+_{\mathbf{k}_3,-,\sigma}\psi^-_{\mathbf{k}_4-\mathbf{p},-,-\sigma}\rho_{\mathbf{p},+,\sigma} >_T +$$
$$\int d\mathbf{p}g_2^p < \psi^+_{\mathbf{k}_1,+,\sigma}\psi^-_{\mathbf{k}_2,+,\sigma}\psi^+_{\mathbf{k}_3,-,\sigma}\psi^-_{\mathbf{k}_4-\mathbf{p},-,-\sigma}\rho_{\mathbf{p},+,-\sigma} >_T +$$
$$\int d\mathbf{p}g_4 < \psi^+_{\mathbf{k}_1,+,\sigma}\psi^-_{\mathbf{k}_2,+,\sigma}\psi^+_{\mathbf{k}_3,+,\sigma}\psi^-_{\mathbf{k}_4-\mathbf{p},+,-\sigma}\rho_{\mathbf{p},-,-\sigma} >_T]\}$$
Similar Dyson equations holds for $< \psi^+_{\mathbf{k}_1,+,\sigma}\psi^-_{\mathbf{k}_2,+,\sigma}\psi^+_{\mathbf{k}_3,-,\sigma}\psi^-_{\mathbf{k}_4,-,\sigma} >_T$ and $< \psi^+_{\mathbf{k}_1,+,\sigma}\psi^-_{\mathbf{k}_2,+,\sigma}\psi^+_{\mathbf{k}_3,+,\sigma}\psi^-_{\mathbf{k}_4,+,\sigma} >_T$. The Renormalization Group analysis of the preceding sections easily implies that, if $|\bar{\mathbf{k}}_i| = \gamma^h, i = 1,2,3,4$
$$< \psi^+_{\mathbf{k}_1,+,\sigma}\psi^-_{\mathbf{k}_2,+,\sigma}\psi^+_{\mathbf{k}_3,-,-\sigma}\psi^-_{\mathbf{k}_4,-,-\sigma} >_T = \gamma^{-4h}Z_h^{-2}[g^o_{2,h} + O(\bar{g}_h^2)] \qquad (13.62)$$
if $\bar{g}_h = \sup_{k \geq h}(|g^o_{2,k}| + |g^p_{2,k}| + |g_{4,k}|)$. Hence by combining the Schwingr-Dyson equation with the WI, and bounding the corrections as in §6.4, we get $g^p_{2,h} = g^p_{2,0} + O(g^2), g^o_{2,h} = g^o_{2,0} + O(g^2), g^o_{4,h} = g^o_{4,0} + O(g^2)$ from which, proceeding as in Lemma 6.1, (13.42) follows.

## 13.5 The effective renormalizations

The properties of the spin and charge density correlations depend ftom $Z_h^{(1),\varepsilon}$, $Z_h^{(2),\varepsilon}$. The flow equation for $Z_{h-1}^{(1),\varepsilon}$

$$\frac{Z_{h-1}^{(i),\varepsilon}}{Z_{h-1}} = \frac{Z_h^{(i),\varepsilon}}{Z_h}[1 + \beta_i^h(\vec{v}_h, \vec{v}_{h+1}, ..., \vec{v}_0)] \qquad (13.63)$$

and, if $b$ is a constant

$$\beta_1^h = ag_{2,h} - ag_{1,h} + O(U^2) + O(U\gamma^{\frac{h}{2}}) \qquad (13.64)$$

so that, using that $g_{2,h} > g_{1,h}$ as $\hat{v}(0) > \hat{v}(2p_F)$

$$\gamma^{-c_1 Uh} \leq Z_h^{(1),\varepsilon} \leq e^{-c_2 Uh} \qquad (13.65)$$

Regarding $Z_{h-1}^{(2),\varepsilon}$ it holds

$$\frac{Z_h^{(2),\varepsilon}}{Z_h} = 1 + O(U) \qquad (13.66)$$

which says that the density renormalization $Z_h^{(2),\varepsilon}$ is proportional to the wave function renormalization $Z_h$. In order to prove we can decompose $\beta_2^h$ in as sum of two terms; defining $\vec{g}_k = (g_{1,k}, g_{2,k}, g_{4,k})$ we have

$$\beta_2^h(\vec{v}_h, \vec{v}_{h+1}, ..., \vec{v}_0) = \beta_{2,a}^h(\vec{g}_h, \vec{g}_{h+1}, ..., \vec{g}_{-1}) + R_2^h(\vec{v}_h, \vec{v}_{h+1}, ..., \vec{v}_0) \qquad (13.67)$$

where we include in $\beta_{1,a}^h$ only the terms contributing to $\beta_1^h$ obtained contracting the quartic part of $\mathcal{L}\mathcal{V}^{(k)}$, $k \leq -1$ with the dominant part of the propagator $g_L^{(k)}(\mathbf{x})$, $k \leq -1$, and in $R_2^{(h)} = O(U\gamma^{\frac{h}{2}})$ are the remining terms. We can write

$$\beta_{2,a}^h(\vec{g}_h, \vec{g}_h, ..., \vec{g}_h) = \sum_{m_1, m_2, m_3} c_{m_1, m_2, m_3}^h (g_{1,h})^{m_1} (g_{2,h})^{m_2} (g_{4,h})^{m_2} \qquad (13.68)$$

and, by the short memory property, $c_{m_1,m_2,m_3}^h = c_{m_1,m_2,m_3} + O(\gamma^{\frac{h}{2}})$. The coefficients $c_{m_1,m_2,m_3}$ are obtained by the truncated expectations of $m_1^p$ interaction $F_{1,\sigma,\sigma}$, $m_1^o$ interaction $F_{1,\sigma,-\sigma}$, $m_2^p$ interaction $F_{2,\sigma,\sigma}$, $m_2^o$ interaction $F_{2,\sigma,-\sigma}$ and $m_3$ interaction $F_{4,\sigma,-\sigma}$ so that we can write

$$c_{m_1,m_2,m_3} = \sum_{m_1^o + m_1^p = m_1} \sum_{m_2^o + m_2^p = m_2} \sum_{m_3} c_{m_1^o, m_1^p, m_2^o, m_2^p, m_3} \qquad (13.69)$$

Note that

$$c_{0,m_2,m_3} = \sum_{m_2^o + m_2^p = m_2} \sum_{m_3} c_{0,0,m_2^o,m_2^p,m_3}$$

$$c_{1,m_2,m_3} = \sum_{m_2^o + m_2^p = m_2} \sum_{m_3} c_{0,1,m_2^o,m_2^p,m_3} \qquad (13.70)$$

The second of (13.70) follows from the fact that there are no possible contributions obtained contracting $\psi^+_{\omega,\sigma}\psi^-_{-\omega,\sigma}\psi^+_{-\omega,-\sigma}\psi^-_{\omega,-\sigma}$ and any number of $F_2, F_4$, as the fields to be contracted would be, if the external lines have index $(\omega,\sigma)$, $n_1+1-2$ fields $(\omega,\sigma)$, $n_2+1$ fields $(\omega,-\sigma)$, $n_3+1$ fields $(-\omega,\sigma)$, $n_4+1$ fields $(-\omega,-\sigma)$, with $n_1, n_2, n_3, n_4$ even, as they are the number of fields coming from the interactions $F_2^k$ and $F_4^k$ which are bilinear in the densities of fermions of label $(\omega', \sigma')$.

Note finally that
$$F^h_{1,\sigma,\sigma} = -F^h_{2,\sigma,\sigma} \tag{13.71}$$
and this implies
$$c_{0,1,m_2^o,m_2^p,m_3} = -c_{0,0,m_2^o,m_2^p+1,m_3} \tag{13.72}$$
By the analysis of the WI of the auxiliary model, for any $m_2, m_3$
$$c_{0,m_2,m_3} = 0; \qquad c_{1,m_2,m_3} = 0 \tag{13.73}$$
so that
$$e^{c_2 U \sum_{k=h}^{0}[\gamma^{\frac{k}{2}}+|k|^{-2}]} \leq \frac{|Z^{(2)\varepsilon}_{h-1}|}{|Z_{h-1}|} \leq e^{c_1 U \sum_{k=h}^{0}[\gamma^{\frac{k}{2}}+|k|^{-2}]} \tag{13.74}$$

## 13.6 Attractive interactions

All the previous analysis is valid only for repulsive interactions, so that the effective coupling $g_{1,h}$ is vanishing as $h \to -\infty$. For repulsive interactions we cannot prove that $g_{1,h}$ remain small, so that the above approach cannot be used; the same considerations hold in the half filled band case. In such cases there is no Luttinger liquid behaviour, as it appears from the analysis of Lieb and Wu in Ref. [60] via *Bethe ansatz*. If only the states of Bethe ansatz form are considered, one sees that the a gap appears at half filling so that the system is an insulating and no Luttiger liquid behaviour is found. The analysis of interacting spinning fermions in the case of attractive interactions by Renormalization Group meythods, and the proof that the Schwinger functions decays exponentially, is still an open problem.

## Chapter 14

# Fermi Liquids in Two Dimensions

## 14.1 Interacting Fermions in $d = 2$

Higher dimensional fermionic systems are more difficult to analyze, as a consequence of the increasing complexity of the Fermi surface, which is not anymore given by two points as in $d = 1$, but is is a line (in $d = 2$) or a surface (in $d = 3$), whose shape and regularity properties affects in a dramatic way the physical properties.

As we discussed in the introduction, an important question for interacting fermionic systems at higher dimensions is to determine if they are or not *Fermi liquids*, at least at temperatures not too low. While a lowest order analysis says that Fermi liquid behaviour is found at $d = 2, 3$, at least if some regularity and convexity properties are assumed, getting non perturbative results is quite complex and at present only result at $d = 2$ have been found.

The two-point Schwinger function of a $d = 2$ non-interacting fermionic system is given by

$$g(\mathbf{x} - \mathbf{y}) = \frac{1}{\beta L^2} \sum_{\mathbf{k} \in \mathcal{D}_{\beta,L}} \frac{e^{-i\mathbf{k} \cdot (\mathbf{x} - \mathbf{y})}}{-ik_0 + \varepsilon_0(\vec{k}) - \mu} \qquad (14.1)$$

where $\mathbf{k} = (k_0, \vec{k})$ and $\mathcal{D}_{\beta,L} = \mathcal{D}_\beta \times \mathcal{D}_L$ and $\mathcal{D}_\beta = \{k_0 = \frac{2\pi}{\beta}(n_0 + \frac{1}{2}) : n_0 \in \mathbb{Z}\}$ and $\mathcal{D}_L = \{\vec{k} = \frac{2\pi}{L}(n_1, n_2) : -[L/2] \le n_1, n_2 \le [(L-1)/2]\}$. In the case of the Hubbard model

$$\varepsilon_0(\vec{k}) = 2 - \cos k_1 - \cos k_2 \qquad (14.2)$$

and for continuum fermions in the absence of fermionic potential (Jellium model)

$$\varepsilon_0(\vec{k}) = \frac{1}{2m}(k_1^2 + k_2^2) \qquad (14.3)$$

Note that $S_0(\mathbf{x})$ is a function of $x_0 \in \mathbb{R}$ antiperiodic of period $\beta$ and that its Fourier transform $\widehat{S}_0(\mathbf{k})$ is well–defined for any $\mathbf{k} \in \mathcal{D}_{\beta,L}$, even in the thermodynamic limit $L \to \infty$, since $|k_0| \geq \frac{\pi}{\beta}$. We shall refer to this last property by saying that the inverse temperature $\beta$ acts as an infrared cutoff for our theory.

In the limit $\beta, L \to \infty$ the propagator $\widehat{S}_0(\mathbf{k})$ becomes singular on the surface $\{k_0 = 0\} \times \Sigma_F^{(0)}$, where $\Sigma_F^{(0)} \equiv \{\vec{k} \,:\, \varepsilon_0(\vec{k}) - \mu = 0\}$ is the free *Fermi surface*. In the case (14.3), the Fermi surface is a circle with radius $\sqrt{2m\mu}$; in the case of the Hubbard model (14.3) if $0 < \mu < 2$, is a smooth convex closed curve, symmetric around the point $\vec{k} = (0,0)$, while it reduces to a square-shaped curve if $\mu = 2$ (half filled band case); if $0 < \mu < 2$ it can be parameterized as $\vec{k} = \vec{p}_F^{(0)}(\theta)$ in terms of the polar angle $\theta \in [0, 2\pi]$. We shall also denote $|\vec{p}_F^{(0)}(\theta)|$ by $u^{(0)}(\theta)$.

In order to make apparent the structure of the pole singularity of $\widehat{S}_0(\mathbf{k})$ at $\{0\} \times \Sigma_F^{(0)}$, it is sometimes convenient to rewrite $\widehat{S}_0(\mathbf{k})$ in the form:

$$\widehat{S}_0(\mathbf{k}) = \frac{1}{Z_0} \frac{1}{-ik_0 + \vec{v}_F^{(0)}(\theta) \cdot (\vec{k} - \vec{p}_F^{(0)}(\theta)) + R(\vec{k})} \qquad (14.4)$$

where $\theta$ is the polar angle of $\vec{k}$, $Z_0 = 1$ is the *free wave function renormalization* and $\vec{v}_F^{(0)}(\theta) = (\partial \varepsilon_0 / \partial \vec{k})|_{\vec{k} = \vec{p}_F(\theta)}$ is the *free Fermi velocity*. Moreover, near the Fermi surface, $|R(\vec{k})| \leq C|\vec{k} - \vec{p}_F^{(0)}(\theta)|^2$, for some positive constant $C$. In the case (14.3), $\vec{p}_F^{(0)}(\theta), \vec{v}_F^{(0)}(\theta)$ are independent from $\theta$.

The interacting two–point function $S(\mathbf{x} - \mathbf{y})$ turns out to have, in the $L = \infty$ limit, the following structure. Let us call $\widehat{S}(\mathbf{k})$ the Fourier transform of $S(\mathbf{x})$ and $\Sigma(\mathbf{k})$ the *self-energy*, defined as usual by the identity

$$\widehat{S}(\mathbf{k}) = \frac{1}{-ik_0 + \varepsilon_0(\vec{k}) - \mu + \Sigma(\mathbf{k})} \qquad (14.5)$$

By simple symmetry arguments, one can see that $\Sigma(k_0, \vec{k}) = \Sigma(k_0, -\vec{k}) = \Sigma^*(-k_0, \vec{k})$; this allows us to introduce the following definitions.

a) The *interacting Fermi surface* $\Sigma_F$ is defined as

$$\Sigma_F = \left\{ \vec{k} \,:\, \varepsilon_0(\vec{k}) + \frac{1}{2} \sum_{j=\pm} \Sigma(j\pi\beta^{-1}, \vec{k}) = \mu \right\}, \qquad (14.6)$$

We shall be able to parameterize $\Sigma_F$ as $\vec{k} = \vec{p}_F(\theta)$ in terms of the polar angle $\theta \in [0, 2\pi]$ and we shall denote $|\vec{p}_F(\theta)|$ by $u(\theta)$.

b) The *wave function renormalization* is the real quantity

$$Z(\theta) = 1 + i\partial_{k_0}\Sigma \qquad (14.7)$$

where $\partial_{k_0}\Sigma = (\beta/2\pi)[\Sigma(\frac{\pi}{\beta}, \vec{p}_F(\theta)) - \Sigma(-\frac{\pi}{\beta}, \vec{p}_F(\theta))]$.

c) The *Fermi velocity* is the real vector

$$\vec{v}_F(\theta) = \frac{1}{Z(\theta)} \left.\frac{\partial(\varepsilon_0 + \frac{1}{2}\sum_{j=\pm}\Sigma(j\pi\beta^{-1}, \vec{k}))}{\partial \vec{k}}\right|_{\vec{k}=\vec{p}_F(\theta)} \tag{14.8}$$

The Jellium and the Hubbard model are Fermi liquids, in the sense of the definition given in the introduction, if the temperature is not too low; this is the content of the following Theorem, proved in Ref [61; 62] in the Jellium case and in Refs [63; 64] and [65] in the case of the Hubbard model.

**Theorem 14.1.** *Let us consider the 2d Hubbard model with $0 < \mu < \mu_0 \equiv \frac{2-\sqrt{2}}{2}$ and $\beta^{-1} \geq e^{-\frac{a}{|U|}}$ where $a > 0$ is a suitable constant. There exists a constant $U_0 > 0$ such that, if $|U| \leq U_0$, the two point Schwinger function $\widehat{S}(\mathbf{k})$ can be written, in the limit $L = \infty$, as*

$$\widehat{S}(\mathbf{k}) = \frac{1}{Z(\theta)} \frac{1}{-ik_0 + \vec{v}_F(\theta) \cdot (\vec{k} - \vec{p}_F(\theta)) + R(\mathbf{k})} \tag{14.9}$$

*with $Z(\theta), \vec{v}_F(\theta)$ and $\vec{p}_F(\theta)$ real and*

$$Z(\theta) = 1 + a(\theta)U^2 + O(U^3) \tag{14.10}$$

$$\vec{v}_F(\theta) = \vec{v}_F^{(0)}(\theta) + \vec{b}(\theta)U^2 + O(U^3) \quad \vec{p}_F(\theta) = \vec{p}_F^{(0)}(\theta) + \vec{c}(\theta)U + O(U^2)$$

*where $a(\theta), |\vec{b}(\theta)|, |\vec{c}(\theta)|$ are bounded above and below by positive $O(1)$ constants in the region $\beta^{-1} \geq e^{-\frac{a}{|U|}}$. Moreover*

$$|R(\mathbf{k})| \leq C[|\vec{k} - \vec{p}_F(\theta)|^2 + k_0^2 + |\vec{k} - \vec{p}_F(\theta)||k_0|] \tag{14.11}$$

*for some constant $C > 0$.*

*In the Jellium case with an ultraviolet cut-off, a similar result hold for any $\mu$, with $\vec{p}_F(\theta), \vec{v}_F(\theta), Z(\theta)$ $\theta$ independent by rotation invariance.*

The above theorem says that the Jellium model and the Hubbard model are *Fermi liquids* (in the sense explained in the introduction) up to exponentially small temperatures; the free and interacting Schwinger functions have a similar behaviour up to a renormalization of the Fermi velocity and the wave function renormalization. The fact that the renormalization of the parameters are essentially *temperature indipendent* is the crucial property of Fermi liquids. Such a property can be true only at $d \geq 2$; we have seen in fact that interacting fermions in $d = 1$ are Luttinger liquids in which the wave function renormalization for $\beta^{-1} \geq e^{-\frac{a}{|U|}}$ has a logarithmic dependence from the temperature

$$Z - 1 \simeq O(U^2 \log \beta) \tag{14.12}$$

Perturbative considerations suggest also that even in $d=2$ the presence of Fermi liquid behaviour depends crucially from certain convexity and regularity properties of the Fermi surface; if the Fermi surface has some flat pieces a logarithmic behaviour like (14.12) is found, and in presence of cusps, like in the Hubbard model in the half filled case, a behaviour of $Z-1 \simeq O(U^2 \log \beta)$ is found.

## 14.2 Multiscale integration

We consider the case of the Hubbard model. As we have seen in the previous chapters, the Schwinger functions can be written in terms of Grassman integrals

$$S(\mathbf{x}_1,\sigma_1,\varepsilon_1;\ldots;\mathbf{x}_n,\sigma_n,\varepsilon_n) = \frac{\int P(d\psi)e^{-V(\psi)}\psi_{\mathbf{x}_1,\sigma_1}^{\varepsilon_1}\cdots\psi_{\mathbf{x}_n,\sigma_n}^{\varepsilon_n}}{\int P(d\psi)e^{-V(\psi)}}. \quad (14.13)$$

where

$$P(d\psi) = \Big[\prod_{\mathbf{k}\in\mathcal{D}_{\beta,L}}^{\sigma=\uparrow\downarrow}(L^2\beta\widehat{g}_\mathbf{k})d\widehat{\psi}_{\mathbf{k},\sigma}^+ d\widehat{\psi}_{\mathbf{k},\sigma}^-\Big]\cdot\exp\Big\{-\sum_{\mathbf{k}\in\mathcal{D}_{\beta,L}}^{\sigma=\uparrow\downarrow}(L^2\beta\widehat{g}_\mathbf{k})^{-1}\widehat{\psi}_{\mathbf{k},\sigma}^+\widehat{\psi}_{\mathbf{k},\sigma}^-\Big\}, \quad (14.14)$$

and

$$V(\psi) = U\int d\mathbf{x}\,\psi_{\mathbf{x},\uparrow}^+\psi_{\mathbf{x},\uparrow}^-\psi_{\mathbf{x},\downarrow}^+\psi_{\mathbf{x},\downarrow}^- \quad (14.15)$$

where the symbol $\int d\mathbf{x}$ must be interpreted as $\int d\mathbf{x} = \int_{-\beta/2}^{+\beta/2}dx_0\sum_{\vec{x}}$.

The integration procedure of the generating function is similar to the one used in 1d; the more relevant difference is in the fact that we include all the quadratic term in the free integration at each integration step, except only its local part (as it was done in 1d). After the integration of $\psi^{(h+1)},\psi^{(h+2)},\ldots$ the partition function can be rewritten as

$$e^{-L^2\beta F_{L,\beta}} = e^{-L^2\beta F_h}\int P_{E_h,C_h}(d\psi^{(\leq h)})e^{-V^{(h)}(\psi^{\leq h})}, \quad (14.16)$$

where: $V^{(h)}$ can be represented as a sumover monomials in $\psi^{(\leq h)}$ with kernels $W_{2l}^{(h)}$; the Grassmann integration $P_{E_h,C_h}(d\psi^{(\leq h)})$ can be represented as

$$P_{E_h,C_h}(d\psi^{(\leq h)}) = \Big[\prod_\mathbf{k}^{\sigma=\uparrow\downarrow}\Big(\frac{L^2\beta C_h^{-1}(\mathbf{k})}{-ik_0 + E_h(\mathbf{k}) - \mu}\Big)d\psi_{\mathbf{k},\sigma}^{(\leq h)+}d\psi_{\mathbf{k},\sigma}^{(\leq h)-}\Big]$$

$$\cdot\exp\Big\{-\frac{1}{L^2\beta}\sum_{\mathbf{k}\in\mathcal{D}_{\beta,L}^*}^{\sigma=\uparrow\downarrow}C_h(\mathbf{k})(-ik_0+E_h(\mathbf{k})-\mu)\widehat{\psi}_{\mathbf{k},\sigma}^{+(\leq h)}\widehat{\psi}_{\mathbf{k},\sigma}^{-(\leq h)}\Big\} \quad (14.17)$$

where $F_h(\mathbf{k})$ is a function to be iteratively constructed below, with $E_0(k_0, \vec{k}) \equiv \varepsilon_0(\vec{k})$. Moreover $C_h(\mathbf{k})^{-1}$ is a compact support function defined as

$$C_h^{-1}(\mathbf{k}) = H_0\left[\gamma^{-h}| - ik_0 + E_h(\mathbf{k}) - \mu|\right] \qquad (14.18)$$

and $\mathcal{D}^*_{\beta,L}$ the restriction of $\mathcal{D}_{\beta,L}$ to the set of momenta in the support of $C_h^{-1}(\mathbf{k})$.

We introduce the *localization operator* as a linear operator acting on the kernels of $\mathcal{V}^{(h)}$ in the following way:

$$\mathcal{L}W_{2l}^{(h)}(\mathbf{x}_1,\ldots,\mathbf{x}_{2l}) = W_{2l}^{(h)}(\mathbf{x}_1,\ldots,\mathbf{x}_{2l}) \quad \text{if} \quad l = 1, 2$$
$$\mathcal{L}W_{2l}^{(h)}(\mathbf{x}_1,\ldots,\mathbf{x}_{2l}) = 0 \qquad\qquad\qquad \text{if} \quad l \geq 3 \qquad (14.19)$$

We also define $\mathcal{R}$ as $\mathcal{R} = 1 - \mathcal{L}$ and rewrite the r.h.s. of (10.20) as

$$e^{-L^2\beta F_h}\int P_{E_h,C_h}(d\psi^{(\leq h)})e^{-\mathcal{L}\mathcal{V}^{(h)}(\psi^{(\leq h)})-\mathcal{R}\mathcal{V}^{(h)}(\psi^{(\leq h)})}, \qquad (14.20)$$

where by definition $\mathcal{L}\mathcal{V}^{(h)}$ can be written as

$$\mathcal{L}\mathcal{V}^{(h)} = \sum_{\sigma=\uparrow\downarrow}\int d\mathbf{x} d\mathbf{y}\, n_h(\mathbf{x}-\mathbf{y})\, \psi_{\mathbf{x},\sigma}^{(\leq h)+}\psi_{\mathbf{y},\sigma}^{(\leq h)-} + \qquad (14.21)$$

$$+ \sum_{\sigma,\sigma'=\uparrow\downarrow}\int d\mathbf{x}_1 d\mathbf{x}_2 d\mathbf{x}_3 d\mathbf{x}_4\, \lambda_h(\mathbf{x}_1,\mathbf{x}_2,\mathbf{x}_3,\mathbf{x}_4)\, \psi_{\mathbf{x}_1,\sigma}^{(\leq h)+}\psi_{\mathbf{x}_2,\sigma}^{(\leq h)-}\psi_{\mathbf{x}_3,\sigma'}^{(\leq h)+}\psi_{\mathbf{x}_4,\sigma'}^{(\leq h)-}$$

Now, calling $\widehat{n}_h(\mathbf{k})$ the Fourier transform of $n_h(\mathbf{x})$ and defining

$$E_{h-1}(\mathbf{k}) = E_h(\mathbf{k}) + C_h^{-1}(\mathbf{k})\widehat{n}_h(\mathbf{k}), \qquad (14.22)$$

we can rewrite (14.20) as

$$e^{-L^2\beta(F_h+t_h)}\int P_{E_{h-1},C_h}(d\psi^{(\leq h)})e^{-\mathcal{L}_4\mathcal{V}^{(h)}(\psi^{\leq h})-\mathcal{R}\mathcal{V}^{(h)}(\psi^{(\leq h)})}, \qquad (14.23)$$

where $t_h$ is a constant which takes into account the change in the renormalization factor of the measure, of size $|U||h|\gamma^{2h}$. Moreover

$$\mathcal{L}_4\mathcal{V}^{(h)} = \sum_{\sigma,\sigma'=\uparrow\downarrow}\int d\mathbf{x}_1 d\mathbf{x}_2 d\mathbf{x}_3 d\mathbf{x}_4\, \lambda_h(\mathbf{x}_1,\mathbf{x}_2,\mathbf{x}_3,\mathbf{x}_4)\, \psi_{\mathbf{x}_1,\sigma}^{(\leq h)+}\psi_{\mathbf{x}_2,\sigma}^{(\leq h)-}\psi_{\mathbf{x}_3,\sigma'}^{(\leq h)+}\psi_{\mathbf{x}_4,\sigma'}^{(\leq h)-}.$$
$$(14.24)$$

We now define $\widehat{\mathcal{V}}^{(h)} \equiv \mathcal{L}_4\mathcal{V}^{(h)} + \mathcal{R}\mathcal{V}^{(h)}$ and again use the addition principle in order to rewrite (14.23) as

$$e^{-L^2\beta(F_h+t_h)}\int P_{E_{h-1},C_{h-1}}(d\psi^{(\leq h-1)})\int P_{E_{h-1},f_h^{-1}}(d\psi^{(h)})e^{-\widehat{\mathcal{V}}^{(h)}(\psi^{(\leq h-1)}+\psi^{(h)})}$$
$$(14.25)$$

with $P_{E_{h-1},f_h^{-1}}(d\psi^{(h)})$ a Grassmann Gaussian integration such that

$$\int P_{E_{h-1},f_h^{-1}}(d\psi^{(h)})\widehat{\psi}^{(h)-}_{\mathbf{k}_1,\sigma_1}\widehat{\psi}^{(h)+}_{\mathbf{k}_2,\sigma_2} = L^2\beta\delta_{\mathbf{k}_1,\mathbf{k}_2}\delta_{\sigma_1,\sigma_2}\widehat{g}^{(h)}(\mathbf{k}_1)$$

$$\text{with } \widehat{g}^{(h)}(\mathbf{k}) = \frac{f_h(\mathbf{k})}{-ik_0 + E_{h-1}(\mathbf{k}) - \mu} \tag{14.26}$$

and

$$f_h(\mathbf{k}) = H_0\left[\gamma^{-h}\left|-ik_0 + E_h(\mathbf{k}) - \mu\right|\right] - H_0\left[\gamma^{-h+1}\left|-ik_0 + E_{h-1}(\mathbf{k}) - \mu\right|\right].$$
$$\tag{14.27}$$

We define an effective dispersion relation at scale $h$ as

$$\varepsilon_h(\vec{k}) = \frac{1}{2}[E_h(\frac{\pi}{\beta},\vec{k}) + E_h(-\frac{\pi}{\beta},\vec{k})] \tag{14.28}$$

and of course $\varepsilon_h(\vec{k}) = \varepsilon_h(-\vec{k})$. We can define interacting Fermi surface at scale $h$ the set $\vec{k} \in \Sigma^{(h)}$ such that $\varepsilon_h(\vec{k}) = 0$. Note that in general the interaction modifies deeply the shape of the Fermi surface, that is the shape of the interacting and free Fermi surface, respectively $\Sigma^{(h)}$ and $\Sigma^{(0)}$, are different. An important exception is in the Jellium case, in which the interacting Fermi surface is a circle, as a consequence of rotation invariance, and the only effect of the interaction is to change its radius.

If we now define

$$e^{-\mathcal{V}^{(h-1)}(\psi^{(\leq h-1)}) - L^2\beta\widetilde{F}_h} = \int P_{E_{h-1},f_h^{-1}}(d\psi^{(h)})e^{-\widehat{\mathcal{V}}^{(h)}(\psi^{(\leq h-1)}+\psi^{(h)})},$$
$$\tag{14.29}$$

it is easy to see that $\mathcal{V}^{(h-1)}$ is of the same form of $\mathcal{V}^{(h)}$. We iterate this procedure up to the first scale $h_\beta$ such that $\gamma^{h_\beta-1} < \min\{|k_0 - \text{Im}E_{h_\beta}(\mathbf{k})|$. By the properties of $E_h(\mathbf{k})$, that will be described and proved below, it will turn out that $h_\beta$ is finite and actually larger than $[\log_\gamma(\pi/2e_0\beta)]$. Moreover, this definition is such that $f_{h_\beta}(\mathbf{k}) = C_{h_\beta}^{-1}(\mathbf{k})$, hence the propagator associated with $P_{E_{h_\beta-1},C_{h_\beta}}(d\psi^{(\leq h_\beta)})$ is given by $\widehat{g}^{(h_\beta)}(\mathbf{k})$.

On scale $h_\beta$ we define

$$e^{-L^2\beta(\widetilde{F}_{h_\beta}+t_{h_\beta})} = \int P_{E_{h_\beta-1},C_{h_\beta}}(d\psi^{(\leq h_\beta)})e^{-\widehat{\mathcal{V}}^{(h_\beta)}(\psi^{(\leq h_\beta)})}, \tag{14.30}$$

so that we have

$$F_{L,\beta} = F_0 + \sum_{h=h_\beta}^{0}(\widetilde{F}_h + t_h). \tag{14.31}$$

Note that the above procedure allows us to rewrite the effective coupling $\lambda_h(\mathbf{x}_1, \mathbf{x}_2, \mathbf{x}_3, \mathbf{x}_4)$ on scale $h$ and the renormalized dispersion relation $E_h(\mathbf{k})$ as functionals of $U$ and $\lambda_j, E_j$, with $h < j \le 0$:

$$E_{h-1}(\mathbf{k}) = E_h(\mathbf{k}) + C_h^{-1}(\mathbf{k})\widehat{\beta}_h^2(\mathbf{k}; E_h, \lambda_h, \ldots, E_0, \lambda_0, U)$$
$$\lambda_{h-1}(\underline{\mathbf{x}}) = \lambda_h(\underline{\mathbf{x}}) + \beta_h^4(\underline{\mathbf{x}}; E_h, \lambda_h, \ldots, E_0, \lambda_0, U) \qquad (14.32)$$

where in the second line we defined $\underline{\mathbf{x}} = \{\mathbf{x}_1, \mathbf{x}_2, \mathbf{x}_3, \mathbf{x}_4\}$. The functionals $\widehat{\beta}_h^2$ and $\beta_h^4$ are called the $E$–component and the $\lambda$–component of the *Beta function*.

## 14.3 Bounds for the Feynman graphs

As in previous chapters the kernels $W_m^{(h)}$ can be written as over trees and Feynnman graphs

$$W_m^{(h)} = \sum_{\tau \in T_h} \sum_G \text{Val}(G) \qquad (14.33)$$

It is easy to bound $\text{Val}(G)$ in momentum space. Let us consider the Jellium case for definiteness, in which the propagator is singular on a circle with radius $p_F^{(h)}$. By expoiting the conservation of momenta associated to each interaction, $(n_v^0 - s_v + 1)$ independent integrations remain for any non trivial vertex $v$ of the tree, if $n_v^0$ are the propagator internal to the cluster $v$ but not to any smaller one ; by the change of variables $(k_1, k_2) \to ((p_F^{(h)} + \rho)\cos\theta, (p_F^{(h)} + \rho)\sin\theta)$, and using that $\rho = O(\gamma^{h_v})$ we get that each integration can be bounded by $O(\gamma^{2h_v})$. Using that $|g^{(h)}(\mathbf{k})| \le C\gamma^{-h}$, we get that each Feynman graph can be bounded by, if $||\lambda_k|| \le \varepsilon$

$$|\text{Val}(G)| \le C\varepsilon^n \prod_v \gamma^{-h_v n_v^0}\gamma^{2h_v(n_v^0-s_v+1)} \prod_v \gamma^{-(h_v-h_{v'})z_v} \qquad (14.34)$$

with $z_v = 1$ if $n_v^e$ has four external lines and $z_v = 0$ otherwise. By proceeding as in §3.3 we can rewrite (14.34) as

$$|\text{Val}(G)| \le C\varepsilon^n \gamma^{-h_{v_0}d_{v_0}} \prod_v \gamma^{-(h_v-h_{v'})(d_v+z_v)} \qquad (14.35)$$

with $d_v = \frac{n_v^e}{2} - 2$, so that the sum over the scales can be performed as $d_v + z_v > 0$ (as there are no vertices with two external lines by constructions). Note that the dimension $d_v$ is identical in $d = 1$ and $d = 2$.

However, here comes a crucial difference between the $d = 1$ and the $d \ge 2$ case, when one is interested to non-perturbative results. As we have seen, in order to get convergent bounds one needs estimates in the coordinate space. In $d = 1$, bounds for the Feynman graphs are the same in the coordinate or momentum space; this is however not true in $d \geq 2$. In the coordinate space, noting that each integration contributes with a factor $\gamma^{-(d+1)h_v(s_v-1)}$ for each vertex of the tree, and that the size of each propagator is $O(\gamma^{h_v})$, one gets the bound

$$|\text{Val}(G)| \leq C\varepsilon^n \prod_v \gamma^{-(d+1)h_v(s_v-1)} \gamma^{h_v n_v^0} \prod_v \gamma^{-(h_v-h_{v'})z_v} \leq$$
$$C\varepsilon^n \gamma^{-h_{v_0} \bar{d}_{v_0}} \prod_v \gamma^{-(h_v-h_{v'})(\bar{d}_v + z_v)} \qquad (14.36)$$

with, if $m_v^4$ is the number of end-points following $v$ in $\tau$,

$$\bar{d}_v = -(d+1) + \frac{n_v^e}{2} - (1-d)m_v^4 \qquad (14.37)$$

so that the bound of each graph in coordinate space becomes greater and greater with the order, except that in $d = 1$.

The point is that in performing bounds in momentum space one incorporates automatically informations from the geometry of the Fermi surface and the momentum conservations, which are apparently lost in the coordinate space; one has then to refine the bounds in the coordinate space using a technique based on the idea of *sectors* in order to take into account the geometry of the Fermi surface also in the coordinate space.

## 14.4 The sector decomposition

In order to improve the bounds we proceed somewhat similarly introducing a further decomposition of the single scale propagators $g^{(h)}(\mathbf{x} - \mathbf{y})$. In $d = 1$, the propagator was written as the sum of two terms, each of them concentrated around one of the two points of the $d = 1$ Fermi surface. In $d = 2$ the Fermi surface is a curve, and we can write the propagator as sum of propagators concentrated on a single part or *sector* in which the Fermi surface can be divided. While this decomposition of the Fermi surface in sectors can be realized in several different ways (or even consider pointlike sectors) it appears technically convenient to consider "anisotropic sectors" which are $O(\gamma^{h/2})$ wide in the direction tangential to the Fermi surface and $O(\gamma^h)$ thick in the normal direction.

Let us still consider the Jellium case. Passing to polar coordinates we can write the propagator as

$$g^{(h)}(\mathbf{x}-\mathbf{y}) = \int dk_0 d\theta \int |\vec{k}|d|\vec{k}| f_h(k_0^2 + E_h(\vec{k})^2) \frac{e^{i\mathbf{k}(\mathbf{x}-\mathbf{y})}}{-ik_0 + E_h(\vec{k})} \quad (14.38)$$

with $\varepsilon_h = C_h(|\vec{k}|^2 - (p_F^{(h)})^2)$, and we can introduce another decomposition over the integration in $\theta$ in the following way. The anolous of radius $\gamma^h$ around the Fermi surface is divided in *sectors* centered at $\theta = \theta_\omega$ and of angular width $\gamma^{h/2}$. Then $1 = \sum_\omega \zeta_{h,\omega}(\theta)$, where $\zeta_{h,\omega}(\theta)$ are compact support functions with support in $\gamma^{h/2-1/2} \leq |\theta - \theta_\omega| \leq \gamma^{h/2+1/2}$, $\sum_\omega 1 = \gamma^{-h/2}$ and, if $p_F(\theta) = p_F(\cos\theta.\sin\theta)$

$$g_\omega^{(h)}(\mathbf{x}-\mathbf{y}) = e^{i\vec{p}_F(\theta_\omega)(\vec{x}-\vec{y})} \bar{g}_\omega^{(h)}(\mathbf{x}-\mathbf{y}) \quad (14.39)$$

with

$$\bar{g}_\omega^{(h)}(\mathbf{x}-\mathbf{y}) = \int dk_0 d\theta \zeta_{h,\omega}(\theta) \int k dk f_h(k_0^2 + E_h^2(\vec{k})) \frac{e^{ik_0(t-s)+i[(\vec{k}-\vec{\omega}p_F^{(h)})(\vec{x}-\vec{y})]}}{-ik_0 + E_h(\vec{k})} \quad (14.40)$$

which is bounded by, for any $N$

$$|g_\omega^h(\mathbf{x}-\mathbf{y})| \leq \gamma^{3h/2} \frac{C_N}{1 + [\gamma^h|t-s| + \gamma^h|(\vec{x}-\vec{y})_r| + \gamma^{h/2}|(\vec{x}-\vec{y})_t|]^N} \quad (14.41)$$

where $(\vec{x}-\vec{y})_r = |\vec{x}-\vec{y}|\cos\theta_\omega$ and $(x-y)_t = |\vec{x}-\vec{y}|\sin\theta_\omega$. We can write

$$\psi_x = \sum_h \sum_\omega e^{i\vec{p}_F^{(h)}(\theta_\omega)\vec{x}} \psi_{\omega,\mathbf{x}}^{(h)} \quad (14.42)$$

where $\psi_{\omega,\mathbf{x}}^{(h)}$ has propagator given by $\bar{g}_\omega^h(\mathbf{x}-\mathbf{y})$. Note that (14.42) is the analogous of (12.39) in the $d=1$ case; note however that the over the quasi-particles indices increases with the scale as $\sum_\omega = \gamma^{-h/2}$.

In the general case, the Fermi surface is not a circle and its shape is not preserved when the interaction is present. We proceed by induction assuming that $\varepsilon_h(\vec{k})$ verifies the following properties.

1) If $|e| \leq \bar{e}$, $\varepsilon_h(\vec{k}) - \mu = e$ defines a convex curve $\Sigma^{(h)}(e)$, encircling the origin and symmetric by reflection with respect to it, which can be represented in polar coordinates as $\vec{p} = u_h(\theta,e)\vec{e}_r(\theta)$. Moreover $u_h(\theta,e) \geq c > 0$ and, if $r_h(\theta,e)$ is the curvature radius $r_h(\theta,e)^{-1} \geq c > 0$

2) If $|e| \leq \bar{e}$ and $\vec{p} = u_h(\theta,e)\vec{e}_r(\theta)$, then

$$0 < c_1 \leq \vec{\nabla}\varepsilon_h(\vec{p}) \cdot \vec{e}_r(\theta) \leq c_2 . \quad (14.43)$$

3) If $\mu < \mu_0 \equiv \frac{2-\sqrt{2}}{2}$ and $|e| \leq \bar{e}$, then

$$n \leq 4, \quad \vec{k}_i \in \Sigma^{(h)}(e), \; i=1,\ldots,2n, \quad \Rightarrow \quad |\sum_{i=1}^{2n} \vec{k}_i| < 2\pi . \quad (14.44)$$

4) The following bounds are true

$$|E_h(\mathbf{k}) - E_0(\mathbf{k})| \leq C'_0|U| \qquad |\partial_{k_i}E_h(\mathbf{k}) - \partial_{k_i}E_0(\mathbf{k})| \leq C'_1|U|\,,$$
$$|\partial_{k_{i_1}}\partial_{k_{i_2}}E_h(\mathbf{k}) - \partial_{k_{i_1}}\partial_{k_{i_2}}E_0(\mathbf{k})| \leq C'_2 c_0 \qquad (14.45)$$
$$|\partial_{k_{i_1}}\cdots\partial_{k_{i_n}}E_h(\mathbf{k}) - \partial_{k_{i_1}}\cdots\partial_{k_{i_n}}E_0(\mathbf{k})| \leq C'_n|U||h|\gamma^{(2-n)h}\,, \qquad n \geq 3$$

with $c_0 \equiv |h_\beta|U_0$ small enough.

From the assumptions (1)–(4) we can deduce the following properties. We call $\Sigma_F^{(h)} \equiv \Sigma^{(h)}(0)$ the *Fermi surface on scale h* and we put $u_h(\theta,0)\vec{e}_r(\theta) = \vec{p}_F^{(h)}(\theta)$ and $u_h(\theta) \equiv u_h(\theta,0) = |\vec{p}_F^{(h)}(\theta)|$. We introduce the angles $\theta_{h,\omega} = \pi(\omega + \frac{1}{2})\gamma^{h/2}$, with $\omega$ an integer in the set $O_h = \{0,1,\ldots,\gamma^{-(h-1)/2} - 1\}$ (choosing $\gamma = 4$). Correspondingly we introduce the functions $\zeta_{h,\omega}(\theta)$ with the properties:

$$||\theta - \theta_{h,\omega}|| < \frac{\pi}{4}\gamma^{h/2} \Rightarrow \zeta_{h,\omega}(\theta) = 1$$
$$||\theta - \theta_{h,\omega}|| > \frac{3\pi}{4}\gamma^{h/2} \Rightarrow \zeta_{h,\omega}(\theta) = 0$$
$$\sum_{\omega \in O_h} \zeta_{h,\omega}(\theta) = 1\,, \qquad \forall \theta \in \mathbb{T}^1 \qquad (14.46)$$

where $||\cdot||$ is the usual distance on $\mathbb{T}^1$.

We also introduce the support function $F_{h,\omega}(\mathbf{k}) = f_h(\mathbf{k})\zeta_{h,\omega}(\theta)$, where, if $\mathbf{k} = (k_0, \vec{k})$, then $\theta$ is the polar angle of $\vec{k}$. We shall call the functions $F_{h,\omega}(\mathbf{k})$ the *anisotropic* support functions and the indices $\omega \in O_h$ the *anisotropic* sector indices. Given any $\mathbf{k}$ belonging to the support of $F_{h,\omega}(\mathbf{k})$, we put

$$\vec{k} = \vec{p}_F^{(h)}(\theta_{h,\omega}) + k'_1\vec{n}_h(\theta_{h,\omega}) + k'_2\vec{\tau}_h(\theta_{h,\omega}) = \vec{p}_F^{(h)}(\theta_{h,\omega}) + \vec{k}' \qquad (14.47)$$

where, putting $\vec{e}_t(\theta) = (-\sin\theta, \cos\theta)$, the vectors $\vec{\tau}_h(\theta)$ and $\vec{n}_h(\theta)$ are defined as

$$\vec{\tau}_h(\theta) = \frac{d\vec{p}_F^{(h)}(\theta)}{d\theta}\left|\frac{d\vec{p}_F^{(h)}(\theta)}{d\theta}\right|^{-1} = \frac{u'_h(\theta)\vec{e}_r(\theta) + u_h(\theta)\vec{e}_t(\theta)}{\sqrt{u'_h(\theta)^2 + u_h(\theta)^2}}\,,$$
$$\vec{n}_h(\theta) = \frac{u_h(\theta)\vec{e}_r(\theta) - u'_h(\theta)\vec{e}_t(\theta)}{\sqrt{u'_h(\theta)^2 + u_h(\theta)^2}} \qquad (14.48)$$

$\vec{n}_h(\theta_{h,\omega})$ and $\vec{\tau}_h(\theta_{h,\omega})$ are the normal and tangential direction to the interacting Fermi surface at scale $h$ in correspondence of the angle $\theta_{h,\omega}$. By using that

$$[\vec{p}_F^{(h)}(\theta_1) - \vec{p}_F^{(h)}(\theta_2)] \cdot \vec{n}^{(h)}(\theta_2) = O(\theta_1 - \theta_2)^2$$
$$[\vec{p}_F^{(h)}(\theta_1) - \vec{p}_F^{(h)}(\theta_2)] \cdot \vec{\tau}^{(h)}(\theta_2) = O(\theta_1 - \theta_2) \qquad (14.49)$$

we obtain $|[\vec{p}_F^{(h)}(\theta) - \vec{p}_F^{(h)}(\theta_{h,\omega})]\vec{n}^{(h)}(\theta_{h,\omega})| \le c\gamma^h$ and $|[\vec{p}_F^{(h)}(\theta) - \vec{p}_F^{(h)}(\theta_{h,\omega})]\vec{\tau}^{(h)}(\theta_{h,\omega})| \le c\gamma^{h/2}$ so that

$$|k_1'| \le C\gamma^h \qquad |k_2'| \le C\gamma^{h/2} \tag{14.50}$$

Moreover if $\mathbf{k} = (k_0, \vec{k})$ belongs to the support of $F_{h,\omega}(\mathbf{k})$, $\omega \in O_h$ then

$$\left|\frac{\partial E_{h-1}(\mathbf{k})}{\partial k_2'}\right| \le C\gamma^{\frac{h}{2}}, \tag{14.51}$$

for some constant $C$. In fact, if $\mathbf{k} = (k_0, \vec{k})$ belongs to the support of $F_{h,\omega}(\mathbf{k})$, we can write $E_{h-1}(\mathbf{k}) = \varepsilon_h(\vec{k}) + (E_{h-1}(\mathbf{k}) - E_h(\mathbf{k})) + (E_h(\mathbf{k}) - \varepsilon_h(\vec{k}))$ where, by the properties described above, $\left|\partial_{k_2'}[(E_{h-1}(\mathbf{k}) - E_h(\mathbf{k})) + (E_h(\mathbf{k}) - \varepsilon_h(\vec{k}))]\right| \le C\gamma^h$. Moreover it is easy to prove that $\partial_{k_2'}\varepsilon_h(\vec{k}) = O(\gamma^{h/2})$. In fact we have $\vec{\nabla}\varepsilon_h(\vec{k}) = |\vec{\nabla}\varepsilon_h(\vec{k})|\vec{n}_h(\theta, e)$ where $|\vec{\nabla}\varepsilon_h(\vec{k})| = O(1)$, $e = \varepsilon_h(\vec{k}) - \mu = O(\gamma^h)$, $\theta$ is the polar angle of $\vec{k}$ and $\vec{n}_h(\theta, e)$ is the outgoing normal vector at $\Sigma^{(h)}(e)$ in $\vec{k}$. Furthermore $\|\theta - \theta_{h,\omega}\| = O(\gamma^{h/2})$ so that

$$\frac{\partial \varepsilon_h(\vec{k})}{\partial k_2'} = |\vec{\nabla}\varepsilon_h(\vec{k})|\vec{n}_h(\theta, e) \cdot \vec{\tau}_h(\theta_{h,\omega}) = |\vec{\nabla}\varepsilon_h(\vec{k})|\vec{n}_h(\theta) \cdot \vec{\tau}_h(\theta_{h,\omega}) + O(\gamma^h) =$$
$$= |\vec{\nabla}\varepsilon_h(\vec{k})|\sin(\theta - \theta_{h,\omega}) + O(\gamma^h) = O(\gamma^{h/2}) \tag{14.52}$$

and (14.51) follows.

Using the decomposition (14.47), we can rewrite

$$\psi_{\mathbf{x},\sigma}^{(h)\pm} \equiv \sum_{\omega \in O_h} e^{\pm i\vec{p}_F^{(h)}(\theta_{h,\omega})\vec{x}} \psi_{\mathbf{x},\sigma,\omega}^{(h)\pm} , \quad P(d\psi^{(h)}) = \prod_{\omega \in O_h} P(d\psi_\omega^{(h)}) , \tag{14.53}$$

where $P(d\psi_\omega^{(h)})$ is the Grassmannian integration with propagator

$$g_\omega^{(h)}(\mathbf{x}) = \tag{14.54}$$

$$\frac{1}{\beta} \sum_{k_0 \in \mathcal{D}_\beta} \int_{-\pi}^{\pi} \int_{-\pi}^{\pi} \frac{d\vec{k}'}{(2\pi)^2} e^{-i(k_0 x_0 + \vec{k}'\vec{x})} \frac{F_{h,\omega}(\mathbf{k}' + \mathbf{p}_F^{(h)}(\theta_{h,\omega}))}{-ik_0 + E_{h-1}(\mathbf{k}' + \mathbf{p}_F^{(h)}(\theta_{h,\omega})) - \mu}$$

where we defined $\mathbf{p}_F^{(h)}(\theta_{h,\omega}) \stackrel{def}{=} (0, \vec{p}_F^{(h)}(\theta_{h,\omega}))$. We insert the decomposition (14.53) into the r.h.s. of (14.20):

$$\int P_{E_{h-1}, f_h^{-1}}(d\psi^{(h)}) \exp\left\{-\widehat{\mathcal{V}}^{(h)}\left(\psi_{\mathbf{x},\sigma}^{(\le h-1)\pm} + \sum_{\omega \in O_h} e^{\pm i\vec{p}_F^{(h)}(\theta_{h,\omega})\vec{x}} \psi_{\mathbf{x},\sigma,\omega}^{(h)\pm}\right)\right\}, \tag{14.55}$$

and in this way we induce a decomposition of the kernels of $\mathcal{V}^{(h-1)}$ into a sum of contributions labelled by the choices of the sector labels of the integrated fields $\psi_{\mathbf{x},\sigma,\omega}^{(h)\pm}$.

The bounds on the (decomposed) kernels of $\mathcal{V}^{(h-1)}$ are based on the following key bound on the asymptotic behavior of $g_\omega^{(h)}(\mathbf{x})$, following by the assumption 1)-4) and obtained as usual integrating by parts:

$$|g_\omega^{(h)}(\mathbf{x})| \leq \frac{C_{M,N}\gamma^{\frac{3}{2}h}}{[1+(\gamma^h|d_\beta(x_0)|+\gamma^h|x_1'|)^M][1+\gamma^{-h}(\gamma^h|x_2'|)^N]} \quad (14.56)$$

where $d_\beta(x_0) = \beta\pi^{-1}\sin(\pi\beta^{-1}x_0)$ and $N, M$ are positive integers. The above bound implies that

$$\int d\mathbf{x}\, |\mathbf{x}|^j \cdot |g_\omega^{(h)}(\mathbf{x})| \leq C_j \gamma^{-(1+j)h}, \quad j \geq 0 \quad (14.57)$$

It is sufficient to consider the integral obtained by substituting $|\mathbf{x}|^j$ with $(\sqrt{|d_\beta(x_0)|^2+|x_1'|^2})^{j_1}|x_2'|^{j_2}$, $j_1+j_2=j$. This integral can be bounded by choosing $M \geq 3+j_1$ and $N = 2+j_2$ and by doing the rescalings $x_0 \to \gamma^{-h}x_0$, $x_1' \to \gamma^{-h}x_1'$ and $x_2' \to \gamma^{-h(j_2+1)/(j_2+2)}x_2'$; one gets a bound proportional to $\gamma^{-j_1 h - \alpha_{j_2} h}$, with $\alpha_j = 1/2 + (j+1)^2/(j+2) \leq j+1$.

## 14.5 The sector lemma

Now that we have introduced the sector fields $\psi_\omega$ we can rewrite $\mathcal{V}^{(h)}(\tau, \psi^{(\leq h)})$ in the r.h.s. of (12.44) as:

$$\mathcal{V}^{(h)}(\tau, \psi^{(\leq h)}) = \sum_{\mathbf{P}\in\mathcal{P}_\tau, \Omega\in\mathcal{O}_\tau} \mathcal{V}^{(h)}(\tau, \mathbf{P}, \Omega)$$

$$\mathcal{V}^{(h)}(\tau, \mathbf{P}, \Omega) = \int d\mathbf{x}_{v_0} \widetilde{\psi}_{\Omega_{v_0}}^{(\leq h)}(P_{v_0}) K_{\tau,\mathbf{P},\Omega}^{(h+1)}(\mathbf{x}_{v_0}) \quad (14.58)$$

where

$$\widetilde{\psi}_{\Omega_v}^{(\leq h)}(P_v) = \prod_{f \in P_v} e^{i\varepsilon(f)\vec{p}_F^{(h)}(\theta_{h,\omega(f)})\vec{x}(f)} \psi_{\mathbf{x}(f),\sigma(f),\omega(f)}^{(\leq h)\varepsilon(f)} \quad (14.59)$$

and $K_{\tau,\mathbf{P},\Omega}^{(h+1)}(\mathbf{x}_{v_0})$ is defined inductively by the equation, valid for any $v \in \tau$ which is not an endpoint,

$$K_{\tau,\mathbf{P},\Omega}^{(h_v)}(\mathbf{x}_v) = \frac{1}{s_v!} \quad (14.60)$$

$$\prod_{i=1}^{s_v}[K_{v_i}^{(h_v+1)}(\mathbf{x}_{v_i})]\, \mathcal{E}_{h_v}^T[\widetilde{\psi}_{\Omega_1}^{(h_v)}(P_{v_1}\backslash Q_{v_1}), \ldots, \widetilde{\psi}_{\Omega_{s_v}}^{(h_v)}(P_{v_{s_v}}\backslash Q_{v_{s_v}})], \quad (14.61)$$

where $\Omega_i = \{\omega(f), f \in P_{v_i}\backslash Q_{v_i}\}$ and $\widetilde{\psi}_{\Omega_i}^{(h_v)}(P_{v_i}\backslash Q_{v_i})$ has a definition similar to (14.61). Moreover, if $v$ is an endpoint and $h_v \leq 0$, $K_v^{(h_v)}(\mathbf{x}_v) =$

$\lambda_{h_v-1}(\mathbf{x}_v)$, while if $h_v = +1$ $K_v^{(1)}$ is equal to one of the kernels of the monomials produced by the ultraviolet integration.

If writing each truncated expectation as in (3.108), we get an expression of the form

$$\mathcal{V}^{(h)}(\tau, \mathbf{P}, \Omega) = \sum_{T \in \mathbf{T}} \int d\mathbf{x}_{v_0} \widetilde{\psi}_{\Omega_{v_0}}^{(\leq h)}(P_{v_0}) W_{\tau, \mathbf{P}, \Omega \backslash \Omega_{v_0}, T}^{(h)}(\mathbf{x}_{v_0})$$

$$= \sum_{T \in \mathbf{T}} \mathcal{V}^{(h)}(\tau, \mathbf{P}, \Omega, T) \qquad (14.62)$$

where $\mathbf{T}$ is a special family of graphs on the set of points $\mathbf{x}_{v_0}$, obtained by putting together an anchored tree graph $T_v$ for each non trivial vertex $v$. Note that any graph $T \in \mathbf{T}$ becomes a tree graph on $\mathbf{x}_{v_0}$, if one identifies all the points in the sets $x_v$, for any vertex $v$ which is also an endpoint. Given $\tau \in \mathcal{T}_{h,n}$ and the labels $\mathbf{P}, \Omega, T$, calling $v_1^*, \ldots, v_n^*$ the endpoints of $\tau$ and putting $h_i = h_{v_i^*}$, the explicit representation of $W_{\tau, \mathbf{P}, \Omega \backslash \Omega_{v_0}, T}^{(h)}(\mathbf{x}_{v_0})$ in (14.62) is

$$W_{\tau, \mathbf{P}, \Omega \backslash \Omega_{v_0}, T}(\mathbf{x}_{v_0}) = \left[ \prod_{i=1}^n K_{v_i^*}^{h_i}(\mathbf{x}_{v_i^*}) \right]$$

$$\cdot \prod_{v \text{ not e.p.}} \frac{1}{s_v!} \int dP_{T_v}(\mathbf{t}_v) \det G^{h_v, T_v}(\mathbf{t}_v) \left[ \prod_{l \in T_v} \delta_{\omega_l^+, \omega_l^-} \delta_{\sigma_l^-, \sigma_l^+} \widetilde{g}_{\omega_l}^{(h_v)}(\mathbf{x}_l - \mathbf{y}_l) \right]$$

We can expand the kernels of the effective potential in Feynman graphs. A bound for each graph is obtained taking into account that we get a factor $\gamma^{-(\frac{5}{2})h_v(s_v-1)}$ for the integration over the coordinates, a factor $\gamma^{\frac{3}{2}h_v n_v^0}$ from the propagators. A rough estimates for the sum over sectors gives a factor $\gamma^{\frac{-h_v}{2}}$ for each propagator so that

$$|\mathrm{Val}| \leq C\varepsilon^n \prod_v \gamma^{\frac{3}{2}h_v n_v^0} \gamma^{-\frac{1}{2}h_v n_v^0} \gamma^{-\frac{5}{2}h_v(s_v-1)} \gamma^{-(h_v - h_{v'})z_v} \leq$$

$$C\varepsilon^n \gamma^{-h\hat{d}_{v_0}} \prod_v \gamma^{-(h_v - h_{v'})(\hat{d}_v + z_v)} \qquad (14.63)$$

with $\hat{d}_v = -5/2 + \frac{n_v^e}{2} + m_v^4/2$, so that again the dimension depend from $m_v^4$ and the size of each Feynman graph depend on the order. In performing the above bounds we have sum over the $\omega$-indices as they were indipendent, but this is not so; if we write the Feynman graphs in momentum space it is clear that the $\omega$ have constraints.

In order to keep memory of such constraints in the coordinate space, we note that, thanks to momentum conservation and compact support properties of propagator Fourier transforms, $\mathcal{V}^{(h)}(\tau, \mathbf{P}, \Omega)$ vanishes for some choices of $\Omega$. We define, for any $h \leq 0$ and $\omega \in O_h$, the *sector* $S_{h,\omega}$ as

$$S_{h,\omega} = \{\vec{k} = \rho \vec{e}_r(\theta) \in \mathbb{R}^2 : |\varepsilon_h(\vec{k})| \leq \gamma^h e_0, \zeta_{h,\omega}(\theta) \neq 0\}. \quad (14.64)$$

Note that the definition of s-sector has the property that the s-sector $S_{h+1,\omega}$ of scale $h+1$ contains the union of two s-sectors of scale $h$: $S_{h+1,\omega} \supseteq \{S_{h,2\omega} \cup S_{h,2\omega+1}\}$. This follows from the the fact that $\gamma = 4$ (so that $\gamma^{1/2} = 2$). Moreover $S_{h,2\omega}$ and $S_{h,2\omega+1}$ are the two only sectors on scale $h$ strictly contained into $S_{h+1,\omega}$.

We now observe that the field variables $\widehat{\psi}_{\mathbf{k}(f),\omega(f),\sigma(f)}^{\leq h_{v_0},\varepsilon(f)}$ have the same supports as the functions $C_{h_{v_0}}^{-1}(\mathbf{k}(f)) \zeta_{h_{v_0},\omega(f)}(\theta(f))$ and $h(f) \leq h_i - 1$, $\forall f \in P_{v_i^*}$; hence in the expression (14.63), we can freely multiply $\widehat{K}_{v_i^*}^{h_i}(\mathbf{k}_{v_i^*})$ by $\prod_{f \in P_{v_i^*}} \widetilde{F}_{h_i-1,\widetilde{\omega}(f)}(\vec{k})$, where $\widetilde{F}_{h,\omega}(\vec{k})$ is a smooth function equal to 1 on $S_{h,\omega}$ and with a support slightly greater than $S_{h,\omega}$, while $\widetilde{\omega}(f) \in O_{h_i-1}$ is the unique sector index such that $S_{h(f),\omega(f)} \subseteq S_{h_i-1,\widetilde{\omega}(f)}$. By passing to the coordinate representation we get that each $K_{v_i^*,\Omega_{v_i^*}}^{h_i}(\mathbf{x}_{v_i^*})$ is replaced by a $\widetilde{K}_{v_i^*,\Omega_{v_i^*}}^{h_i}(\mathbf{x}_{v_i^*})$ given by the convolution of $K_{v_i^*,\Omega_{v_i^*}}^{h_i}(\mathbf{x}_{v_i^*})$ with the functions $\widetilde{F}_{h_i-1,\widetilde{\omega}(f)}(\vec{k})$

In order to explicitly keep track of the constraints satisfied by the sector indices $\Omega$, which is crucial for performing the dimensional bounds on the free energy, we introduce the following constraint functions. Given a tree $\tau \in \mathcal{T}_{h,n}$ with all its labels, a vertex $v \in \tau$ and the set of anisotropic sector indices $\Omega_v = \{\omega(f) \in O_{h(f)}, f \in P_v\}$ labelled by $P_v$, we define

$$\chi(\Omega_v) = \chi\Big(\forall f \in P_v, \exists \vec{k}(f) \in S_{h(f),\omega(f)} : \sum_{f \in P_v} \varepsilon(f)\vec{k}(f) = 0\Big) \quad (14.65)$$

if $|P_v| \leq 8$, and $= 1$ if $|P_v| \geq 10$. In free energy and the effective potential can be controlled by bounding the following function

$$J_{h,n}^{(F)}(\tau,\mathbf{P},T) = \sum_{\Omega \setminus \Omega_{ext}^{(F)}} \Big[\prod_v \chi(\Omega_v)\Big] \int d(\mathbf{x}_{v_0} \setminus \mathbf{x}^*) |W_{\tau,\mathbf{P},\Omega,T}(\mathbf{x}_{v_0})| \quad (14.66)$$

where $\mathbf{x}^*$ is an arbitrary point in $\mathbf{x}_{v_0}$, $\Omega \in \mathcal{O}_\tau$ and, if $2l_0 = |P_{v_0}| > 0$ and $0 < F \leq 2l_0$, $\Omega_{ext}^{(F)} \subset \Omega_{v_0}$ is an arbitrary subset of the sector indices in $\Omega_{v_0}$ of cardinality $F$, $|\Omega_{ext}^{(F)}| = F$: in particular, if $l_0 = 0$ or $F = 0$, $\sum_{\Omega \setminus \Omega_{ext}^{(F)} \in \mathcal{O}_\tau}$ coincides with $\sum_{\Omega \in \mathcal{O}_\tau}$.

Note that we could freely insert $\prod_v \chi(\Omega_v)$ in (14.66), because of the constraints following from momentum conservation and the compact support properties of propagator Fourier transforms. In particular in momentum space we see that the compact support properties of the cut-off functions and the constraints coming from the momentum conservation imposes constraints on the $\omega$-indices associated to the external fields associated to each vertex $v$; such constraints remain of course also in the coordinate space, and we keep memory of them by multiplying by the $\chi(\Omega_v)$ functions.

The crucial property of the functions $\chi(\Omega_v)$ is the following, also called *Sector Lemma*

$$\sum_{\Omega_{\tilde{v}}^{(h)} \prec \Omega_{\tilde{v}}^{(h_0)}}^{*} \chi(\Omega_{\tilde{v}}^{(h)}) \leq c\gamma^{(h_0-h)\left[\frac{1}{2}(|P_{\tilde{v}}|-3)\chi(4\leq|P_{\tilde{v}}|\leq 8)+\frac{1}{2}(|P_{\tilde{v}}|-1)\chi(|P_{\tilde{v}}|\geq 10)\right]} \quad (14.67)$$

where $\Omega_{\tilde{v}}^{(h)} \prec \Omega_{\tilde{v}}^{(h_0)}$ means that the indices in $\Omega_{\tilde{v}}^{(h)}$ satisfy the following constraint: given $f \in P_v$ and the corresponding index $\omega_f^{(h_0)} \in O_{j_f^{(h_0)}}$ of $\Omega_{\tilde{v}_0}^{(h_0)}$, then the index $\omega_f^{(h)} \in O_{j_f^{(h)}}$ of $\Omega_{\tilde{v}_0}^{(h)}$ is such that $S_{j_f^{(h)},\omega_f^{(h)}} \subset S_{j_f^{(h_0)},\omega_f^{(h_0)}}$. The symbol $*$ on the sums in the second line means that the sector index in $\Omega_{ext}^{(1)}$, associated to one of the fields in $P_{\tilde{v}_0}$ (say to the field $f_0 \in P_{\tilde{v}_0}$), is not summed over.

Let us consider for definiteness $|P_{\tilde{v}}| = 4$ and suppose that we have fixed the $\omega$ on some larger sectors at scale $h_0$; then the sum over sectors is not bounded by $\gamma^{2(h_0-h)}$, as one would find proceeding naively as in the derivation of (14.63), but by $\gamma^{1/2(h_0-h)}$.

The proof of (14.67) was given in [61] (see also [63] with a version with the notations followed here). It is related to well known parallelogram lemma, saying that, given four vectors on a circle with vanishing sum, the angles of two of them fix the angles of the other two.

## 14.6 Bounds for the tree expansion

By using the constraints due to the sector lemma, we can finally obtain a bound in which the dimension associated to each vertex are indeed positive and independent from $m_v^4$.

**Theorem 14.2.** *Given* $h_\beta \leq h \leq 0$, $\tau \in \mathcal{T}_{h,n}$, $\mathbf{P} \in \mathcal{P}_\tau$, $T \in \mathbf{T}$, *under the assumptions 1)–4) and if* $|\tilde{\lambda}_{h_{v^*}-1,\Omega_{v^*}}| \leq C|U|$, *then*

$$J_{h,n}^{(1)}(\tau, \mathbf{P}, T) \leq (c|U|)^n \gamma^{h(\frac{5}{2}-\frac{3}{4}|P_{v_0}|)} \prod_{v \text{ not e.p.}} \frac{1}{s_v!} \gamma^{\delta(|P_v|)} \quad (14.68)$$

where
$$\delta(p) = 1 - \frac{p}{4} + \chi(p \geq 10). \tag{14.69}$$

*Proof*: We shall describe the proof in the case that all the endpoints are of type $\lambda$. A posteriori, it will be clear that the possible presence of endpoints of scale $+1$ with $p \geq 6$ external legs does not qualitatively change the argument.

(14.68) follows by (14.56) and using that, by the Gram–Hadamard inequality

$$|\det G_\alpha^{h_v,T_v}(\mathbf{t}_v)| \leq c^{\sum_{i=1}^{s_v}|P_{v_i}|-|P_v|-2(s_v-1)} \cdot \gamma^{h_v \frac{3}{4}\left(\sum_{i=1}^{s_v}|P_{v_i}|-|P_v|-2(s_v-1)\right)}. \tag{14.70}$$

and noting that, by the bounds for $g_{\omega_l}^{(h_v)}$

$$\prod_{v \text{ not e.p.}} \frac{1}{s_v!} \int \prod_{l \in T_v} d(\mathbf{x}_l - \mathbf{y}_l) |g_{\omega_l}^{(h_v)}(\mathbf{x}_l - \mathbf{y}_l)| \leq c^n \prod_{v \text{ not e.p.}} \frac{1}{s_v!} \gamma^{-h_v(s_v-1)}. \tag{14.71}$$

Moreover, by suitably taking into account the constraint functions $\chi(\mathcal{S}(P_v))$, the sum over the choices of the sector indices gives

$$\sum_{\Omega \backslash \Omega_{ext}^{(1)}} \prod_{v \in \tau} \left( \chi(\Omega_v) \prod_{l \in T_v} \delta_{\omega_l^+, \omega_l^-} \right) \leq \tag{14.72}$$

$$\leq c^n \gamma^{-\frac{1}{2}hn} \prod_{v \text{ not e. p.}} \gamma^{\left[-\frac{1}{2}m_v^4 + \frac{1}{2}(|P_v|-3)\chi(4 \leq |P_v| \leq 8) + \frac{1}{2}(|P_v|-1)\chi(|P_v| \geq 10)\right]}$$

where $m_v^4$ denotes the number of endpoints (all of type $\lambda$, by hypothesis) following $v$ on $\tau$. It is straightforward to check that Theorem 14.2 follows by combining the bounds (14.70), (14.71) and (14.72).

The meaning of (14.72) is quite clear. It says that the sum over the $\omega$ associated to the external fields of the end-points can be done fixing the scale of the sectors to the larger one; then we get factor $\gamma^{-\frac{1}{2}\bar{m}_v^4 h_v}$, where $\bar{m}_v^4$ are the end-points following $v$ on $\tau$ but not following any subtrees of $\tau$ (see also (8.17)). Then $\prod_v \gamma^{-\frac{1}{2}\bar{m}_v^4 h_v} = \gamma^{-\frac{1}{2}hn} \prod_{v \text{ not e. p.}} \gamma^{-\frac{1}{2}m_v^4}$. The remaining sum over sectors, associated to the fields coming out from the end-points not with the larger scale, are bounded by (14.67). It is important to use the sector lemma only with a fixed number of external lines, as the constant in the lemma have a bad dependence from the number of external lines.

In order to prove (14.72), let us first note that, by the definition of sector $S_{h,\omega}$ and by the properties of $E_h(\mathbf{k})$, the following crucial property is true: *given* $\omega \in O_h$ *and the sector* $S_{h,\omega}$, *for any* $j > h$ *there is a unique* $\omega'(j;\omega) \in O_j$ *such that* $S_{j,\omega'(j;\omega)} \supset S_{h,\omega}$.

This property allows us to give a meaning to the following definition. Given $\tau \in \mathcal{T}_{h,n}$, $\mathbf{P} \in \mathcal{P}_\tau$ and $v \in \tau$, we introduce the symbol $\Omega_v^{(j)}$ to denote the set

$$\Omega_v^{(j)} = \{\omega(f), f \in P_v \; : \; h(f) \geq j\} \cup \{\omega'(j;\omega(f)), f \in P_v \; : \; j > h(f)\} \equiv$$
$$\equiv \{\omega_f^{(j)} \in O_{j_f^{(j)}}, f \in P_v\} \tag{14.73}$$

where the last identity *defines* the scales $j_f^{(j)}$ and the sector indices $\omega_f^{(j)}$, $f \in P_v$. If $\Omega_v = \Omega_v^{(h)}$ in the following we shall also denote:

$$\chi(\Omega_v^{(j)}) = \chi\left(\forall f \in P_f, \exists \vec{k}(f) \in S_{j_f^{(j)},\omega_f^{(j)}} \; : \; \sum_{f \in P_f} \varepsilon(f)\vec{k}(f) = \vec{0}\right), \tag{14.74}$$

Note in particular that:

$$\chi(\Omega_v^{(j)}) \leq \chi(\Omega_v^{(k)}), \quad \text{if} \quad j \leq k. \tag{14.75}$$

Given $\tau \in \mathcal{T}_{h,n}$, we define the set $V_c(\tau)$ of $c$-vertices of $\tau$ as the set of vertices $v$ of $\tau$ which either are endpoints or have the property that their set $\mathcal{I}_v$ of internal lines is non empty; in the following we shall often drop the dependence on $\tau$ (when it is clear from the context). Note that by definition, if $\tau \in \mathcal{T}_{h,n}$, then $|V_c(\tau)| = O(n)$ and it holds $\prod_{v \in \tau} \chi(\Omega_v) = \prod_{v \in V_c} \chi(\Omega_v)$. Moreover, using that $\Omega_v = \Omega_v^{(h)}$, we see that we can replace the product $\prod_{v \in \tau} \chi(\Omega_v)$ in the l.h.s. of (14.72) by $\prod_{v \in V_c} \chi(\Omega_v^{(h)})$.

We now begin to inductively bound the l.h.s. of (14.72); first of all we shall bound the sum corresponding to the first $c$-vertex following the root; then we will iteratively enter its structure. After each step we will be left with a product of sector sums of the same form of the initial one, but on larger scales.

We call $\tilde{v}_0$ the first $c$-vertex following the root and $h_0 \equiv h_{\tilde{v}_0}$ (the scale label of the legs belonging to $T_{\tilde{v}_0}$); using (14.75), we find

$$\prod_{v \in V_c} \chi(\Omega_v^{(h)}) \leq \chi(\Omega_{\tilde{v}_0}^{(h)}) \prod_{v > \tilde{v}_0, v \in V_c} \chi(\Omega_v^{(h_0)}). \tag{14.76}$$

Substituting (14.76) into the l.h.s. of (14.72), we find

$$\sum_{\Omega \backslash \Omega_{ext}^{(1)}}^* \prod_{v \in V_c} \left(\chi(\Omega_v^{(h)}) \prod_{l \in T_v} \delta_{\omega_l^+,\omega_l^-}\right) \leq$$

$$\leq \sum_{\Omega_{\tilde{v}_0}^{(h_0)}}^* \left[\sum_{\Omega_{\tilde{v}_0}^{(h)} \prec \Omega_{\tilde{v}_0}^{(h_0)}}^* \chi(\Omega_{\tilde{v}_0}^{(h)})\right] \sum_{\Omega \backslash \Omega_{\tilde{v}_0}} \prod_{v > \tilde{v}_0, v \in V_c} \chi(\Omega_v^{(h_0)}) \prod_{l \in T} \delta_{\omega_l^+,\omega_l^-} \tag{14.77}$$

where the symbol $*$ on the sums in the second line means that the sector index in $\Omega_{ext}^{(1)}$, associated to one of the fields in $P_{\tilde{v}_0}$ (say to the field $f_0 \in P_{\tilde{v}_0}$), is not summed over.

The r.h.s. of (14.77) can be bounded by the r.h.s. of (14.67) times

$$\sum_{\cup_v \Omega_v^{(h_0)}}^{*} \Big[ \prod_{v > \tilde{v}_0, v \in V_c} \chi(\Omega_v^{(h_0)}) \Big] \Big[ \prod_{l \in T} \delta_{\omega_l^+, \omega_l^-} \Big], \qquad (14.78)$$

where the $*$ on the sum recalls again that we are not summing over the sector index of $f_0 \in P_{\tilde{v}_0}$.

We will now prove that (14.78) can be reduced to a product of contributions analogue to the l.h.s. of (14.77), with $h_0$ replacing $h$. In fact, calling $\underline{\tilde{v}}_0 = \{v_1, \ldots, v_{s_{\tilde{v}_0}}\}$ the set of $c$-vertices immediately following $\tilde{v}_0$ on $\tau$ and $\Omega_{\underline{\tilde{v}}_0}^{(h_0)} = \cup_{v \in \underline{\tilde{v}}_0} \Omega_v^{(h_0)}$, we can rewrite (14.78) as

$$\sum_{\Omega_{\underline{\tilde{v}}_0}^{(h_0)}}^{*} \prod_{v \in \underline{\tilde{v}}_0} \Big[ \sum_{\cup_{w > v} \Omega_w^{(h_0)} \setminus \Omega_v^{(h_0)}} \prod_{w \geq v, w \in V_c} \chi(\Omega_w^{(h_0)}) \prod_{l \in \cup_{w \geq v} T_w} \delta_{\omega_l^+, \omega_l^-} \Big] \prod_{l \in T_{\tilde{v}_0}} \delta_{\omega_l^+, \omega_l^-}$$

$$\equiv \sum_{\Omega_{\underline{\tilde{v}}_0}^{(h_0)}}^{*} \prod_{v \in \underline{\tilde{v}}_0} F_v(\Omega_v^{(h_0)}) \prod_{l \in T_{\tilde{v}_0}} \delta_{\omega_l^+, \omega_l^-} \qquad (14.79)$$

The function $F_v(\Omega_v^{(h_0)})$, defined by (14.79), is the sum over the "internal sector indices" of the product of the constraint functions corresponding to the vertex $v$. Note that also the l.h.s. of (14.77) could have been written in terms of one of this functions; the l.h.s. of (14.77) is in fact equal to $\sum_{\Omega_{v_0}^{(h)}}^{*} F_{v_0}(\Omega_{v_0}^{(h)})$.

We now choose as the root of $T_{\tilde{v}_0}$ the vertex $v_i \in \underline{\tilde{v}}_0$ such that $f_0 \in P_{v_i}$; then we select a leaf $v^*$ of $T_{\tilde{v}_0}$ and we call $l^*$ the branch of $T_{\tilde{v}_0}$ anchored to $v^*$. Calling $\Omega_{\underline{\tilde{v}}_0 \setminus v^*}^{(h_0)} = \cup_{v \in \underline{\tilde{v}}_0 \setminus v^*} \Omega_v^{(h_0)}$, we denote by $\mathcal{F}_{\underline{\tilde{v}}_0 \setminus v^*}(\Omega_{\underline{\tilde{v}}_0 \setminus v^*})$ the product of the constraint functions corresponding to the set of vertices $\underline{\tilde{v}}_0 \setminus v^*$:

$$\mathcal{F}_{\underline{\tilde{v}}_0 \setminus v^*}(\Omega_{\underline{\tilde{v}}_0 \setminus v^*}^{(h_0)}) \stackrel{def}{=} \prod_{v \in \underline{\tilde{v}}_0 \setminus v^*} F_v(\Omega_v^{(h_0)}) \prod_{l \in T_{\tilde{v}_0} \setminus l^*} \delta_{\omega_l^+, \omega_l^-} \qquad (14.80)$$

so that we can rewrite (14.79) as

$$\sum_{\omega_{l^*}^+, \omega_{l^*}^-} \delta_{\omega_{l^*}^+, \omega_{l^*}^-} \sum_{\Omega_{\underline{\tilde{v}}_0 \setminus v^*}}^{**} \mathcal{F}_{\underline{\tilde{v}}_0 \setminus v^*}(\Omega_{\underline{\tilde{v}}_0 \setminus v^*}^{(h_0)}) \sum_{\Omega_{v^*}^{(h_0)}}^{*} F_{v^*}(\Omega_{v^*}^{(h_0)}), \qquad (14.81)$$

where the ∗∗ on the second sum means that we are not summing neither on $\omega_{f_0}$ nor on $\omega_{l^*}$ and the ∗ on the third sum recalls that we are not summing over $\omega_{l^*}$. Bounding the last sum by $\sup_{\omega_{l^*}} \sum^*_{\Omega^{(h_0)}_{v^*}} F_{v^*}(\Omega^{(h_0)}_{v^*})$, we see that the last sum can be factorized out:

$$(14.81) \leq \sum_{\Omega_{\widetilde{\underline{v}}_0 \setminus v^*}}^{*} \mathcal{F}_{\widetilde{\underline{v}}_0 \setminus v^*}(\Omega^{(h_0)}_{\widetilde{\underline{v}}_0 \setminus v^*}) \cdot \sup_{\omega_{l^*}} \Big[ \sum_{\Omega^{(h_0)}_{v^*}}^{*} F_{v^*}(\Omega^{(h_0)}_{v^*}) \Big]. \qquad (14.82)$$

It is now clear that we can iterate the same procedure by choosing another leaf of $T_{\widetilde{v}_0} \setminus l^*$ and by factorizing out the corresponding contribution. At the end of the procedure we reach the root of $T_{\widetilde{v}_0}$ and we finally find that (14.78) can be bounded by

$$\prod_{v \in \widetilde{\underline{v}}_0} \sup_{\omega^*_v} \sum_{\Omega^{(h_0)}_v}^{*} F_v(\Omega^{(h_0)}_v), \qquad (14.83)$$

where $\omega^*_v$ is the sector index corresponding to the line of $T_{\widetilde{v}_0}$ entering $v \in \widetilde{\underline{v}}_0$, if $v$ is not the root, or to $l_0$, otherwise; the ∗ on the sum $\sum^*_{\Omega^{(h_0)}_v}$ means that we are not summing over $\omega^*_v$.

Now, if $v \in \widetilde{\underline{v}}_0$ is an endpoint, then the corresponding contribution in (14.83) can be easily bounded by

$$\sum_{\Omega^{(h_0)}_v}^{*} F_v(\Omega^{(h_0)}_v) \leq c\gamma^{-\frac{h_0}{2}}, \qquad (14.84)$$

where we used again the sector counting Lemma.

If $v \in \widetilde{\underline{v}}_0$ is not an endpoint, the corresponding factor in (14.83) has exactly the same form as the l.h.s. of (14.77), and we can bound it by repeating the same procedure described above; then, proceeding by induction, we find (14.72). ∎

## 14.7 Flow of runing coupling constants

Of course the bound (14.68) depend on the assumption that the effective coupling is small and the properties (1)–(4).

In general the flow of $\lambda_h$ is quite complex, especially in the non rotational invariant case. By a second order analysis it seems that the effective coupling increases without limit, and this is generally seen as signaling a phase transition. However, at finite temperature, we can take advantage from the fact that all the scales from $h_\beta = O(\log \beta)$ to $-\infty$ can be integrated in a single step, that is the temperature acts as an infrared cut-off. This implies that we can prove, by induction

$$\int d\mathbf{x}_2 d\mathbf{x}_3 d\mathbf{x}_4 |\tilde{\lambda}_{j,\Omega_4}(\mathbf{x})| \leq C|U|, \qquad j \geq h \qquad (14.85)$$

In order to prove (14.85) for $j < 0$, we get from (14.68)

$$\lambda_{j,\Omega_4}(\mathbf{x}) = \lambda_0(\mathbf{x}) + \sum_{j'=j+1}^{0} \beta^4_{j'}(\mathbf{x}) \qquad (14.86)$$

which implies, using that $|\beta^4_{j',\Omega_4}(\mathbf{x})| \leq c|U|^2$,

$$|\tilde{\lambda}_{j,\Omega_4}(\mathbf{x})| \leq C_0|U| + c|j||U|^2 \leq C|U|, \qquad (14.87)$$

where in the last passage we used that $|j||U| \leq c_0$.

More subtle is the proof of the assumptions 1)-4), as they do not follow simply from (14.68). Indeed from (14.68) we know that $|\partial^n \beta^2_j| \leq C|U|\gamma^j \gamma^{-jn}$, and proceeding as in (14.86) such bound is not enough to control the second derivative of the effective dispersion relation. This seems to say that all our procedure is not consistent; our analysis was based under the assumption that the interacting Fermi surface verify certain regularity assumptions, but the bounds we got at the end seems not strong enough to derive such regularity properties.

It is however possible to improve the bound (14.68) in the case $|P_{v_0}| = 2$. We can expect such improvment is possible becouse we could extract some of the "loop propagators" from the determinants, which would imply additional constraints on the $\omega$ variables. Let us consider for instance the second order contribution which is given by the graph if Fig. 14.1

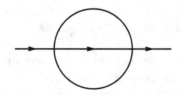

Fig. 14.1   Second order contribution to $\beta^2_h$.

We can bound its value by a factor $\gamma^{-\frac{5}{2}h}$ from the integration, a factor $\gamma^{\frac{9}{2}h}$ from the propagators and a factor $\gamma^{-h}$ from the sector lemma (one for each vertex), so that the final bound is $\gamma^h$, in agreement with the bound (14.68). On the other hand the bound can be easily improved noting that, fixed the $\omega$ of a vertex, the other $\omega$ variables are fixed. In this way we get a bound $\gamma^{\frac{3h}{2}}$ which is better than the one in (14.68) but still not enough to prove the assumptions 1)-4).

An essential optimal bound for the second order graph can be obtained by using a decomposition based on isotropic sector (with side $O(\gamma^h)$ either in the normal or parallel direction). It is easy to see, by purely dimensional reasons, that the propagator is bounded in such case by $\gamma^{2h}/(1+(\gamma^h|\mathbf{x}|)^N)$, and that the sector lemma for isotropic sectors gives $\gamma^{-h}|h|$, when applied to the vertices, so that we get the bound $CU^2\gamma^{6h}\gamma^{-3h}\gamma^{-h}|h|$ from which the needed regularity property follow.

Isotropic sectors are not really suitable to perform a non-perturbative analysis like the previous ones (as the reader can easily check); however we extract some loop propagators (not too many in order to avoid factorials) and further expand them in isotropic sectors, leaving the other propagators in the determinats with the anisotropic ones. We refer for such analyisis to Ref.[64]; here we simply state the result

$$\int d\mathbf{x} |\beta_{h,\bar{\omega}}^2(\mathbf{x})| \leq C_0 |U| |h| \gamma^{2h}$$

$$\int d\mathbf{x} |\mathbf{x}|^n |\beta_{h,\bar{\omega}}^2(\mathbf{x})| \leq C_n |U|^2 |h| \gamma^{(2-n)h}, \qquad n \geq 1 \qquad (14.88)$$

Given $h_\beta \leq h \leq 0$, we can write $\varepsilon_h(\vec{k}) = \varepsilon_0(\vec{k}) + \sum_{j=h+1}^{0}(\varepsilon_{j-1}(\vec{k}) - \varepsilon_j(\vec{k}))$. From this identity and the inductive assumption (11.20) we soon find, if $c_0 \leq 1$,

$$|\varepsilon_h(\vec{k}) - \varepsilon_0(\vec{k})| \leq C_0 |U| \sum_{j=h_{\vec{k}}}^{0} |j| \gamma^{2j} \leq C_0' |U|,$$

$$|\vec{\nabla}\varepsilon_h(\vec{k}) - \vec{\nabla}\varepsilon_0(\vec{k})| \leq 2C_1 |U| \sum_{j=h_{\vec{k}}}^{0} \gamma^j |j| \leq C_1' |U| \qquad (14.89)$$

$$|\partial^2_{k_i k_r}\varepsilon_h(\vec{k}) - \partial^2_{k_i k_r}\varepsilon_0(\vec{k})| \leq C_2 |U| \left[ |U| \sum_{j=h_{\vec{k}}+3}^{0} |j| + \sum_{j=h+1}^{h_{\vec{k}}+3} |j| \right] \leq 4C_2 c_0$$

where $h_{\vec{k}} = \min\{j \geq h+1 : C_j^{-1}(\pi\beta^{-1}, \vec{k}) > 0\}$ and we used that $C_j^{-1}(\pi\beta^{-1}, \vec{k}) = 1$ if $j \geq h_{\vec{k}} + 3$ and that $\sum_{j=h+1}^{h_{\vec{k}}+3} \leq 3$. The first two bounds in (14.89) show that $\varepsilon_h(\vec{k})$ and $\vec{\nabla}\varepsilon_h(\vec{k})$ are close within $O(U)$ and $O(U^2)$ to $\varepsilon_0(\vec{k})$ and $\vec{\nabla}\varepsilon_0(\vec{k})$ respectively. This properties and the validity of properties (1)–(4) in the unperturbed case $h = 0$ guarantee that the equation $\varepsilon_h(\vec{k}) - \mu = e$ can be inverted for $h < 0$ if $e$ is small enough (as it follows from an application of implicit function Theorem). Then the set of vectors $\Sigma^{(h)}(e)$ defines a closed curve enclosing the origin, close to $\Sigma^{(0)}(e)$ within $O(U)$. Also, the third bound in (14.89) implies that the second derivatives of $\varepsilon_h(\vec{k})$ are close to the second derivatives of $\varepsilon_0(\vec{k})$, if $c_0$ small enough. This means that $\Sigma^{(h)}(e)$ is convex (since so $\Sigma^{(0)}(e)$ is) and $r_h(\theta, e) = r_0(\theta, e) + O(c_0^2)$ (because the curvature radius $r_h(\theta, e)$ is computed in terms of the first two derivatives of $\varepsilon_h(\vec{k})$).

Finally it is a straigthforward consequence of (14.89) the fact that the wave function renormaliztion $Z(\theta)$ is essentially independent fom $\beta$, that is the fact that the Jellium and Hubbard model not at half-filling are Fermi liquids.

## 14.8 Other results in $d = 2$

Interacting fermionic systems in $d = 2$ have been the subject of several studies in recent times. As we have seen, the property to be a Fermi liquid is crucially related to improvment in the naive bounds due to the sector lemma (so that one gets an $O(\gamma^{2h}|h|)$ instead of $O(\gamma^h)$), whose validity rely on the convexity of the Fermi surface.

One can then expect that Fermi liquid behavious indeed breaks down when the Fermi surface has some flat pieces, when parallelogram rule cannot be applied. In the Hubbard model in the half-filled case the Fermi surface is a square, and it is not changed by the interaction; it has been proved in Refs.[66],[67] that such a system is not a Fermi liquid.

Theorem 15.1 holds under the condition $\mu < \mu_0$, but it could be extended to any $\mu < 2$, under the smallness condition $|U| \leq U_0(\mu)$, with $\lim_{\mu \to 2^-} U_0(\mu) = 0$. Such a result, combined with the result of Refs.[66; 67] discussed above, implies that the 2D Hubbard model shows a transition from Fermi to non–Fermi liquid behavior, in the above range of temperatures, depending on the choice of $\mu$. It would be very interesting to explicitly investigate the crossover between the two regimes.

If the temperature is low enough, it is expected that Fermi liquid behavior breaks down, as a consequence of quantum instabilities present in the systems. It is possible to destroy such instabilities by choosing properly an highly *non symmetric* dispersion relation, for instance by introducing an external magnetic field; indeed for such a system in Ref.[68] a proof of Fermi liquid behavior was given up to *zero temperature* in 2d.

## Chapter 15

# BCS Model with Long Range Interaction

## 15.1 BCS model

As we have seen in the previous chapter, the finite temperature acts as an infrared cut-off so that the flow of the effective coupling is trivially bounded. Much more complex is to get informations at zero temperature; the effective coupling $\lambda_h(\mathbf{k})$ is a function of the momenta and the flow equations are very complex to study. In the Jellium case some analytical insights on the flow of the $\lambda_h(\mathbf{k})$ can be obtained, see Refs. [15; 35; 69], and it emerges that the relevant effective interactions in (14.24) are the ones involving the momenta in a *Cooper pair* configuration, namely $\vec{k}_1 \simeq -\vec{k}_3, \vec{k}_2 \simeq -\vec{k}_4$, $\sigma = -\sigma'$. The flow equations suggest that the running couplings reach the domain of convergence of the expansions so that the approach seen in the previous chapter fails at temperatures very low; this is not surprising as one expects the appearance of phase transitions at very low temperatures.

The analytic difficulties for studying a model with a generic interaction at very low temperatures and the above considerations motivate the search of simpler models. Bardeen, Cooper and Schreiffer, see Ref. [70], developed their theory of superconductivity by introducing and studying the *BCS model*, describing a system of electrons with an instantaneous infinite range interaction involving only electrons of opposite momentum and spin (*Cooper pairs*), with hamiltonian

$$H_{BCS} = \sum_{\sigma=\pm} \int_V d\vec{x} a^+_{\vec{x},\sigma}(-\frac{\nabla^2}{2}-\mu)a^-_{\vec{x},\sigma} - \frac{\lambda}{V}[\int_V d\vec{x} a^+_{\vec{x},+} a^+_{\vec{x},-}][\int_V d\vec{y} a^-_{\vec{y},-} a^-_{\vec{y},+}]$$

(15.1)

where $\nabla$ is the Laplacian, $\mu$ is the chemical potential, $a^\pm_{\vec{x},\sigma}$ are creation or annihilation spin $\frac{1}{2}$ fermionic field operators in a $d$-dimensional box with side $L$ and $V = L^d$, $m$ is the electron effective mass and $\lambda > 0$ is the (attractive) coupling. The model is not solvable but it was shown in Ref. [70] that a superconducting phase is energetically favorable with respect to a normal phase.

The Schwinger functions can be written in terms of a Grassman integral with interaction

$$\mathcal{V}_{BCS} = -\frac{\lambda}{(\beta V)^3} \sum_{\mathbf{k},\mathbf{k}'} \sum_{\substack{\vec{p}=0, p_0=\frac{2\pi}{\beta}n_0 \\ n_0 \in \mathbb{Z}}} \widehat{\psi}^+_{\mathbf{k},+} \widehat{\psi}^+_{-\mathbf{k}+\mathbf{p},-} \widehat{\psi}^-_{\mathbf{k}',-} \widehat{\psi}^-_{-\mathbf{k}'+\mathbf{p},+} \quad (15.2)$$

Note that the interaction (15.2) cannot be factorized in a simple way, as the sum over $p_0$ has the effect that it is not the square of the total number of Cooper pairs. Hence, even if the Hamiltonian (15.1) seems much more tractable than the full model, this lack of factorization has prevented, despite many attempts, the rigorous computation of correlations of the BCS model.

A simpler model is obtained considering the following interaction

$$\mathcal{V} = -\frac{\lambda}{(\beta V)^3} \sum_{\mathbf{k},\mathbf{k}'} \sum_{\substack{\vec{p}=0, p_0=\frac{2\pi}{\beta}n_0 \\ n_0 \in \mathbb{Z}}} \widehat{v}(p_0) \widehat{\psi}^+_{\mathbf{k},+} \widehat{\psi}^+_{-\mathbf{k}+\mathbf{p},-} \widehat{\psi}^-_{\mathbf{k}',-} \widehat{\psi}^-_{-\mathbf{k}'+\mathbf{p},+} \quad (15.3)$$

or, in coordinate space

$$\mathcal{V} = -\frac{\lambda}{V} \int d\mathbf{x} \int d\mathbf{y} v(x_0 - y_0) \psi^+_{\mathbf{x},+} \psi^+_{\mathbf{x},-} \psi^-_{\mathbf{y},-} \psi^-_{\mathbf{y},+} \quad (15.4)$$

where $\int d\mathbf{x}$ stands for $\int dx_0 \sum_{\vec{x}}$, $v(x_0 - y_0)$ is a *Kac potential* with a *long but finite* range potential $\kappa^{-1}$ given by

$$v(x_0 - y_0) = \frac{1}{\beta} \sum_{\substack{p_0 = \frac{2\pi n_0}{\beta} \\ n_0 \in \mathbb{Z}}} e^{-ip_0(x_0-y_0)} \frac{\kappa^2}{\kappa^2 + p_0^2} \quad (15.5)$$

Moreover

$$P(d\psi) = \mathcal{N}^{-1} \mathcal{D}\psi \cdot \exp\left\{ -\frac{1}{V\beta} \sum_{\sigma=\pm} \sum_{\mathbf{k} \in \mathcal{D}} (-ik_0 + \varepsilon(\vec{k}) - \mu) \widehat{\psi}^+_{\mathbf{k},\sigma} \widehat{\psi}^-_{\mathbf{k},\sigma} \right\},$$

(15.6)

with $\varepsilon(\vec{k}) = \sum_{i=1}^{d}(1 - \cos k_i)$, $\mathcal{N} = \prod_{\mathbf{k} \in \mathcal{D}} [(V\beta)^{-2}(-k_0^2 - (\varepsilon(\vec{k}) - \mu)^2)]$. The two point correlation functions are given by the following Grassmann integrals

$$<\widehat{\psi}^\varepsilon_{\mathbf{k},\sigma} \widehat{\psi}^{-\varepsilon'}_{\mathbf{k}',\sigma'}>_{L,\beta,h} = \lim_{M\to\infty} \frac{\int P(d\psi) e^{-\mathcal{V}-h\sum_{\sigma=\pm}\int d\mathbf{x}\psi^\sigma_{\mathbf{x},\sigma}\psi^\sigma_{\mathbf{x},-\sigma}} \psi^\varepsilon_{\mathbf{k},\sigma}\psi^{-\varepsilon'}_{\mathbf{k}',\sigma'}}{\int P(d\psi) e^{-\mathcal{V}-h\sum_{\sigma=\pm}\int d\mathbf{x}\psi^\sigma_{\mathbf{x},\sigma}\psi^\sigma_{\mathbf{x},-\sigma}}}$$

(15.7)

aand the following theorem holds, see [71]

**Theorem 15.1.** *Assume $\mu < d$ and $\lambda > 0$; there exists $\beta_c(\lambda) > 0$ such that for $\beta \geq \beta_c(\lambda)$ and $0 < \kappa < \kappa_0 = C^{-1}\lambda^{-\frac{1}{2}}\beta^{-d/2-2}$ for a suitable constant $C$ the Schwinger functions (15.7) with $v(x_0 - y_0)$ given by (15.5) are such that*

$$\lim_{h \to 0^+} \lim_{L \to \infty} < \widehat{\psi}^-_{\mathbf{k},\sigma} \widehat{\psi}^+_{\mathbf{k},\sigma} >_{L,\beta,h} = \frac{ik_0 + (\varepsilon(\vec{k}) - \mu)}{k_0^2 + (\varepsilon(\vec{k}) - \mu)^2 + \Delta(\beta)^2} \quad (15.8)$$

$$\lim_{h \to 0^+} \lim_{L \to \infty} < \widehat{\psi}^+_{\mathbf{k},+} \widehat{\psi}^+_{-\mathbf{k},-} >_{L,\beta,h} = \frac{\Delta(\beta)}{k_0^2 + (\varepsilon(\vec{k}) - \mu)^2 + \Delta(\beta)^2} \quad (15.9)$$

*where $\Delta(\beta)$ is the real negative solution of the BCS gap equation*

$$1 = \lambda \frac{1}{V\beta} \sum_{k \in \mathcal{D}} \sum_{\mathbf{k}} \frac{\tanh\left(\frac{\beta}{2}\sqrt{(\varepsilon(\vec{k}) - \mu)^2 + \Delta(\beta)^2}\right)}{2\sqrt{(\varepsilon(\vec{k}) - \mu)^2 + \Delta(\beta)^2}} \quad (15.10)$$

*and $\beta_c(\lambda)$ is the minimal $\beta$ such that (15.10) admits a solution.*

Eq.(15.10) is the well known *gap equation* found in Ref.[70] and the r.h.s. of (15.8), (15.9) are the Schwinger functions of the BCS mean field model. It is well known that, if $\beta = T^{-1}$, (15.10) has a solution for $T \leq T_c$ with $T_c = Ae^{-\frac{a}{\lambda}}$, with $A, a$ suitable positive constants, and, for $T$ close to $T_c$, $\Delta \simeq B(T_c - T)^{\frac{1}{2}}$ for a suitable constant $B$. The above theorem then implies that, if $\lambda, \kappa, \beta$ are chosen so that $\beta \geq \beta_c$ and $0 < \kappa < C^{-1}\lambda^{-\frac{1}{2}}\beta^{-d/2-2}$ then the phenomenon of spontaneous mass generation is present as $<\psi^+_{\mathbf{k},+}\psi^+_{-\mathbf{k},-}>$ is different from zero and the Schwinger functions in coordinate space have an exponential decay proportional to $\Delta(\beta)$.

Note that we can prove convergence only for small $\kappa$, as it turns out that $\kappa \leq C^{-1}\lambda^{-\frac{1}{2}}\beta^{-d/2-2}$ for a suitable constant $C$; we have not tryed to optimize the power of $\beta^{-1}$ in the above bound and small improvements could be easily obtained using the techniques in this paper, which however would not change qualitatively our results. Of course it would be interesting the prove the same theorem up to $\kappa = \infty$, so obtaining a real solution of the BCS model with instantaneous interaction, or at least up to $\kappa$ independent from $\lambda$ and $\beta$.

## 15.2 Partial Hubbard-Stratonovich transformation

We start the analysis of (15.7) by splitting the interaction $\mathcal{V}$ (15.3) as sum over two terms, one with $n_0 = 0$ and the other with $n_0 \neq 0$

$$\mathcal{V} = \bar{\mathcal{V}} + \widehat{\mathcal{V}} \tag{15.11}$$

$$\bar{\mathcal{V}} = -\frac{\lambda}{(\beta V)^3} \sum_{\mathbf{k},\mathbf{k}'} \widehat{\psi}^+_{\mathbf{k},+} \widehat{\psi}^+_{-\mathbf{k},-} \widehat{\psi}^-_{\mathbf{k}',-} \widehat{\psi}^-_{-\mathbf{k}',+} = -2N^+ N^- \tag{15.12}$$

$$\widehat{\mathcal{V}} = -\frac{\lambda}{(\beta V)^3} \sum_{\substack{\mathbf{k},\mathbf{k}' \\ \vec{p}=0, |p_0| \geq \frac{2\pi}{\beta}}} \widehat{v}(p_0) \widehat{\psi}^+_{\mathbf{k},+} \widehat{\psi}^+_{-\mathbf{k}+\mathbf{p},-} \widehat{\psi}^-_{\mathbf{k}',-} \widehat{\psi}^-_{-\mathbf{k}'+\mathbf{p},+} \tag{15.13}$$

with $N^\sigma = \int d\mathbf{x} \psi^\sigma_{\mathbf{x},\sigma} \psi^\sigma_{\mathbf{x},-\sigma}$. Let us consider the *generating function* $\mathcal{S}_{L,\beta,h}(J)$

$$e^{\mathcal{S}_{L,\beta,h}(J)} = \tag{15.14}$$
$$\int P(d\psi) e^{2N^+ N^- - \widehat{\mathcal{V}}} e^{-h\frac{\sqrt{2\beta V}}{\sqrt{\lambda}} N^+ - h\frac{\sqrt{2\beta V}}{\sqrt{\lambda}} N^-} e^{\int d\mathbf{x} \sum_\sigma [J^+_{\mathbf{x},\sigma} \psi^-_{\mathbf{x},\sigma} + \psi^+_{\mathbf{x},\sigma} J^-_{\mathbf{x},\sigma}]}$$

where $J^\pm$ are external Grassmann fields, so that

$$< \psi^\varepsilon_{\mathbf{x},\sigma} \psi^{\varepsilon'}_{\mathbf{y},\sigma'} > = \frac{\partial^2}{\partial J^{-\varepsilon}_{\mathbf{x},\sigma} \partial J^{-\varepsilon'}_{\mathbf{y},\sigma'}} \mathcal{S}_{L,\beta,h}(J)|_{J=0} \tag{15.15}$$

By using the identity (*Hubbard-Stratonovich transformation*) ($\phi = u + iv$, $\bar{\phi} = u - iv$, $u,v \in \mathbb{R}^2$, $a,b \in \mathbb{R}^2$)

$$e^{2ab} = \frac{1}{2\pi} \int_{\mathbb{R}^2} du dv e^{-\frac{1}{2}|\phi|^2} e^{a\phi + b\bar{\phi}} \tag{15.16}$$

we can rewrite the above expression as

$$e^{\mathcal{S}_{L,\beta,h}(J)} = \frac{1}{2\pi} \int_{\mathbb{R}^2} du dv e^{-\frac{1}{2}|\phi|^2} \int P(d\psi) e^{-\widehat{\mathcal{V}}}$$
$$e^{(\phi - h\frac{\sqrt{2\beta V}}{\sqrt{\lambda}})N^+ + (\bar{\phi} - h\frac{\sqrt{2\beta V}}{\sqrt{\lambda}})N^-} e^{\int d\mathbf{x} \sum_\sigma [J^+_{\mathbf{x},\sigma} \psi^-_{\mathbf{x},\sigma} + \psi^+_{\mathbf{x},\sigma} J^-_{\mathbf{x},\sigma}]} \tag{15.17}$$

Performing the change of variables $(u,v) \to \sqrt{2\beta V}(u + \frac{h}{\sqrt{\lambda}}, v)$ and defining $N^\varepsilon \equiv \frac{\sqrt{\lambda}}{(2\beta V)^{1/2}} \mathcal{D}^\varepsilon$ we obtain

$$e^{\mathcal{S}_{L,\beta,h}(J)} = \frac{\beta V}{\pi} \int_{\mathbb{R}^2} du dv e^{-\beta V(v^2 + (u + \frac{h}{\sqrt{\lambda}})^2)} e^{-\beta V \mathcal{F}_{L,\beta,h}(u,v) + \mathcal{B}_{L,\beta,h}(u,v,J)}$$
$$\tag{15.18}$$

where

$$e^{-\beta V \mathcal{F}_{L,\beta,h}(u,v) + \mathcal{B}_{L,\beta,h}(u,v,J)} = \int P(d\psi) e^{-\widehat{\mathcal{V}}} e^{\sqrt{\lambda}\phi \mathcal{D}^+ + \sqrt{\lambda}\bar{\phi}\mathcal{D}^-} e^{\int d\mathbf{x} \sum_\sigma [J^+_{\mathbf{x},\sigma}\psi^-_{\mathbf{x},\sigma} + \psi^+_{\mathbf{x},\sigma}J^-_{\mathbf{x},\sigma}]} \quad (15.19)$$

and (by definition) $\mathcal{B}_{L,\beta,h}(u,v,J)$ vanishes for $J=0$ so that $\mathcal{F}_{L,\beta,h}(u,v)$ is given by

$$e^{-\beta V \mathcal{F}_{L,\beta,h}(u,v)} = \int P(d\psi) e^{-\widehat{\mathcal{V}}} e^{\sqrt{\lambda}\phi \mathcal{D}^+ + \sqrt{\lambda}\bar{\phi}\mathcal{D}^-} \quad (15.20)$$

The conclusion is that (15.7) can be written as

$$< \psi^\varepsilon_{\mathbf{k},\sigma} \psi^{\varepsilon'}_{-\varepsilon\varepsilon'\mathbf{k},-\varepsilon\varepsilon'\sigma} >= \quad (15.21)$$

$$\frac{1}{Z_{L,\beta,h}} \int_{\mathbb{R}^2} du dv e^{-\beta V(v^2 + (u + \frac{h}{\sqrt{\lambda}})^2)} e^{-\beta V \mathcal{F}_{L,\beta}(u,v)} \widehat{S}^{\varepsilon,\varepsilon'}_{L,\beta}(\mathbf{k},u,v)$$

where $S^{\varepsilon,\varepsilon'}_{L,\beta}(u,v) = \partial_{J^{-\varepsilon}_{\mathbf{k},\sigma}} \partial_{J^{-\varepsilon'}_{-\varepsilon\varepsilon'\mathbf{k},-\varepsilon\varepsilon'\sigma}} \mathcal{B}(u,v,J)|_{J=0}$ ($\sigma$-independent) and

$$Z_{L,\beta,h} = \int_{\mathbb{R}^2} du dv e^{-\beta V(v^2 + (u + \frac{h}{\sqrt{\lambda}})^2)} e^{-\beta V \mathcal{F}_{L,\beta}(u,v)} \quad (15.22)$$

We can write

$$\mathcal{F}_{L,\beta}(u,v) = t_{BCS}(u,v) + \bar{\mathcal{F}}_{L,\beta}(u,v) \quad (15.23)$$

where, if $E(\vec{k}) = \varepsilon(\vec{k}) - \mu$ and $\phi = u + iv$

$$t_{BCS}(u,v) = -\frac{1}{V} \sum_{\vec{k}} \frac{2}{\beta} \log \frac{\cosh\left(\frac{\beta}{2}\sqrt{E^2(\vec{k}) + \lambda|\phi|^2}\right)}{\cosh \frac{\beta}{2} E(\vec{k})} \quad (15.24)$$

is the free energy in the mean field BCS model and $\bar{\mathcal{F}}_{L,\beta}(u,v)$ is the rest. The following lemma is true.

**Lemma 15.1.** *There exist constants $C, C_1$ such that, if $0 < \kappa < \kappa_0 = C^{-1}\lambda^{-\frac{1}{2}}\beta^{-d/2-2}$, then*

$$|\bar{\mathcal{F}}_{L,\beta}(u,v)| \leq C_1 \frac{\lambda}{V} \kappa^2 \beta^{d+2} \log \beta \quad (15.25)$$

The above Lemma says that the correction to mean field behaviour *vanishes* in the thermodynamic limit, if the range is long enough.

Calling $V\bar{\mathcal{F}}_{L,\beta}(u,v) \equiv \widehat{\mathcal{F}}_{L,\beta}(u,v)$, we can write the two point Schwinger functions as

$$\frac{1}{Z_{L,\beta,h}} \int_{\mathbb{R}^2} du dv e^{-\beta V[v^2 + (u + \frac{h}{\sqrt{\lambda}})^2 + t_{BCS}(u,v)]} e^{-\beta \widehat{\mathcal{F}}_{L,\beta}(u,v)} S^{\varepsilon,\varepsilon'}_{L,\beta}(\mathbf{k},u,v)$$

$$(15.26)$$

By the saddle point theorem, for $\beta$ large enough

$$\lim_{L\to\infty} \frac{e^{-\beta V(v^2+(u+\frac{h}{\sqrt{\lambda}})^2+t_{BCS}(u,v))}}{\int dudv e^{-\beta V(v^2+(u+\frac{h}{\sqrt{\lambda}})^2+t_{BCS}(u,v))}} = \delta(u-u_0)\delta(v) \qquad (15.27)$$

where $u_0$ is given by the negative (for $h > 0$) solution of

$$u_0\left[\lambda \int \frac{d\vec{k}}{(2\pi)^d} \frac{\tanh\left(\frac{\beta}{2}\sqrt{E^2(\vec{k})+\lambda u_0^2}\right)}{2\sqrt{E^2(\vec{k})+\lambda u_0^2}} - 1\right] = \frac{2h}{\sqrt{\lambda}} \qquad (15.28)$$

which in the limit $h \to 0$ reduces to the BCS equation (15.10). Moreover there exist constants $C, C_2$ such that, if $0 < \kappa < \kappa_0 = C^{-1}\lambda^{-\frac{1}{2}}\beta^{-d/2-2}$ then

$$S_{L,\beta}^{-,+}(\mathbf{k},u,v) = \frac{ik_0+(\varepsilon(\vec{k})-\mu)}{k_0^2+(\varepsilon(\vec{k})-\mu)^2+\lambda|\phi|^2} + R_{L,\beta}^{-,+}(\mathbf{k},u,v) \qquad (15.29)$$

$$S_{L,\beta}^{+,+}(\mathbf{k},u,v) = \frac{\sqrt{\lambda}\phi}{k_0^2+(\varepsilon(\vec{k})-\mu)^2+\lambda|\phi|^2} + R_{L,\beta}^{+,+}(\mathbf{k},u,v) \qquad (15.30)$$

with

$$|R_{L,\beta}^{-,+}(\mathbf{k},u,v)|, |R_{L,\beta}^{+,+}(\mathbf{k},u,v)| \leq C_2\lambda\kappa^2\beta^{3d+5}V^{-1} \qquad (15.31)$$

Hence, by inserting (15.29)-(15.30) in (15.26) and using (15.25)-(15.31)-(15.27), Theorem 15.1 is proved.

## 15.3 Corrections to the mean field

In order to prove Lemma 15.1, recalling that $\phi = u + iv$, we write

$$\int P(d\psi)e^{\sqrt{\lambda}\phi\mathcal{D}^+ + \sqrt{\lambda}\bar{\phi}\mathcal{D}^-}e^{-\widehat{\mathcal{V}}(\psi)} =$$
$$e^{-\beta V t_{BCS}(u,v)}\int P_\phi(d\psi)e^{-\widehat{\mathcal{V}}(\psi)} = e^{-\beta V t_{BCS}(u,v) - \beta V \bar{\mathcal{F}}_{L,\beta}(u,v)} \qquad (15.32)$$

where

$$\widehat{\mathcal{V}}(\psi) = -\frac{\lambda}{V}\int d\mathbf{x}d\mathbf{y}\widetilde{v}(x_0-y_0)\psi_{\mathbf{x},+}^+\psi_{\mathbf{x},-}^+\psi_{\mathbf{y},-}^-\psi_{\mathbf{y},+}^- \qquad (15.33)$$

where

$$\tilde{v}(x_0 - y_0) = \frac{1}{\beta} \sum_{\substack{p_0 = \frac{2\pi n_0}{\beta} \\ n_0 \neq 0}} e^{-ip_0(x_0-y_0)} \frac{\kappa^2}{\kappa^2 + p_0^2} \,. \tag{15.34}$$

Moreover

$$P_\phi(d\psi) = \prod_{\mathbf{k}} \frac{d\widehat{\psi}_{\mathbf{k}}^+ d\widehat{\psi}_{\mathbf{k}}^-}{\mathcal{N}(\mathbf{k})} \left\{ -\frac{1}{V\beta} \sum_{\mathbf{k}} \sum_{\varepsilon,\varepsilon'=\pm} \widehat{\psi}_{\varepsilon\mathbf{k},\varepsilon}^\varepsilon T_{\varepsilon,\varepsilon'} \widehat{\psi}_{\varepsilon'\mathbf{k},\varepsilon'}^{-\varepsilon'} \right\} \tag{15.35}$$

where $\mathcal{N}(\mathbf{k})$ is the normalization of $P_\phi(d\psi)$,

$$t_{BCS}(u,v) = -\frac{1}{V\beta} 2 \sum_{\mathbf{k}} \log \frac{k_0^2 + E(\vec{k})^2 + \lambda|\phi|^2}{k_0^2 + E(\vec{k})^2} \tag{15.36}$$

and the $2 \times 2$ matrix $T(\mathbf{k})$ is given by

$$T(\mathbf{k}) = \begin{pmatrix} -ik_0 + E(\vec{k}) & \sqrt{\lambda}\phi \\ \sqrt{\lambda}\bar{\phi} & -ik_0 - E(\vec{k}) \end{pmatrix}$$

We can write $t_{BCS}(u,v)$ as (15.24) by explicitly performing the sums over $k_0$ and of course $|t_{BCS}(u,v)| \leq \sqrt{|\lambda|}C[1+|\phi|]$. If $\varepsilon, \varepsilon' = \pm$, the propagator of $P_\phi(d\psi)$ is given by

$$\int P_\phi(d\psi) \psi_{\mathbf{x},\varepsilon}^\varepsilon \psi_{\mathbf{y},\varepsilon'}^{-\varepsilon'} \equiv g_{\varepsilon,\varepsilon'}(\mathbf{x}-\mathbf{y}) = \frac{1}{V\beta} \sum_{\mathbf{k}} e^{-i\mathbf{k}(\mathbf{x}-\mathbf{y})}[T^{-1}(\mathbf{k})]_{\varepsilon,\varepsilon'} \,. \tag{15.37}$$

We decompose the propagator $g_{\varepsilon,\varepsilon'}(\mathbf{x}-\mathbf{y})$ into a sum of two propagators supported in the regions of $k_0$ "large" and "small", respectively. The regions of $k_0$ large and small are defined in terms of a smooth compact support function $H_0(t)$, $t \in \mathbb{R}^+$, such that $H_0(t) = 1$ if $t < 1/\gamma$ and $= 0$ $t > 1$, with $\gamma > 1$. We define $h(k_0) = H_0(|k_0|)$ so that we can rewrite $g_{\varepsilon,\varepsilon'}(\mathbf{x}-\mathbf{y})$ as:

$$g_{\varepsilon,\varepsilon'}(\mathbf{x}-\mathbf{y}) = g_{\varepsilon,\varepsilon'}^{(u.v.)}(\mathbf{x}-\mathbf{y}) + g_{\varepsilon,\varepsilon'}^{(i.r.)}(\mathbf{x}-\mathbf{y}) \tag{15.38}$$

where

$$g_{\varepsilon,\varepsilon'}^{(i.r.)}(\mathbf{x}-\mathbf{y}) = \frac{1}{V\beta} \sum_{\mathbf{k}} e^{-i\mathbf{k}(\mathbf{x}-\mathbf{y})} h(k_0)[T^{-1}(\mathbf{k})]_{\varepsilon,\varepsilon'} \tag{15.39}$$

$$g_{\varepsilon,\varepsilon'}^{(u.v.)}(\mathbf{x}-\mathbf{y}) = \frac{1}{V\beta} \sum_{\mathbf{k}} e^{-i\mathbf{k}(\mathbf{x}-\mathbf{y})} (1-h(k_0))[T^{-1}(\mathbf{k})]_{\varepsilon,\varepsilon'} \tag{15.40}$$

We can write then $P_\phi(d\psi) = P_\phi(d\psi^{(u.v.)})P_\phi(d\psi^{(i.r.)})$ and

$$\int P_\phi(d\psi^{(u.v.)}) e^{-\widehat{V}(\psi^{(i.r.)} + \psi^{(u.v.)})} = \mathcal{N}_0 e^{-V^0(\psi^{(i.r.)})} \tag{15.41}$$

The ultraviolet integration can be done as in §12.4; on the other hand the infrared part of the propagator verifies the following bound

$$|g^{(i.r.)}_{\varepsilon,\varepsilon}(\mathbf{x}-\mathbf{y})| \leq \frac{C_N}{1+[\beta^{-1}|\mathbf{x}-\mathbf{y}|]^N} \qquad (15.42)$$

The integration over all the coordinates can be done integrating over all the $n-1$ coordinate differences of the extreme points of the lines in $T$, using that each integration contributes to the final bound for $\mathcal{E}^T(\mathcal{V}^0;n)$ a factor $\beta^{d+1}$; the integration over the other coordinate differences can be performed by using that

$$|\widetilde{v}(x_0-y_0)| = |\frac{1}{\beta}\sum_{\substack{k_0\neq 0 \\ k_0=\frac{2\pi}{\beta}n_0}} e^{ik_0(x_0-y_0)}\frac{\kappa^2}{\kappa^2+k_0^2}| \leq \qquad (15.43)$$

$$\frac{1}{(2\pi)^2\beta}\sum_{n_0\neq 0}\frac{\kappa^2\beta^2}{n_0^2} \leq \beta^{-1}C(\kappa\beta)^2 \qquad (15.44)$$

which implies

$$\int d\mathbf{x} V^{-1}|\widetilde{v}(x_0)| \leq C(\kappa\beta)^2 \qquad (15.45)$$

if $C$ is a suitable constant. By using the Gram inequality and that $detG$ in the truncated expectation is $O(\log \beta^s)$, is $s$ is the order of $G$, we obtain

$$|\mathcal{E}^T(\mathcal{V}^0;n)| \leq (\beta V)n!C^n\lambda^n(\kappa^2\beta^2)^n \qquad (15.46)$$

$$|\log\beta|^{n+1}\beta^{(d+1)n}\beta^{-(d+1)} \leq (\beta V)C^n\lambda^n(\kappa^2\beta^{d+3+\eta})^n\beta^{-(d+1)}n!$$

for a constant $0 \leq \eta < 1$. Hence, by assuming $\kappa \leq C^{-1}\lambda^{-\frac{1}{2}}\beta^{\frac{-d-3-\eta}{2}} \equiv \kappa_0$, it follows

$$|\bar{\mathcal{F}}_{L,\beta}(u,v)| \leq C\lambda\kappa^2\beta^{2+\eta} \qquad (15.47)$$

The above analysis says that $\bar{\mathcal{F}}_{L,\beta,h}$, which is the correction to mean field, is given by a convergent expansion if the interaction range is long enough. A closer look to the Feynman graphs shows, however, that each graph obeys to a much better bound as it vanishes as $V \to \infty$. Expanding in terms of Feynman graphs corresponds to evaluting the determinants as sums of $n!$ terms and give up the combinatorial better bound based on the Gram-Hadamard inequality. However an extra factor $V^{-1}$ can be gained in the estimates without expanding in graphs (i.e. without loosing convergence); we proceed essentially as in chapt.3.

We have to bound

$$\int d\mathbf{x}_1 \int d\mathbf{y}_1 \int d\underline{\mathbf{x}} \int d\underline{\mathbf{y}} [\prod_{i=1}^n \frac{\lambda}{V}v(x_{0,i}-y_{0,i})]\mathcal{E}^T(\widetilde{\psi}(\mathbf{x}_1\cup\mathbf{y}_1)...\widetilde{\psi}(\mathbf{x}_n\cup\mathbf{y}_n))$$

$$(15.48)$$

where $\int d\mathbf{x} = \int d\mathbf{x}_2 \ldots \int d\mathbf{x}_n$, $\int d\mathbf{y} = \int d\mathbf{y}_2 \ldots \int d\mathbf{y}_n$ and we defined $\widetilde{\psi}(\mathbf{x}) = \psi^+_{\mathbf{x},+}\psi^+_{\mathbf{x},-}$, $\widetilde{\psi}(\mathbf{y}) = \psi^-_{\mathbf{y},-}\psi^-_{\mathbf{y},+}$ and $\widetilde{\psi}(\mathbf{x} \cup \mathbf{y}) = \psi^+_{\mathbf{x},+}\psi^+_{\mathbf{x},-}\psi^-_{\mathbf{y},-}\psi^-_{\mathbf{y},+}$.

By (3.74) we can write $\mathcal{E}^T(\mathcal{V}^0;n)$ as (see fig.15.1)

$$\mathcal{E}^T(\mathcal{V}^0;n) = \int d\mathbf{x}_1 d\mathbf{y}_1 \frac{\lambda}{V}\widetilde{v}(x_{0,1}-y_{0,1}) \sum_{\substack{K_1,K_2;K_1 \cap K_2=0 \\ K_1 \cup K_2=2,\ldots,n}} (-1)^{\pi} H_2(\mathbf{x}_1;K_1)H_2(\mathbf{y}_1;K_2) \tag{15.49}$$

$$+ \int d\mathbf{x}_1 d\mathbf{y}_1 \frac{\lambda}{V}\widetilde{v}(x_{0,1} - y_{0,1})H_4(\mathbf{x}_1,\mathbf{y}_1) \tag{15.50}$$

 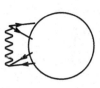

Fig. 15.1 Decomposition of $\mathcal{E}^T$

The crucial point is that, by translation invariance, $H_2(\mathbf{x};K_1)$ and $H_2(\mathbf{y};K_2)$ are $\mathbf{x},\mathbf{y}$ indipendent so that the first addend in (15.50) vanishes (because $\widetilde{v}(p_0) = 0$ if $p_0 = 0$)

$$\frac{1}{\beta} \sum_{\substack{p_0 \neq 0 \\ p_0 = \frac{2\pi}{\beta}n_0}} \frac{\kappa^2}{\kappa^2 + p_0^2}\delta_{p_0,0}H_2(0;K_1)H_2(0;K_2) = 0 \tag{15.51}$$

On the other hand we can write

$$\frac{1}{\beta V}\int d\mathbf{x}_1 d\mathbf{y}_1 \frac{\lambda}{V}|\widetilde{v}(x_{0,1} - y_{0,1})H_4(\mathbf{x}_1,\mathbf{y}_1)| \leq \tag{15.52}$$

$$\frac{1}{\beta V}\lambda V^{-1}\beta^{-1}(\kappa\beta)^2 \int d\mathbf{x}_1 d\mathbf{y}_1 |H_4(\mathbf{x}_1,\mathbf{y}_1)| \tag{15.53}$$

Moreover for $n \geq 1$

$$\frac{1}{V\beta}\frac{\lambda}{V}(\kappa\beta)^2\beta^{-1}\int d\mathbf{x}\int d\mathbf{y}|H_4(\mathbf{x},\mathbf{y})| \leq \tag{15.54}$$

$$\frac{1}{\beta}\frac{1}{V}C^n(\lambda\kappa^2\beta^2)^n\beta^{(d+1)n}|\log\beta|^n n! \tag{15.55}$$

so that for $\kappa \leq \kappa_0$ and summing over $n$ (dividing by $n!$) we get the bound (15.25). Note that, with respect to the previous bound, we have a fermionic integration, giving an extra $\beta^{d+1} \log \beta^{-1}$, replacing an integration over $\widetilde{v}$, giving an extra $V^{-1}\beta^{-1}$.

# Appendix A

# The Ising Model Fermionic Representation

## A.1 The Grassmann representation of the 2d Ising model with open boundary conditions

In order to represent the sum over multipolygons in (9.5) as a Grassmann integral, we first associate to each site $\mathbf{x} \in \Lambda$, a set of four Grassmann variables, $\overline{H}_{\mathbf{x}}, H_{\mathbf{x}}, \overline{V}_{\mathbf{x}}, V_{\mathbf{x}}$, that must be thought as associated to four new sites drawn very near to $\mathbf{x}$ and to its right, left, up side, down side respectively, see Fig A.1. We shall denote these sites by $R_{\mathbf{x}}, L_{\mathbf{x}}, U_{\mathbf{x}}, D_{\mathbf{x}}$ respectively.

Fig. A.1 The four Grassmann fields associated to the sites $\mathbf{x}$ and $\mathbf{y}$.

If $t = \tanh \beta J$, we consider the action

$$S(t) = t \sum_{\mathbf{x} \in \Lambda} \left[ \overline{H}_{\mathbf{x}} H_{\mathbf{x}+\hat{e}_1} + \overline{V}_{\mathbf{x}} V_{\mathbf{x}+\hat{e}_0} \right]$$

$$+ \sum_{\mathbf{x} \in \Lambda} \left[ \overline{H}_{\mathbf{x}} H_{\mathbf{x}} + \overline{V}_{\mathbf{x}} V_{\mathbf{x}} + \overline{V}_{\mathbf{x}} \overline{H}_{\mathbf{x}} + V_{\mathbf{x}} \overline{H}_{\mathbf{x}} + H_{\mathbf{x}} \overline{V}_{\mathbf{x}} + V_{\mathbf{x}} H_{\mathbf{x}} \right] \quad (A.1)$$

where $\hat{e}_1, \hat{e}_0$ are the coordinate versors in the horizontal and vertical directions, respectively. Open boundary conditions are assumed.

The following identity holds:

$$\frac{Z}{2^{L^2}(\cosh\beta J)^B} = (-1)^{L^2}\int \prod_{\mathbf{x}\in\Lambda}\mathrm{d}\overline{H}_{\mathbf{x}}\mathrm{d}H_{\mathbf{x}}\mathrm{d}\overline{V}_{\mathbf{x}}\mathrm{d}V_{\mathbf{x}}e^{S(t)} \qquad (A.2)$$

where $\Xi_I$ in the l.h.s. is calculated using open boundary conditions; below is given the proof of this classical result (following the exposition in [72]).

In order to prove (A.2) we expand the exponential in the r.h.s., we integrate term by term the Grassmann variables, and we get a summation over terms that we want to put in correspondence with the terms in the summation over mutipolygons of (9.5).

We represent, as in Chapt.2, each Grassmann field as an half line so that every term obtained by the contraction of the Grassmann variables is represented by the union of the lines. The figure one obtains (called a dimer) exactly coincide with a multipolygon if one shrinks the sites $R_{\mathbf{x}}, L_{\mathbf{x}}, U_{\mathbf{x}}, D_{\mathbf{x}}$ to let them coincide with $\mathbf{x}$. This graphical construction allows to put in correspondence each dimer with a unique multipolygon. We then have to show that the total weight of the dimer corresponding to the same multipolygon $\gamma$ is exactly $(-1)^{L^2}t^{|\gamma|}$, where $(-1)^{L^2}$ is the same factor appearing in the r.h.s. of (A.2) and, if $|\gamma|$ is the length of $\gamma$, $t^{|\gamma|}$ is the weight (9.5) assigns to $\gamma$.

We first note that the correspondence between dimers and multipolygons is not one to one, because an empty site $\mathbf{x}$ in the multipolygon representation corresponds to three different contractions of Grassmann fields, that is either to $\int \mathrm{d}\overline{H}_{\mathbf{x}}\mathrm{d}H_{\mathbf{x}}\mathrm{d}\overline{V}_{\mathbf{x}}\mathrm{d}V_{\mathbf{x}}\ \overline{H}_{\mathbf{x}}H_{\mathbf{x}}\overline{V}_{\mathbf{x}}V_{\mathbf{x}}$, or to $\int \mathrm{d}\overline{H}_{\mathbf{x}}\mathrm{d}H_{\mathbf{x}}\mathrm{d}\overline{V}_{\mathbf{x}}\mathrm{d}V_{\mathbf{x}}\ V_{\mathbf{x}}\overline{H}_{\mathbf{x}}H_{\mathbf{x}}\overline{V}_{\mathbf{x}}$, or to $\int \mathrm{d}\overline{H}_{\mathbf{x}}\mathrm{d}H_{\mathbf{x}}\mathrm{d}\overline{V}_{\mathbf{x}}\mathrm{d}V_{\mathbf{x}}\ V_{\mathbf{x}}H_{\mathbf{x}}\overline{V}_{\mathbf{x}}\overline{H}_{\mathbf{x}}$. The total contribution of these three contractions is:

$$\int \mathrm{d}\overline{H}_{\mathbf{x}}\mathrm{d}H_{\mathbf{x}}\mathrm{d}\overline{V}_{\mathbf{x}}\mathrm{d}V_{\mathbf{x}} \qquad (A.3)$$
$$(\overline{H}_{\mathbf{x}}H_{\mathbf{x}}\overline{V}_{\mathbf{x}}V_{\mathbf{x}} + V_{\mathbf{x}}\overline{H}_{\mathbf{x}}H_{\mathbf{x}}\overline{V}_{\mathbf{x}} + V_{\mathbf{x}}H_{\mathbf{x}}\overline{V}_{\mathbf{x}}\overline{H}_{\mathbf{x}}) = 1 - 1 - 1 = -1$$

as wanted.

It is easy to realize that, unless for the above ambiguity, the correspondence between dimers and multipolygons is unique. And, since each side of a dimer is weighted by a factor $t$ and each empty site is weighted by $(-1)$, the weights of the corresponding figures are the same, at least *in absolute value*. From now on we shall extract from the weight of $\gamma$ the contribution of the empty sites together with the trivial factor $t^{|\gamma|}$ (that is we redefine the weight of $\gamma$ by dividing it by $(-1)^{L^2-n_\gamma}t^{|\gamma|}$, where $n_\gamma$ is the number of sites belonging to $\gamma$, possibly different from $|\gamma|$, if $\gamma$ has self intersections).

We are then left with proving that the weight of a dimer (as just redefined) is exactly $(-1)^{n_\gamma}$; in this way the sign of every configuration of dimers together with the minus signs of the empty sites, (A.4), would reproduce exactly the factor $(-1)^{L^2}$ in (A.2).

We start with considering the simplest dimer, that is the square with unit side. Let us denote its corner sites with $(0,0) \equiv \mathbf{x}_1$, $(1,0) \equiv \mathbf{x}_2$, $(1,1) \equiv \mathbf{x}_3$, $(0,1) \equiv \mathbf{x}_4$ and let us prove that its weight is $(-1)^4 = 1$. The explicit expression of its weight in terms of Grassmann integrals, as generated by the expansion of the exponent in (A.2) is:

$$\int \prod_{i=1}^{4} \mathrm{d}\overline{H}_{\mathbf{x}_i} \mathrm{d}H_{\mathbf{x}_i} \mathrm{d}\overline{V}_{\mathbf{x}_i} \mathrm{d}V_{\mathbf{x}_i} \Big[ \overline{H}_{\mathbf{x}_1} H_{\mathbf{x}_2} \cdot \quad\quad (A.4)$$

$$V_{\mathbf{x}_2}\overline{H}_{\mathbf{x}_2} \cdot \overline{V}_{\mathbf{x}_2}V_{\mathbf{x}_3} \cdot \overline{V}_{\mathbf{x}_3}\overline{H}_{\mathbf{x}_3} \cdot (-H_{\mathbf{x}_3}\overline{H}_{\mathbf{x}_4}) \cdot H_{\mathbf{x}_4}\overline{V}_{\mathbf{x}_4} \cdot (-V_{\mathbf{x}_4}\overline{V}_{\mathbf{x}_1}) \cdot V_{\mathbf{x}_1}H_{\mathbf{x}_1}\Big]$$

In the previous equation, we wrote the different binomials corresponding to the segments of the dimer following the anticlockwise order, starting from $\overline{H}_{\mathbf{x}_1}$. By collecting the minus signs and by permutating the position of $\overline{H}_{\mathbf{x}_1}$ from the first to the last position, we find that (A.4) is equal to

$$-\int \prod_{i=1}^{4} \mathrm{d}\overline{H}_{\mathbf{x}_i} \mathrm{d}H_{\mathbf{x}_i} \mathrm{d}\overline{V}_{\mathbf{x}_i} \mathrm{d}V_{\mathbf{x}_i} \cdot \quad\quad (A.5)$$

$$\Big[ H_{\mathbf{x}_2}V_{\mathbf{x}_2}\overline{H}_{\mathbf{x}_2}\overline{V}_{\mathbf{x}_2} \cdot V_{\mathbf{x}_3}\overline{V}_{\mathbf{x}_3}\overline{H}_{\mathbf{x}_3}H_{\mathbf{x}_3} \cdot \overline{H}_{\mathbf{x}_4}H_{\mathbf{x}_4}\overline{V}_{\mathbf{x}_4}V_{\mathbf{x}_4} \cdot \overline{V}_{\mathbf{x}_1}V_{\mathbf{x}_1}H_{\mathbf{x}_1}\overline{H}_{\mathbf{x}_1}\Big]$$

where now we wrote separated from a dot the contributions corresponding to the same site. The explicit computation of (A.5) gives $-[(-1)(-1)(+1)(-1)] = +1$, as desired.

Let us now consider a generic dimer $\gamma$ not winding up the lattice and without self intersections, and let us prove by induction that its weight is $(-1)^{n_\gamma}$. We will then assume that the dimers with number of sites $k \leq n_\gamma$ have weights $(-1)^k$. The first step from which the induction starts is the case $k = 4$, that we have just considered.

Let us consider the smallest rectangle $R$ containing $\gamma$. Necessarily, each side of $R$ has non empty intersection with $\gamma$. Let us enumerate the corners of $\gamma$ which are also extremes of straight segments belonging to the sides of $R$, starting from the leftmost among the lowest of these points (possibly coinciding with the lower left corner of $R$) and proceeding in anticlockwise order; call $\mathbf{x}_j$ the site with label $j$. Note that two consecutive indeces $j, j+1$ could represent the same site $\mathbf{x}_j \equiv \mathbf{x}_{j+1} \in \Lambda_M$; in that case $\mathbf{x}$ would be a corner of $R$. Call $2N$ the cardinality of the set of the enumerated points (it is even by construction) and let us identify the label $2N + 1$ with the label 1.

Let us denote with the symbol $(2j - 1 \to 2j)$, $j = 1, \ldots, N$, the product of Grassmann fields corresponding to the straight line connecting the point $2j - 1$ with $2j$ (not including the fields located in $2j - 1$ and in $2j$), written in the anticlockwise order and with the sign induced by the expansion of the exponential in (A.2). That is, if the two fields belonging to a binomial appearing in (A.1), written following the anticlockwise order, are in the same order as they appear in (A.1), we will assign a $+$ sign to the second of those two fields (of course, second w.r.t. the anticlockwise order); otherwise a $-$ sign. As an example, if $2j - 1$ and $2j$ are two points on the upper horizontal side of $R$, $(2j - 1 \to 2j)$ would be equal to

$$(-\overline{H}_{\mathbf{x}_{2j-1}-\hat{e}_1})\overline{V}_{\mathbf{x}_{2j-1}-\hat{e}_1} V_{\mathbf{x}_{2j-1}-\hat{e}_1} H_{\mathbf{x}_{2j-1}-\hat{e}_1}$$
$$\cdots\cdots (-\overline{H}_{\mathbf{x}_{2j}+\hat{e}_1})\overline{V}_{\mathbf{x}_{2j}+\hat{e}_1} V_{\mathbf{x}_{2j}+\hat{e}_1} H_{\mathbf{x}_{2j}+\hat{e}_1} \qquad (A.6)$$

With a small abuse of notation, in the following we shall also denote with the symbol $(2j - 1 \to 2j)$ the straight line connecting $2j - 1$ with $2j$ on the polygon (*i.e.* the geometric object, not only the algebraic one).

Moreover, let us denote with the symbol $[2j \to 2j + 1]$, $j = 1, \ldots, N$, the product of Grassmann fields corresponding to the *non* straight line connecting the point $2j$ with $2j + 1$ (including the fields located in $2j$ and in $2j + 1$) in the order induced by the choice of proceeding in anticlockwise order and with the sign induced by the expansion of the exponential in (A.2). With a small abuse of notation we shall also denote with the same symbol $[2j \to 2j + 1]$ the corresponding line connecting $2j$ with $2j + 1$ on the polygon $\gamma$. The sites $2j$ and $2j + 1$ could either coincide (in that case $2j$ is a corner of $R$) or, if they do not, they could belong to the same side of $R$ or to different adjacent sides of $R$. Let us denote with $\gamma_j$ the union of $[2j \to 2j + 1]$ with the shortest path on $R$ connecting $2j$ with $2j + 1$. The key remark is that $n_{\gamma_j} < n_\gamma$ so that, by the inductive hypothesis, the weight of $\gamma_j$ is $(-1)^{n_{\gamma_j}}$.

With these notations and remarks, let us calculate the weight of $\gamma$. We write the weight in terms of a Grassmann integral as follows:

$$-\int \prod_{\mathbf{x} \in \gamma} \mathrm{d}\overline{H}_{\mathbf{x}} \mathrm{d}H_{\mathbf{x}} \mathrm{d}\overline{V}_{\mathbf{x}} \mathrm{d}V_{\mathbf{x}} \quad (1 \to 2)[2 \to 3] \cdots (2N{-}1 \to 2N)[2N \to 1]$$
$$(A.7)$$

The minus sign in front of the integral, appearing for the same reason why it appears in (A.7), is due to the permutation of the field $\overline{H}_{\mathbf{x}_1}$ from the first position (that is the one one gets by expanding the exponential in (A.2), writing the Grassmann binomials starting from site 1 and proceeding in anticlockwise order) to the last one (that is the position it appears into the product $[2N \to 1]$).

By a simple explicit calculation, it is straightforward to verify that the integral of the "straight line" $(2j-1 \to 2j)$ gives a contribution $(-1)^{\ell_{2j-1}-1}$, where $\ell_{2j-1}$ is the length of the segment $(2j-1 \to 2j)$ (note that $\ell_{2j-1}-1$ is the number of sites belonging to $(2j-1 \to 2j)$, excluding the extremes). We are left with computing the integral of the "non straight line" $[2j \to 2j+1]$. We must distinguish 12 different cases, which we shall now study in detail.

1) $j$ and $j+1$ are distinct and they belong to the low side of $R$. In this case

$$[j \to j+1] = \int H_{\mathbf{x}_j} \cdot V_{\mathbf{x}_j} \overline{H}_{\mathbf{x}_j} \cdot \{\overline{V}_{\mathbf{x}_j} \cdots (-\overline{V}_{\mathbf{x}_{j+1}})\} \cdot V_{\mathbf{x}_{j+1}} H_{\mathbf{x}_{j+1}} \cdot \overline{H}_{\mathbf{x}_{j+1}},$$
(A.8)

as it follows from the rules explained above. In order to compute (A.8) we use the inductive hypothesis, telling us that the weight of $\gamma_j$ is $(-1)^{n_{\gamma_j}}$, that is, explicitely:

$$(-1)^{D_j+d_j} = \int V_{\mathbf{x}_j} H_{\mathbf{x}_j} \cdot \overline{H}_{\mathbf{x}_j} (j \to j+1) H_{\mathbf{x}_{j+1}} \cdot V_{\mathbf{x}_{j+1}} \overline{H}_{\mathbf{x}_{j+1}} \cdot \{\overline{V}_{\mathbf{x}_j} \cdots (-\overline{V}_{\mathbf{x}_{j+1}})\}$$
(A.9)

In the last equation we called $D_j$ the length of the non straight line $[j \to j+1]$ (note that $D_j + 1$ is the number of sites belonging to $[j \to j+1]$, including both extremes), we denoted by the symbol $(j \to j+1)$ the product of Grassmanian fields corresponding to the straight line on $R$ connecting $\mathbf{x}_j$ with $\mathbf{x}_{j+1}$ and by $d_j$ its length (note that $d_j - 1$ is the number of sites belonging to $(j \to j+1)$, excluding both extremes). By performing the integration over the fields in $(j \to j+1)$, we find:

$$(-1)^{D_j+1} = \int V_{\mathbf{x}_j} H_{\mathbf{x}_j} \overline{H}_{\mathbf{x}_j} H_{\mathbf{x}_{j+1}} V_{\mathbf{x}_{j+1}} \overline{H}_{\mathbf{x}_{j+1}} \{\overline{V}_{\mathbf{x}_j} \cdots (-\overline{V}_{\mathbf{x}_{j+1}})\} =$$
$$= \int V_{\mathbf{x}_j} H_{\mathbf{x}_j} \overline{H}_{\mathbf{x}_j} \{\overline{V}_{\mathbf{x}_j} \cdots (-\overline{V}_{\mathbf{x}_{j+1}})\} H_{\mathbf{x}_{j+1}} V_{\mathbf{x}_{j+1}} \overline{H}_{\mathbf{x}_{j+1}} \qquad (A.10)$$

and the last line is clearly equal to the r.h.s. of (A.8).

2) $j$ and $j+1$ coincide with the low right corner of $R$. In this case

$$[j \to j+1] = \int H_{\mathbf{x}_j} \cdot V_{\mathbf{x}_j} \overline{H}_{\mathbf{x}_j} \cdot \overline{V}_{\mathbf{x}_j} = -1. \qquad (A.11)$$

3) $j$ and $j+1$ are distinct and they belong to the low and the rights sides of $R$, respectively. In this case

$$[j \to j+1] = \int H_{\mathbf{x}_j} \cdot V_{\mathbf{x}_j} \overline{H}_{\mathbf{x}_j} \cdot \{\overline{V}_{\mathbf{x}_j} \cdots H_{\mathbf{x}_{j+1}}\} \cdot V_{\mathbf{x}_{j+1}} \overline{H}_{\mathbf{x}_{j+1}} \cdot \overline{V}_{\mathbf{x}_{j+1}} . \quad (A.12)$$

Calling $\vec{0}$ the lower right corner of $R$, the inductive hypothesis tells us that:

$$(-1)^{D_j+d_j} = \int V_{\mathbf{x}_j} H_{\mathbf{x}_j} \cdot \overline{H}_{\mathbf{x}_j} (j \to \vec{0}) H_{\vec{0}} \cdot V_{\vec{0}} \overline{H}_{\vec{0}} \cdot$$

$$\overline{V}_{\vec{0}}(\vec{0} \to j+1) V_{\mathbf{x}_{j+1}} \cdot \overline{V}_{\mathbf{x}_{j+1}} \overline{H}_{\mathbf{x}_{j+1}} \cdot \{\overline{V}_{\mathbf{x}_j} \cdots H_{\mathbf{x}_{j+1}}\} . \quad (A.13)$$

In the last equation we called $d_j$ the length of the shortest path on $R$ connecting $j$ with $j+1$ that is the sum of the lengths of $(j \to \vec{0})$ and $(\vec{0} \to j+1)$. By performing the integration over the fields in $(j \to \vec{0})$, in $\vec{0}$ and in $(\vec{0} \to j+1)$ we find:

$$(-1)^{D_j+1} = \int V_{\mathbf{x}_j} H_{\mathbf{x}_j} \overline{H}_{\mathbf{x}_j} V_{\mathbf{x}_{j+1}} \overline{V}_{\mathbf{x}_{j+1}} \overline{H}_{\mathbf{x}_{j+1}} \{\overline{V}_{\mathbf{x}_j} \cdots H_{\mathbf{x}_{j+1}}\} =$$

$$= \int V_{\mathbf{x}_j} H_{\mathbf{x}_j} \overline{H}_{\mathbf{x}_j} \{\overline{V}_{\mathbf{x}_j} \cdots H_{\mathbf{x}_{j+1}}\} V_{\mathbf{x}_{j+1}} \overline{V}_{\mathbf{x}_{j+1}} \overline{H}_{\mathbf{x}_{j+1}} \quad (A.14)$$

and the last line is clearly equal to the r.h.s. of (A.12).

4) $j$ and $j+1$ are distinct and they belong to the right side of $R$. In this case

$$[j \to j+1] = \int V_{\mathbf{x}_j} \cdot \overline{V}_{\mathbf{x}_j} \overline{H}_{\mathbf{x}_j} \cdot \{H_{\mathbf{x}_j} \cdots H_{\mathbf{x}_{j+1}}\} \cdot V_{\mathbf{x}_{j+1}} \overline{H}_{\mathbf{x}_{j+1}} \cdot \overline{V}_{\mathbf{x}_{j+1}} . \quad (A.15)$$

The inductive hypothesis tells us that:

$$(-1)^{D_j+d_j} = \int V_{\mathbf{x}_j} \overline{H}_{\mathbf{x}_j} \cdot \overline{V}_{\mathbf{x}_j} (j \to j+1) V_{\mathbf{x}_{j+1}} \cdot \overline{V}_{\mathbf{x}_{j+1}} \overline{H}_{\mathbf{x}_{j+1}} \cdot \{H_{\mathbf{x}_j} \cdots H_{\mathbf{x}_{j+1}}\} . \quad (A.16)$$

By performing the integration over the fields in $(j \to j+1)$ we find:

$$(-1)^{D_j+1} = \int V_{\mathbf{x}_j} \overline{H}_{\mathbf{x}_j} \overline{V}_{\mathbf{x}_j} V_{\mathbf{x}_{j+1}} \overline{V}_{\mathbf{x}_{j+1}} \overline{H}_{\mathbf{x}_{j+1}} \{H_{\mathbf{x}_j} \cdots H_{\mathbf{x}_{j+1}}\} =$$

$$= \int V_{\mathbf{x}_j} \overline{H}_{\mathbf{x}_j} \overline{V}_{\mathbf{x}_j} \{H_{\mathbf{x}_j} \cdots H_{\mathbf{x}_{j+1}}\} V_{\mathbf{x}_{j+1}} \overline{V}_{\mathbf{x}_{j+1}} \overline{H}_{\mathbf{x}_{j+1}} \quad (A.17)$$

and the last line is clearly equal to the r.h.s. of (A.15).

5) $j$ and $j+1$ coincide with the upper right corner of $R$. In this case

$$[j \to j+1] = \int V_{\mathbf{x}_j} \cdot \overline{V}_{\mathbf{x}_j} \overline{H}_{\mathbf{x}_j} \cdot H_{\mathbf{x}_j} = -1 . \quad (A.18)$$

6) $j$ and $j+1$ are distinct and they belong to the right and upper sides of $R$, respectively. In this case

$$[j \to j+1] = \int V_{\mathbf{x}_j} \cdot \overline{V}_{\mathbf{x}_j} \overline{H}_{\mathbf{x}_j} \cdot \{H_{\mathbf{x}_j} \cdots V_{\mathbf{x}_{j+1}}\} \cdot \overline{V}_{\mathbf{x}_{j+1}} \overline{H}_{\mathbf{x}_{j+1}} \cdot H_{\mathbf{x}_{j+1}} . \quad (A.19)$$

Calling $\vec{0}$ the upper right corner of $R$, the inductive hypothesis tells us that:

$$(-1)^{D_j+d_j} = \int V_{\mathbf{x}_j} \overline{H}_{\mathbf{x}_j} \cdot \overline{V}_{\mathbf{x}_j} (j \to \vec{0}) V_{\vec{0}} \cdot \overline{V}_{\vec{0}} \overline{H}_{\vec{0}} \cdot$$

$$H_{\vec{0}}(\vec{0} \to j+1)(-\overline{H}_{\mathbf{x}_{j+1}}) \cdot H_{\mathbf{x}_{j+1}} \overline{V}_{\mathbf{x}_{j+1}} \cdot \{H_{\mathbf{x}_j} \cdots V_{\mathbf{x}_{j+1}}\} . \quad (A.20)$$

By performing the integration over the fields in $(j \to \vec{0})$, in $\vec{0}$ and in $(\vec{0} \to j+1)$ we find:

$$(-1)^{D_j+1} = \int V_{\mathbf{x}_j} \overline{H}_{\mathbf{x}_j} \overline{V}_{\mathbf{x}_j} (-\overline{H}_{\mathbf{x}_{j+1}}) H_{\mathbf{x}_{j+1}} \overline{V}_{\mathbf{x}_{j+1}} \{H_{\mathbf{x}_j} \cdots V_{\mathbf{x}_{j+1}}\} =$$

$$= \int V_{\mathbf{x}_j} \overline{H}_{\mathbf{x}_j} \overline{V}_{\mathbf{x}_j} \{H_{\mathbf{x}_j} \cdots V_{\mathbf{x}_{j+1}}\} (-\overline{H}_{\mathbf{x}_{j+1}}) H_{\mathbf{x}_{j+1}} \overline{V}_{\mathbf{x}_{j+1}} \quad (A.21)$$

and the last line is clearly equal to the r.h.s. of (A.19).

7) $j$ and $j+1$ are distinct and they belong to the upper side of $R$. In this case

$$[j \to j+1] = \int (-\overline{H}_{\mathbf{x}_j}) \cdot H_{\mathbf{x}_j} \overline{V}_{\mathbf{x}_j} \cdot \{V_{\mathbf{x}_j} \cdots V_{\mathbf{x}_{j+1}}\} \cdot \overline{V}_{\mathbf{x}_{j+1}} \overline{H}_{\mathbf{x}_{j+1}} \cdot H_{\mathbf{x}_{j+1}}$$
$$(A.22)$$

The inductive hypothesis tells us that:

$$(-1)^{D_j+d_j} = \int \overline{V}_{\mathbf{x}_j} \overline{H}_{\mathbf{x}_j} \cdot H_{\mathbf{x}_j} (j \to j+1)(-\overline{H}_{\mathbf{x}_{j+1}}) \cdot H_{\mathbf{x}_{j+1}} \overline{V}_{\mathbf{x}_{j+1}} \cdot \{V_{\mathbf{x}_j} \cdots V_{\mathbf{x}_{j+1}}\}$$
$$(A.23)$$

By performing the integration over the fields in $(j \to j+1)$ we find:

$$(-1)^{D_j+1} = \int \overline{V}_{\mathbf{x}_j} \overline{H}_{\mathbf{x}_j} H_{\mathbf{x}_j} (-\overline{H}_{\mathbf{x}_{j+1}}) H_{\mathbf{x}_{j+1}} \overline{V}_{\mathbf{x}_{j+1}} V_{\mathbf{x}_j} V_{\mathbf{x}_{j+1}} =$$

$$= \int \overline{V}_{\mathbf{x}_j} \overline{H}_{\mathbf{x}_j} H_{\mathbf{x}_j} V_{\mathbf{x}_j} \cdots V_{\mathbf{x}_{j+1}} (-\overline{H}_{\mathbf{x}_{j+1}}) H_{\mathbf{x}_{j+1}} \overline{V}_{\mathbf{x}_{j+1}} \quad (A.24)$$

and the last line is clearly equal to the r.h.s. of (A.22).

8) $j$ and $j+1$ coincide with the upper left corner of $R$. In this case

$$[j \to j+1] = \int (-\overline{H}_{\mathbf{x}_j}) \cdot H_{\mathbf{x}_j} \overline{V}_{\mathbf{x}_j} \cdot V_{\mathbf{x}_j} = -1 . \quad (A.25)$$

9) $j$ and $j+1$ are distinct and they belong to the upper and left sides of $R$, respectively. In this case

$$[j \to j+1] = \int (-\overline{H}_{\mathbf{x}_j}) \cdot H_{\mathbf{x}_j} \overline{V}_{\mathbf{x}_j} \cdot \{V_{\mathbf{x}_j} \cdots$$
$$(-\overline{H}_{\mathbf{x}_{j+1}})\} \cdot H_{\mathbf{x}_{j+1}} \overline{V}_{\mathbf{x}_{j+1}} \cdot V_{\mathbf{x}_{j+1}} . \quad (A.26)$$

Calling $\vec{0}$ the upper left corner of $R$, the inductive hypothesis tells us that:

$$(-1)^{D_j+d_j} = \int \overline{V}_{\mathbf{x}_j} \overline{H}_{\mathbf{x}_j} \cdot H_{\mathbf{x}_j}(j \to \vec{0})(-\overline{H}_{\vec{0}}) \cdot H_{\vec{0}} \overline{V}_{\vec{0}} \cdot$$
$$V_{\vec{0}}(\vec{0} \to j+1)(-\overline{V}_{\mathbf{x}_{j+1}}) \cdot V_{\mathbf{x}_{j+1}} H_{\mathbf{x}_{j+1}} \cdot \{V_{\mathbf{x}_j} \cdots (-\overline{H}_{\mathbf{x}_{j+1}})\}. \quad \text{(A.27)}$$

By performing the integration over the fields in $(j \to \vec{0})$, in $\vec{0}$ and in $(\vec{0} \to j+1)$ we find:

$$(-1)^{D_j+1} = \int \overline{V}_{\mathbf{x}_j} \overline{H}_{\mathbf{x}_j} H_{\mathbf{x}_j}(-\overline{V}_{\mathbf{x}_{j+1}}) V_{\mathbf{x}_{j+1}} H_{\mathbf{x}_{j+1}} \{V_{\mathbf{x}_j} \cdots (-\overline{H}_{\mathbf{x}_{j+1}})\} =$$
$$= \int \overline{V}_{\mathbf{x}_j} \overline{H}_{\mathbf{x}_j} H_{\mathbf{x}_j} \{V_{\mathbf{x}_j} \cdots (-\overline{H}_{\mathbf{x}_{j+1}})\} (-\overline{V}_{\mathbf{x}_{j+1}}) V_{\mathbf{x}_{j+1}} H_{\mathbf{x}_{j+1}} \quad \text{(A.28)}$$

and the last line is clearly equal to the r.h.s. of (A.26).

10) $j$ and $j+1$ are distinct and they belong to the left side of $R$. In this case

$$[j \to j+1] = \int (-\overline{V}_{\mathbf{x}_j}) \cdot V_{\mathbf{x}_j} H_{\mathbf{x}_j} \cdot \{\overline{H}_{\mathbf{x}_j} \cdots (-\overline{H}_{\mathbf{x}_{j+1}})\} \cdot H_{\mathbf{x}_{j+1}} \overline{V}_{\mathbf{x}_{j+1}} \cdot V_{\mathbf{x}_{j+1}}.$$
(A.29)

The inductive hypothesis tells us that:

$$(-1)^{D_j+d_j} = \int H_{\mathbf{x}_j} \overline{V}_{\mathbf{x}_j} \cdot V_{\mathbf{x}_j}(j \to j+1)(-\overline{V}_{\mathbf{x}_{j+1}}) \cdot V_{\mathbf{x}_{j+1}} H_{\mathbf{x}_{j+1}} \cdot \{\overline{H}_{\mathbf{x}_j} \cdots (-\overline{H}_{\mathbf{x}_{j+1}})\}.$$
(A.30)

By performing the integration over the fields in $(j \to j+1)$ we find:

$$(-1)^{D_j+1} = \int H_{\mathbf{x}_j} \overline{V}_{\mathbf{x}_j} V_{\mathbf{x}_j}(-\overline{V}_{\mathbf{x}_{j+1}}) V_{\mathbf{x}_{j+1}} H_{\mathbf{x}_{j+1}} \{\overline{H}_{\mathbf{x}_j} \cdots (-\overline{H}_{\mathbf{x}_{j+1}})\} =$$
$$= \int H_{\mathbf{x}_j} \overline{V}_{\mathbf{x}_j} V_{\mathbf{x}_j} \{\overline{H}_{\mathbf{x}_j} \cdots (-\overline{H}_{\mathbf{x}_{j+1}})\} (-\overline{V}_{\mathbf{x}_{j+1}}) V_{\mathbf{x}_{j+1}} H_{\mathbf{x}_{j+1}} \quad \text{(A.31)}$$

and the last line is clearly equal to the r.h.s. of (A.29).

11) $j$ and $j+1$ coincide with the lower left corner of $R$. In this case it is necessarily $j \equiv 2N$ and $j+1 \equiv 1$ and we have:

$$[2N \to 1] = \int (-\overline{V}_{\mathbf{x}_1}) \cdot V_{\mathbf{x}_1} H_{\mathbf{x}_1} \cdot \overline{H}_{\mathbf{x}_1} = +1. \quad \text{(A.32)}$$

Note that this time the result is $+1$. This "wrong" sign exactly compensates the minus sign appearing in the r.h.s. of (A.7).

12) $j$ and $j+1$ are distinct and they belong to the left and lower sides of

$R$, respectively. In this case it is necessarely $j \equiv 2N$ and $j + 1 \equiv 1$ and we have

$$[2N \to 1] = \int (-\overline{V}_{\mathbf{x}_{2N}}) \cdot V_{\mathbf{x}_{2N}} H_{\mathbf{x}_{2N}} \cdot \{\overline{H}_{\mathbf{x}_{2N}} \cdots (-\overline{V}_{\mathbf{x}_1})\} \cdot V_{\mathbf{x}_1} H_{\mathbf{x}_1} \cdot \overline{H}_{\mathbf{x}_1} \,. \tag{A.33}$$

Calling $\vec{0}$ the lower left corner of $R$, the inductive hypothesis tells us that:

$$(-1)^{D_N + d_N} = \int H_{\mathbf{x}_{2N}} \overline{V}_{\mathbf{x}_{2N}} \cdot \tag{A.34}$$

$$V_{\mathbf{x}_{2N}}(2N \to \vec{0})(-\overline{V}_{\vec{0}}) \cdot V_{\vec{0}} H_{\vec{0}} \cdot \overline{H}_{\vec{0}}(\vec{0} \to 1) H_{\mathbf{x}_1} \cdot V_{\mathbf{x}_1} \overline{H}_{\mathbf{x}_1} \cdot \{\overline{H}_{\mathbf{x}_{2N}} \cdots (-\overline{V}_{\mathbf{x}_1})\}$$

By performing the integration over the fields in $(2N \to \vec{0})$, in $\vec{0}$ and in $(\vec{0} \to 1)$ we find:

$$(-1)^{D_N} = \int H_{\mathbf{x}_{2N}} \overline{V}_{\mathbf{x}_{2N}} V_{\mathbf{x}_{2N}} H_{\mathbf{x}_1} V_{\mathbf{x}_1} \overline{H}_{\mathbf{x}_1} \{\overline{H}_{\mathbf{x}_{2N}} \cdots (-\overline{V}_{\mathbf{x}_1})\} =$$

$$= \int H_{\mathbf{x}_{2N}} \overline{V}_{\mathbf{x}_{2N}} V_{\mathbf{x}_{2N}} \{\overline{H}_{\mathbf{x}_{2N}} \cdots (-\overline{V}_{\mathbf{x}_1})\} H_{\mathbf{x}_1} V_{\mathbf{x}_1} \overline{H}_{\mathbf{x}_1} \tag{A.35}$$

and the last line is clearly equal to the r.h.s. of (A.26). It follows that $[2N \to 1] = -(-1)^{D_N+1}$, consistently with the result in item (11) above. Also in this case, the appearently "wrong" sign exactly compensates the minus sign appearing in the r.h.s. of (A.7).

Combining the results of previous items, we can simply say that the integration of $(2j - 1 \to 2j)$ contributes to the weight of $\gamma$ with $(-1)^{\ell_{2j-1}-1}$; the integration of $[2j \to 2j+1]$, with $j < N$, contributes with $(-1)^{L_{2j}+1}$ (here we defined $L_{2j}$ to be the length of $[2j \to 2j+1]$), while $[2N \to 1]$ with $(-1)^{L_{2N}}$. Substituting these results into (A.7), we find that the weight of $\gamma$ is equal to $(-1)^{n_\gamma}$, as desired.

The above discussion concludes the proof in the case of polygons without self intersections. Let us call *simple* a polygon without self intersections. If $\gamma$ is not simple, calling $\nu_\gamma$ the number of its self intersections, we can easily prove that its weight is equal to $(-1)^{\nu_\gamma}$ times the product of the weights of a number of simple polygons, defined as follows. We draw with two colors, white and black, both the disconnetted interiors of the polygon and its exterior, call them $A_1, \ldots, A_n$ and $A_0$ respectively. The drawing is done in such a way that $A_0$ is white and two adjacent sets $A_i$ and $A_j$, $0 \le i < j \le n$, have different colors (we call $A_i$ and $A_j$ adjacent if their boundaries have a common side). Then we consider the set $\mathcal{P}$ of simple polygons obtained as the boundaries of the black sets, thought as completely disconnetted one from the other. The "disconnection" of the boundaries of the black regions (which originally could touch each other through the corners) is realized by the elementary disconnetion of the intersection elements described in Fig.A2

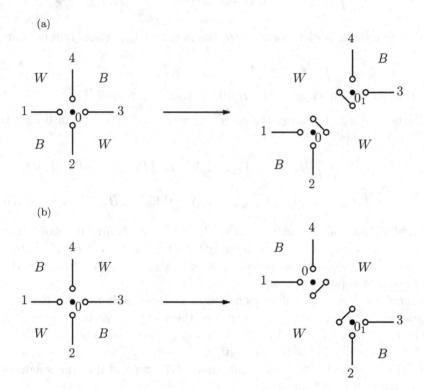

Fig. A.2 The two elementary operations of disconnecting an intersection. The labels $W$ and $B$ mean that the corresponding regions must be coloured white and black respectively. Note that the operation of disconnecting an intersection involves the doubling of the site 0 at the center of the intersection: in the figure we call 0 and $0_1$ its two copies after the disconnection.

We claim that the weight of $\gamma$ is $(-1)^{\nu_\gamma} \prod_{\gamma' \in \mathcal{P}} (-1)^{n_{\gamma'}}$, which is the desired result (recall that $\mathcal{P}$ is the set of polygons obtained as boundaries of the black sets, *after* the disconnection described in Fig.A2). Note that the factor $(-1)^{\nu_\gamma}$ in front of the product of the weights of the disconnected simple polygons is due to the doubling of the centers of the intersections, implied by our definition of disconnection.

In order to prove the claim we explicitely write the contribution from the intersection in both cases (a) and (b) of Fig. A2, and we show that it is equal to the contribution of the two corner elements on the r.h.s. of Fig. A2, unless for a minus sign, to be associated to the new site $0_1$.

The contribution of the left hand side of case $(a)$ in Fig. A2 is:

$$\int d\overline{H}_{\mathbf{x}_0} dH_{\mathbf{x}_0} d\overline{V}_{\mathbf{x}_0} dV_{\mathbf{x}_0} \left[ \overline{H}_{\mathbf{x}_1} H_{\mathbf{x}_0} \cdot \overline{H}_{\mathbf{x}_0} H_{\mathbf{x}_3} \cdot \overline{V}_{\mathbf{x}_2} V_{\mathbf{x}_0} \cdot \overline{V}_{\mathbf{x}_0} V_{\mathbf{x}_4} \right]. \quad (A.36)$$

Multiplying (A.36) by

$$-\int d\overline{H}_{\mathbf{x}_{0_1}} dH_{\mathbf{x}_{0_1}} d\overline{V}_{\mathbf{x}_{0_1}} dV_{\mathbf{x}_{0_1}} [\overline{V}_{\mathbf{x}_{0_1}} \overline{H}_{\mathbf{x}_{0_1}} \cdot V_{\mathbf{x}_{0_1}} H_{\mathbf{x}_{0_1}}] = +1, \quad (A.37)$$

we see that it can be equivalently rewritten as

$$-\int (d\overline{H}_{\mathbf{x}_0} dH_{\mathbf{x}_0} d\overline{V}_{\mathbf{x}_0} dV_{\mathbf{x}_0})(d\overline{H}_{\mathbf{x}_{0_1}} dH_{\mathbf{x}_{0_1}} d\overline{V}_{\mathbf{x}_{0_1}} dV_{\mathbf{x}_{0_1}}) \cdot \quad (A.38)$$

$$\cdot \left[ \overline{H}_{\mathbf{x}_1} H_{\mathbf{x}_0} \cdot \overline{V}_{\mathbf{x}_{0_1}} \overline{H}_{\mathbf{x}_{0_1}} \cdot \overline{V}_{\mathbf{x}_2} V_{\mathbf{x}_0} \right] \cdot \left[ \overline{H}_{\mathbf{x}_0} H_{\mathbf{x}_3} \cdot V_{\mathbf{x}_{0_1}} H_{\mathbf{x}_{0_1}} \cdot \overline{V}_{\mathbf{x}_0} V_{\mathbf{x}_4} \right]$$

Exchanging the names of the fields $\overline{V}_{\mathbf{x}_0} \longleftrightarrow \overline{V}_{\mathbf{x}_{0_1}}$ and $\overline{H}_{\mathbf{x}_0} \longleftrightarrow \overline{H}_{\mathbf{x}_{0_1}}$, we easily recognize that (14.31) is equal to $(-1)$ times the contribution of the r.h.s. of case $(a)$ in Fig. A2. The minus sign compensate the fact that after the doubling the new polygon has a site more than the original one.

The argument can be repeated in case $(b)$, so that the proof of the claim is complete.

This concludes the proof of (A.2) in the case of open boundary conditions (*i.e.* in the case where polygons winding up over the lattice are not allowed).

## A.2 The Grassmann representation of the 2d Ising model with periodic boundary conditions

In the case periodic boundary conditions are assumed, the representation in terms of multipolygons is the same, except for the fact that also polygons winding up over the lattice are allowed. In order to construct a Grassmann representation for the multipolygon expansion of Ising with p.b.c., let us start with considering the following expression:

$$\int \prod_{\mathbf{x} \in \Lambda_M} d\overline{H}_\mathbf{x} dH_\mathbf{x} d\overline{V}_\mathbf{x} dV_\mathbf{x} e^{S_{\varepsilon,\varepsilon'}(t)}, \quad (A.39)$$

where $\varepsilon, \varepsilon' = \pm$ and

$$\overline{H}_{\mathbf{x}+M\hat{e}_0} = \varepsilon \overline{H}_\mathbf{x} \quad , \quad \overline{H}_{\mathbf{x}+M\hat{e}_1} = \varepsilon' \overline{H}_\mathbf{x}$$
$$H_{\mathbf{x}+M\hat{e}_0} = \varepsilon H_\mathbf{x} \quad , \quad H_{\mathbf{x}+M\hat{e}_1} = \varepsilon' H_\mathbf{x} \quad , \quad \varepsilon, \varepsilon' = \pm \quad (A.40)$$

where we recall that $M$ is the side of the lattice $\Lambda_M$. Identical definitions are set for the variables $V, \overline{V}$. We shall say that $\overline{H}, H, \overline{V}, V$ satisfy $\varepsilon$–periodic ($\varepsilon'$–periodic) boundary conditions in vertical (horizontal) direction. Note that, unless for a sign and for the replacement $S(t) \to S_{\varepsilon,\varepsilon'}(t)$, (A.39) is the same as the r.h.s. of (A.4).

Clearly, by expanding the exponential in (A.39) and by integrating the Grassmann fields as described in previous section, we get a summation over dimers very similar to the one seen above. In particular the weights assigned to the closed polygons not winding up the lattice are exactly the same as those calculated in previous section. In this case, however, also Grassmann polygons winding up the lattice are allowed. Let us calculate the weight that (A.39) assigns to these polygons (as above we define the weight by descarding the "trivial" factors $t^{|\gamma|}$ and $(-1)^{M^2 - n_\gamma}$).

As an example, let us first calculate the contribution from the simplest polygon $\gamma$ winding up the lattice, the horizontal straight line winding once in horizontal direction. Its weight is given by:

$$\int \overline{V}_{\vec{0}} V_{\vec{0}} \cdot \overline{H}_{\vec{0}} H_{\hat{e}_1} \cdot \overline{V}_{\hat{e}_1} V_{\hat{e}_1} \cdot \overline{H}_{\hat{e}_1} H_{2\hat{e}_1} \cdots \overline{H}_{(M-1)\hat{e}_1} H_{M\hat{e}_1}. \qquad (A.41)$$

Now, we can rewrite $\overline{H}_{M\hat{e}_1}$ as $\varepsilon' \overline{H}_{\vec{0}}$. Also, permutating the field $H_{\vec{0}}$ from the last position to the third one, we see that (A.41) is equal to:

$$(-\varepsilon') \int \overline{V}_{\vec{0}} V_{\vec{0}} H_{\vec{0}} \overline{H}_{\vec{0}} \cdot \overline{V}_{\hat{e}_1} V_{\hat{e}_1} H_{\hat{e}_1} \overline{H}_{\hat{e}_1} \cdots \overline{V}_{(M-1)\hat{e}_1} V_{(M-1)\hat{e}_1} H_{(M-1)\hat{e}_1} \overline{H}_{(M-1)\hat{e}_1}$$

$$= (-\varepsilon')(-1)^M = (-\varepsilon')(-1)^{n_\gamma} \qquad (A.42)$$

where, in the last identity, we used that the length of the straight polygon $\gamma$ is exactly $M$. Repeating the lengthy construction of previous section, it can be (straightforwardly) proven that a generic polygon $\gamma$ winding up once in horizontal direction has a weight (as assigned by (A.39)) equal to $(-\varepsilon')(-1)^{n_\gamma}$. Analogously a polygon $\gamma$ winding up once in horizontal direction has a weight (as assigned by (A.39)) equal to $(-\varepsilon)(-1)^{n_\gamma}$.

Let us now consider the simplest polygon $\gamma$ winding up $h$ times in horizontal direction and $v$ times in vertical direction, that is the union of $h$ distinct horizontal lines and $v$ distinct vertical lines each of them winding once over the lattice in horizontal or vertical direction, respectively. Repeating the same simple calculation of (A.41)–(11.20), we easily see that the weight assigned by (A.39) to $\gamma$ is $(-\varepsilon')^h(-\varepsilon)^v(-1)^{M(h+v)}$. Note that $\gamma$ has $(-1)^{h \cdot v}$ self intersections, so that $n_\gamma = M(h+v) - h \cdot v$ and the weight can be rewritten as $(-\varepsilon')^h(-\varepsilon)^v(-1)^{h \cdot v}(-1)^{n_\gamma}$. Again, repeating the lengthy construction of previous section, it can be (straightforwardly) proven that a generic polygon $\gamma$ winding up $h$ times in horizontal direction and $v$ times in vertical direction has a weight (as assigned by (A.39)) equal to $(-\varepsilon')^h(-\varepsilon)^v(-1)^{h\cdot v}(-1)^{n_\gamma}$.

Since the weight assigned to a generic polygon is the one just computed, which is in general different from $(-1)^{n_\gamma}$, it is clear that there exists *no* choice of $\varepsilon, \varepsilon' = \pm 1$ such that (A.39) is equal to $(-1)^{M^2}(2\cosh^2 \beta J)^{-M^2}$ times $\Xi_I$, where now $\Xi_I$ is the Ising model partition function in the volume $\Lambda_M$ with periodic boundary conditions. However it is easy to realize that $(-1)^{M^2}\Xi_I(2\cosh^2 \beta J)^{-M^2}$ is equal to a suitable linear combination of the expressions in (A.39), with different choices of $\varepsilon, \varepsilon' = \pm 1$: it holds that

$$(-1)^{M^2}\frac{Z}{(2\cosh^2\beta J)^{M^2}} = \frac{1}{2}\sum_{\varepsilon,\varepsilon'=\pm 1}\int\prod_{\mathbf{x}\in\Lambda_M}\mathrm{d}\overline{H}_{\mathbf{x}}\mathrm{d}H_{\mathbf{x}}\mathrm{d}\overline{V}_{\mathbf{x}}\mathrm{d}V_{\mathbf{x}}(-1)^{\delta_{(\varepsilon,\varepsilon')}}e^{S_{\varepsilon,\varepsilon'}(t)},$$
(A.43)

where $\delta_{+,-} = \delta_{-,+} = \delta_{-,-} = 0$ and $\delta_{+,+} = 1$. In order to verify the last identity it is sufficient to verify that the weight assigned to each polygon $\gamma$ is exactly $(-1)^{n_\gamma}$. If $\gamma$ winds up the lattice $h$ times in horizontal direction and $v$ times in vertical direction, from the calculation above it follows that the weight is:

$$\frac{1}{2}\sum_{\varepsilon,\varepsilon'=\pm 1}(-1)^{\delta_{(\varepsilon,\varepsilon')}}(-\varepsilon')^h(-\varepsilon)^v(-1)^{h\cdot v}(-1)^{n_\gamma}$$
$$= \frac{1}{2}(-1)^{n_\gamma}\Big[(-1)^{h+v+hv+\delta_{+,+}} + (-1)^{v+hv+\delta_{+,-}}$$
$$+ (-1)^{h+hv+\delta_{-,+}} + (-1)^{hv+\delta_{-,-}}\Big]$$
(A.44)

The expression between square brackets on the last line is equal to $(-1)^{hv}[-(-1)^{h+v}+(-1)^v+(-1)^h+1]$. Now, if $h$ and $v$ are both even, this is equal to $(+1)[-1+1+1+1] = 2$; if $h$ is even and $v$ is odd (or viceversa), it is equal to $(+1)[+1-1+1+1] = 2$; if they are both odd, it is equal to $(-1)[-1-1-1+1] = 2$. That is, (A.44) is identically equal to $(-1)^{n_\gamma}$, as wanted, and (A.43) is proven.

# Bibliography

[1] Itzykson C. and Zuber J. *The quantum theory of fields*, McGraw-Hill,New York, 1985.
[2] Weinberg, S. *The Quantum Theory of Fields*, Cambridge University Press, 1995.
[3] Glimm J. and Jaffe A. *Quantum Physics: a Functional Integralpoint of wiew*, Springer, 1987.
[4] Streater R.F., Wightman A.S. *PCT,spin and statistics and all that*. Benjiamin Publishing, 1964.
[5] Osterwalder K., Schrader R.: Axioms for Euclidean Green's Functions. *Comm. Math. Phys. 31*, 83–112, 1973; *Comm. Math. Phys. 42*, 281–305, 1975.
[6] Wilson K.G. The Renormalization Group and critical phenomena *Rev. Mod.Phys.* 55, 583, 1983.
[7] Polchinski,J. *Nucl. Phys. B* 231 , 269, 1984.
[8] Gallavotti G.: Renormalization theory and ultraviolet stability for scalar fields via renormalization group methods. *Rev.Mod.Phys.* 57, 471–562, 1985.
[9] Rivasseau V.: *From perturbative to constructive renormalization*, Princeton University Press, Princeton, 1994.
[10] Benfatto G., Gallavotti G. *Renormalization group*. Physics Notes, 1. Princeton Paperbacks. Princeton University Press, 1995.
[11] Salmhofer M.*Renormalization. An introduction*. Texts and Monographs in Physics. Springer-Verlag, Berlin, 1999.
[12] Itzykson C., Drouffe, J. Statistical field theory, Cambridge monographs in Mathematical Notes, Cambr. Univ. Press, 1989.
[13] Gallavotti G. :Statistical Mechanics. Springer, 1999.
[14] Baxter R.J. *Exactly solved models in statistical mechanics*. Academic Press, 1982.
[15] Abrikosov A.A., Gorkov L.P., Dzyaloshinski I.Y.: *Methods of Quantum Field Theory in Statistical Physics*, Dover Publications, New York, 1965.
[16] Negele J.W., H.Orland H. *Quantum many particle systems* Addison-Wesley, 1987.
[17] Anderson P.W.: *The theory of superconductivity in high-$T_c$ cuprates*, Princeton Press, Princeton, 1997.

[18] Mattis D. C., Lieb E. H.: Exact Solution of a Many–Fermion System and its Associated Boson Field. *J. Math. Phys. 6*, 304–312, 1965.

[19] Feldman, J, Knoerrer H. and Trubowitz E. *Fermion functional integral and the Renormalization Group* American Math. Soc. (2000).

[20] Gallavotti G., Nicoló F.: Renormalization theory for four dimensional scaler fields, I, *Comm. Math. Phys 100* (1985), 545–590; II, *Comm. Math. Phys. 101* (1985), 471–562.

[21] Caianiello E.R.: Number of Feynman graphs and convergence, *Nuovo Cimento 3* (1956), No. 1, 223–225.

[22] Battle G., Federbush P. A phase cell cluster expansion for Euclidean field theories. Ann.Phys. 142,95, 1982.

[23] Brydges D.: A short course on Cluster Expansions, Les Houches 1984, K. Osterwalder, R. Stora eds., North Holland Press, (1986).

[24] Gawedzki K., Kupiainen A.: Gross–Neveu model through convergent perturbation expansions. *Comm.Math.Phys. 102*, 1–30, 1985.

[25] Lesniewski A.: Effective action for the Yukawa$_2$ quantum field theory. *Comm. Math. Phys. 108*, 437–467, 1987.

[26] Feldman J., Magnen J., Rivasseau V, Sénéor R.: Massive Gross–Neveu Model: A renormalizable field theory in two dimensions. *Comm. Math. Phys. 103*, 67–103, 1986.

[27] Mastropietro V.: Non-perturbative Adler-Bardeen Theorem. *J. Math. Phys.* (2007).

[28] Benfatto G., Mastropietro V.: Renormalization group, hidden symmetries and approximate Ward identities in the $XYZ$ model. *Rev. Math. Phys. 13*, 1323–1435, 2001.

[29] Benfatto G., Mastropietro V.: On the density–density critical indices in interacting Fermi systems. *Comm. Math. Phys. 231*, 97–134, 2002.

[30] Benfatto G., Mastropietro V.: Ward identities and vanishing of the Beta function for $d=1$ interacting Fermi systems. *J. Stat. Phys. 115*, 143–184, 2004.

[31] Benfatto G., Mastropietro V.: Ward identities and chiral anomaly in the Luttinger liquid. *Comm. Math. Phys. 258*, 609–655, 2005.

[32] Adler S. L., Bardeen W.A.: Absence of higher order corrections in the anomalous axial vector divergence equation. *Phys. Rev.* **182**, 1517-1536, 1969.

[33] Georgi H., Rawls J.M.: Anomalies of the Axial–Vector Current in Two dimensions. *Phys. Rev.D 3*, 874–879, 1971.

[34] Gomes M., Lowenstein J.H.: Asymptotic scale invariance in a massive Thirring model. *Nucl. Phys. B 45*, 252–266, 1972.

[35] Metzner W., Di Castro C.:Conservation laws and correlation functions in the Luttinger liquid. *Phys. Rev. B* **47**, 16107 (1993).

[36] Benfatto G., Gallavotti G.: Perturbation theory of the Fermi surface in a quantum liquid. A general quasiparticle formalism and one dimensional systems, *J. Statist. Phys. 59* (1990), No. 3–4, 541–664.

[37] Benfatto G., Gallavotti G., Procacci A., Scoppola B.: Beta functions and Schwinger functions for a many fermions system in one dimension, *Comm. Math. Phys. 160* (1994), 93–171.

[38] Mastropietro V.: Non-perturbative aspects of chiral anomalies. *J. Phys. A* **40**, 33, 10349–10365 (2007).
[39] Benfatto G., Falco P., Mastropietro V.: Functional Integral Construction of the Massive Thirring model: Verification of Axioms and Massless Limit, *Comm. Math. Phys.* **273**, 67–118, (2007).
[40] Osterwalder K., Seiler E.: Gauge Field Theories on a Lattice. *Ann.Phys.* **110**, 440–471, 1978.
[41] Mastropietro V.. Renormalization Group and Ward Identities for Infrared QED4. *J. Math. Phys.* **48**, 102303 (2007).
[42] Kasteleyn P.W.: Dimer Statistics and phase transitions, *J. Math.Phys.* **4**,287 (1963).
[43] Hurst C.: New approach to the Ising problem, *J.Math. Phys.* **7**,2, 305-310 (1966).
[44] Samuel S., The use of anticommuting variable integrals in statistical mechanics, *J. Math. Phys.* **21** (1980) 2006.
[45] Spencer T.: A mathematical approach to universality in two dimensions. *Physica A* **279**, 250-259 (2000).
[46] Pinson P., Spencer T.: Universality in 2D critical Ising model. To appear in *Comm. Math. Phys.*
[47] Mastropietro V.: *Non Universality in Ising models with four spin interaction.* *Journ. Stat. Phys.*, **111**, 201-259 (2003).
[48] Mastropietro V.: *Ising models with four spin interaction at criticality.* *Comm. Math. Phys.* **244**, 3 (2004), 595–642.
[49] Giuliani A., Mastropietro V.: Anomalous Universality in the Ashkin-Teller model. *Comm. Math. Phys.* 2003.
[50] Giuliani A., Mastropietro V.: *Anomalous critical exponents in the anisotropic Ashkin–Teller model* **93** (2005) *Phys. Rev. Lett.* 190603-190607.
[51] Tomonaga S.I.: *Progress of Theoretical Physics.* **5**, 544 (1950).
[52] Bonetto F.,Mastropietro,V.: Beta Function and anomaly of the Fermi surface for a d=1 system of interacting fermions in a periodic potential F. Bonetto, V.Mastropietro. *Comm. Math. Phys.* **172**, 1, 57–93 (1992).
[53] Bonetto F., Mastropietro V.: Critical indeces in a $d = 1$ filled band Fermi systems, *Phys. Rev. B* **56**,3,1296-1308 (1997).
[54] Bonetto F., V. Mastropietro V.: Critical Indices for the Yukawa2 model, *Nucl.Phys.B* **497**, 541-554 (1997).
[55] Mastropietro V.:Small denominators and anomalous behaviour in the incommensurate Holstein-Hubbard model, *Commun. Math. Phys.*, **201**, 81-115 (1999).
[56] Mastropietro V.: Incommensurate Charge Density Waves in the adiabatic Hubbard-Holstein model, *Phys. Rev.B 65*, 75113, (2002).
[57] Gentile G., Mastropietro V.: Renormalization group for fermions: a review on mathematical results, *Physics Reports* **352** (2001), no. 4-6, 273–437.
[58] Mastropietro V.: Rigorous proof of Luttinger liquid behavior in the 1d Hubbard model, *J. Stat. Phys.* **121** (2005), no. 3–4, 373–432
[59] Mastropietro V.: The absence of Logarithmic corrections in the 1d Hubbard model. *J. Phys. A* **40**, 13, 3347 (2007).

[60] Lieb E.H., Wu F.Y.: *Phys. Rev. Lett.* 20, 1445–1449 (1968).
[61] Feldman J., Magnen J., Rivasseau V., Trubowitz E.: An infinite volume expansion for many fermions Green functions, *Helv. Phys. Acta* 65 (1992), 679–721.
[62] Disertori M., Rivasseau, V.: Interacting Fermi Liquid in Two Dimensions at Finite Temperature. *Comm. Math. Phys. 215*, 251–290, 2000; 215, 251–290, 2000.
[63] Benfatto G., Giuliani A., Mastropietro V.: Low temperature analysis of two dimensional Fermi systems with symmetric Fermi surface, *Annales Henry Poincare'* n.4 (2003) 137-193.
[64] Benfatto G., Giuliani A., Mastropietro V.: Proof of Fermi liquid behavior in the 2D Hubbard model. *Annales Henri Poincare'* 7 (2006), no. 5, 809–898.
[65] Pedra S., Salmhofer M.: Proceeding of ICMP 2003.
[66] Rivasseau V.: *J. Stat. Phys. 106*, 693–722 (2002).
[67] Afchain S., Magnen J., Rivasseau V.: *Ann. Henri Poincare'* 6, 399–448; 449–483 (2005).
[68] Feldman J., Knoerrer H., Trubowitz E.: *Comm. Math. Phys. 247*, 1–320 (2004).
[69] Shankar, R. Renormalization Grouyp approach to interacting fermions. *Rev. Math. Phys.* 66,129 (1994).
[70] Bardeen J, Cooper L N, Schrieffer J R. *Phys. Rev.* 108,1175 (1957).
[71] Mastropietro V.: Mass generation in a fermionic model. *Comm. Math. Phys.* 269 (2007), no. 2, 401–424
[72] Giuliani A. PHD thesis, Rome University (2005).